Plant Biochemistry II

Publisher's Note

The *International Review of Biochemistry* remains a major force in the education of established scientists and advanced students of biochemistry throughout the world. It continues to present accurate, timely, and thorough reviews of key topics by distinguished authors charged with the responsibility of selecting and critically analyzing new facts and concepts important to the progress of biochemistry from the mass of information in their respective fields.

Following the successful format established by the earlier volumes in this series, new volumes of the *International Review of Biochemistry* will concentrate on current developments in the major areas of biochemical research and study. New volumes on a given subject generally appear at two-year intervals, or according to the demand created by new developments in the field. The scope of the series is flexible, however, so that future volumes may cover areas not included earlier.

University Park Press is honored to continue publication of the *International Review of Biochemistry* under its sole sponsorship beginning with Volume 13. The following is a list of volumes published and currently in preparation for the series:

Volume 1: **CHEMISTRY OF MACROMOLECULES** (H. Gutfreund)
Volume 2: **BIOCHEMISTRY OF CELL WALLS AND MEMBRANES** (C.F. Fox)
Volume 3: **ENERGY TRANSDUCING MECHANISMS** (E. Racker)
Volume 4: **BIOCHEMISTRY OF LIPIDS** (T.W. Goodwin)
Volume 5: **BIOCHEMISTRY OF CARBOHYDRATES** (W.J. Whelan)
Volume 6: **BIOCHEMISTRY OF NUCLEIC ACIDS** (K. Burton)
Volume 7: **SYNTHESIS OF AMINO ACIDS AND PROTEINS** (H.R.V. Arnstein)
Volume 8: **BIOCHEMISTRY OF HORMONES** (H.V. Rickenberg)
Volume 9: **BIOCHEMISTRY OF CELL DIFFERENTIATION** (J. Paul)
Volume 10: **DEFENSE AND RECOGNITION** (R.R. Porter)
Volume 11: **PLANT BIOCHEMISTRY** (D.H. Northcote)
Volume 12: **PHYSIOLOGICAL AND PHARMACOLOGICAL BIOCHEMISTRY** (H. Blaschko)
Volume 13: **PLANT BIOCHEMISTRY II** (D.H. Northcote)
Volume 14: **BIOCHEMISTRY OF LIPIDS II** (T.W. Goodwin)

(Series numbers for the following volumes will be assigned in order of publication)
BIOCHEMISTRY OF NUCLEIC ACIDS II (B.F.C. Clarke)
BIOCHEMISTRY OF CELL DIFFERENTIATION II (J. Paul)
AMINO ACID AND PROTEIN BIOSYNTHESIS II (H.R.V. Arnstein)
BIOCHEMISTRY OF CARBOHYDRATES II (D.J. Manners)
BIOCHEMISTRY AND MODE OF HORMONES II (H.V. Rickenberg)
DEFENSE AND RECOGNITION II (E. S. Lennox)
BIOCHEMISTRY OF NUTRITION (A. Neuberger)
BIOCHEMISTRY OF CELL WALLS AND MEMBRANES II (J.C. Metcalfe)
MICROBIAL BIOCHEMISTRY (J.R. Quayle)

INTERNATIONAL REVIEW OF BIOCHEMISTRY

Volume 13

Plant Biochemistry II

Edited by

D. H. Northcote, Ph.D., F.R.S.
Department of Biochemistry
University of Cambridge
Cambridge, England

UNIVERSITY PARK PRESS
Baltimore · London · Tokyo

UNIVERSITY PARK PRESS
International Publishers in Science and Medicine
Chamber of Commerce Building
Baltimore, Maryland 21202

Typeset by The Composing Room of Michigan, Inc.

Manufactured in the United States of America by Universal Lithographers, Inc., and The Optic Bindery Incorporated

Library of Congress Cataloging in Publication Data

Main entry under title:

Plant Biochemistry II.

(International review of biochemistry; v. 13)
Includes index.
1. Botanical Chemistry. I. Northcote, Donald Henry. II. Series.
QP 501.B527 Vol.13 [QK861] 574.1 92 08s [581.1 92]
ISBN 0-8391-1077-4 77-3149

Consultant Editors' Note

The MTP *International Review of Biochemistry* was launched to provide a critical and continuing survey of progress in biochemical research. In order to embrace even barely adequately so vast a subject as "progress in biochemical research," twelve volumes were prepared. They range in subject matter from the classical preserves of biochemistry—the structure and function of macromolecules and energy transduction—through topics such as defense and recognition and cell differentiation, in which biochemical work is still a relatively new factor, to those territories that are shared by physiology and biochemistry. In dividing up so pervasive a discipline, we realized that biochemistry cannot be confined to twelve neat slices of biology, even if those slices are cut generously: every scientist who attempts to discern the molecular events that underlie the phenomena of life can legitimately parody the cry of *Le Bourgeois Gentilhomme,* "Pas ma foi! il y a plus de quarante ans que je dis de la Biochimie sans que j'en susse rien!" We therefore make no apologies for encroaching even further, in this second series, on areas in which the biochemical component has, until recently, not predominated.

However, we repeat our apology for being forced to omit again in the present collection of articles many important matters, and we also echo our hope that the authority and distinction of the contributions will compensate for our shortcomings of thematic selection. We certainly welcome criticism—we thank the many readers and reviewers who have so helpfully criticized our first series of volumes—and we solicit suggestions for future reviews.

It is a particular pleasure to thank the volume editors, the chapter authors, and the publishers for their ready cooperation in this venture. If it succeeds, the credit must go to them.

H. L. Kornberg
D. C. Phillips

Contents

Preface ix

1
Chemistry, Function, and Evolution of Plastocyanin 1
D. Boulter, B.G. Haslett, D. Peacock, J.A.M. Ramshaw, and M.D. Scawen

2
Electron and Proton Transfer in Chloroplasts 41
D.S. Bendall

3
Riddle of Sucrose 79
H.G. Pontis

4
Biochemistry of Osmotic Regulation 119
H. Kauss

5
Biochemistry of Plant Pathogens 141
J. Friend

6
Glycoproteins 183
R.G. Brown and W.C. Kimmins

7
Functions of Ion Transport in Plant Cells and Tissues 211
E.A.C. MacRobbie

Index 248

Preface

The use of plant tissues to investigate some of the fundamental problems in biology and biochemistry is described in the various articles in this volume. The subjects discussed range from the intriguing details of the electron and proton transport systems present in chloroplasts to the use of amino acid sequences of proteins as a basis for the study of taxonomy and molecular evolution.

Reviews are presented on the transport processes that take place across membranes and on the metabolic changes that occur which control the turgor pressure of the cell. These are problems that have been continuously investigated by plant physiologists and biochemists and that now have additional general interest in the light of modern ideas on the structure and function of cell membranes.

The above topics are related to those that are discussed in the chapters on the advances in the chemistry and role of plant glycoproteins and on the biochemical changes associated with the invasion of plant cells by fungal and other parasites. Both these subjects are concerned with the outer surfaces of the cell and raise important fundamental ideas about the biochemistry of cell defense mechanisms and the basis of cell resistance and susceptibility to attack. The work described here has considerable practical importance for agriculture and plant breeding.

One of the important sources of carbohydrate in food is sucrose. It is used by the plant as the principal material for transporting the fixed carbon resulting from photosynthesis and it may have other specific actions for plant development. The formation and role of this key disaccharide in plant cell metabolism is reviewed so that the various enzyme systems associated with its formation and breakdown can be assessed and compared.

I am grateful to the contributors to this volume for their collaboration which has made my job simple. I feel sure that the volume will be of considerable interest to biochemists in general and plant biochemists in particular.

D.H. Northcote

International Review of Biochemistry
Plant Biochemistry II, Volume 13
Edited by D. H. Northcote

1
Chemistry, Function, and Evolution of Plastocyanin

D. BOULTER,[1] B. G. HASLETT,[1] D. PEACOCK,[2]
J. A. M. RAMSHAW,[3] AND M. D. SCAWEN[4]

[1] Department of Botany, University of Durham, Durham, England;
[2] Department of Biochemistry, The University of Leicester, Leicester, England;
[3] Department of Inorganic Chemistry, The University of Sydney, Sydney, Australia;
and [4] Microbial Research Establishment, Porton, Salisbury, Wiltshire, England

---

PURIFICATION 3
- Purification Procedures 3
- Criteria of Purity 7

PROPERTIES 8
- Oxidation and Reduction 8
- Spectral Properties 9
- Stability, Effects of Denaturing Agents, and Surface Properties 14
- Preparation of Apoplastocyanin 16

STRUCTURE 16
- General Characteristics 16
- Amino Acid Sequence 19
  - *Amino Acid Composition* 19
  - *Amino Acid Sequence Studies* 19
- Secondary and Tertiary Structure 23
- Copper Binding Site 23
- Structural Homologies with Other Type I Copper Proteins 27

BIOLOGICAL ROLE 28
- Localization of Plastocyanin 28
- Role in Photosynthetic Electron Transport 29

BIOSYNTHESIS 31

MOLECULAR EVOLUTION 32
Evolutionary Rate 32
Relative Usefulness of Partial, Compared with Complete, Sequences 32
General Phylogenetic Implications 33
Higher Plants 34
Polymorphism 35
Serological Comparisons 35
Conclusions 35

In compiling this review, particular attention has been paid to the structure and properties of plastocyanin, as these have not been reviewed previously, whereas there are many reviews of photosynthesis (e.g., Trebst (1)) which include the function of plastocyanin in that process. The molecular evolution of plastocyanin cannot be presented in detail at this stage, because most of the data are as yet unpublished.

Copper was first shown to be an essential element for the growth of tomato and sunflower plants by Somner (2). In the same year, Green et al. (3) demonstrated that photosynthetic carbon dioxide assimilation by the green alga *Chlorella ellipsoidea* was inhibited by copper-specific chelating agents, although respiration in the dark was unaffected; the inhibition was reversed by dialyzing the cell suspension against a chelate-free medium, and the effect was considered as evidence that copper was an essential element requirement in photosynthesis. Neish (4) found that 74.6% of the total copper of clover leaves was concentrated in the chloroplast fraction, and Whatley et al. (5) showed that, of the metallic micronutrients, only copper and iron were significantly concentrated in the chloroplast. Initial work on the importance of copper in photosynthesis centered around the possible functions of ascorbate oxidase or polyphenol oxidase in that process, as these were the only copper-containing enzymes that had been isolated from plant material (6).

It was not until 1960 that Katoh and Takamiya described a blue copper protein from *C. ellipsoidea,* which was later shown to be a natural Hill oxidant (7); the name "plastocyanin" was first introduced by Katoh and Takamiya (7) when they isolated a similar blue protein from spinach chloroplasts.

Nieman and Vennesland (8) reported the presence of a soluble factor in spinach chloroplasts that was essential for the photo-oxidation of reduced cytochrome *c*. They later purified this material and, although no mention was made of any blue color or of a copper content, the data would suggest that this protein was plastocyanin (9).

Later work by Katoh and his coworkers showed that plastocyanin was widely distributed among plants and was associated with a chlorophyll-containing subchloroplast fragment (10), and De Kouchkovsky and Fork (11) showed its presence in green, but not white, parts of variegated leaves. More recently, the localization of plastocyanin in the chloroplast has been demonstrated by electron paramagnetic resonance spectroscopy (12); the EPR spectrum of plastocyanin in situ is identical with that of purified plastocyanin.

Plastocyanin has now been purified from a wide range of vascular plants and also from several green and blue-green algae (see Table 1). It has not been isolated from red algae, although Visser et al. (13) have detected an EPR signal attributable to this protein in *Porphyridium aerugineum;* there are no reports of it being found in brown algae. The presence of plastocyanin in *Euglena* has been suggested from spectrophotometric evidence (14), although plastocyanin is reported to be absent in the yellow-green alga *Bumilleriopsis* where it is replaced by cytochrome 553 (15). Katoh et al. (10) did not detect plastocyanin in the photosynthetic bacterium *Rhodopseudomonas palustris,* and its presence has not been reported in any bacterium. However, the similarity between the photosynthetic electron transport system of red and brown algae with that of higher plants suggests that plastocyanin will be isolated from these organisms. In bacteria, in which the mechanisms of photosynthesis are different, it may well be absent.

PURIFICATION

Purification Procedures

The purification of plastocyanin from vascular plants or algae can be divided into five main steps: 1) breaking the cell walls and extracting the protein from the lipoprotein complex of the chloroplast; 2) initial fractionation and concentration; 3) ion exchange chromatography on DEAE-cellulose; 4) ammonium sulfate fractionation (not always useful, depending on source material); 5) final purification either by ion exchange chromatography or gel filtration, or a combination of both.

Step 1 has been carried out by extracting an acetone-dried leaf or algal powder with water (27, 28), by slow freezing and thawing of leaf material, followed by homogenization in an aqueous buffer, pH 7.8 (29), or by blending either fresh or frozen leaves in an aqueous buffer (pH 8) containing Triton X-100 (30) (Boulter et al., unpublished results).

The type of extraction procedure required depends on the nature of the leaf material being used. Thus, plastocyanin could be extracted from leaves of French bean or marrow by freezing and thawing, followed by extraction in an aqueous buffer (29, 31), although this procedure was ineffective when used with spinach leaves (29). In general, it appears that a method which uses either acetone-dried leaves or detergent is likely to be effective with a wide variety of sources (32). Recently, plastocyanin has been extracted from frozen potato and broad bean leaves and fresh cauliflower leaf by treatments with detergent containing buffer (23, 24).

Table 1. Sources of plastocyanin and sequence data[a]

Common English name	Botanical name	Family	No. of residues sequenced	Reference
Blue-green algae *Anabaena*	*Anabaena variabilis*	Cyanophyceae	105[b]	16
Green algae				
Chlorella	*Chlorella fusca*	Chlorophyceae	98[b]	17
Enteromorpha	*Enteromorpha intestinalis*	Chlorophyceae	84	
Scenedesmus	*Scenedesmus acutus*	Chlorophyceae	0	15
Ferns				
Bracken	*Pteridium aquiliaceae*	Pteridaceae	30	
Hard fern	*Blechnum spicant*	Blechnaceae	0	
Male fern	*Dryopteris filix-mas*	Aspidiaceae	40	
Horsetail	*Equisetum arvense*	Equisetaceae	35	
Gymnosperms				
Cycad	*Cycas revoluta*	Cycadaceae	40	
Yew	*Taxus baccata*	Taxaceae	39	
Angiosperms				
Goosegrass	*Galium aparine*	Rubiaceae	30	
Parsnip	*Pastinaca sativa*	Umbelliferae	40	
Cow parsley	*Anthriscus sylvestris*	Umbelliferae	40	
Hogweed	*Heracleum sphondylium*	Umbelliferae	46	
Giant hogweed	*Heracleum mantegazzianum*	Umbelliferae	46	
Ground elder	*Aegopodium podagraria*	Umbelliferae	32	
Dock	*Rumex acetosella*	Polygonaceae	99[b]	18
Buckwheat	*Fagopyrum esculentum*	Polygonaceae	40	
Viburnum	*Viburnum tinus*	Caprifoliaceae	38	
Elder	*Sambucus nigra*	Caprifoliaceae	99[b]	19
Lonicera	*Lonicera periclymenum*	Caprifoliaceae	40	
Fat hen	*Chenopodium album*	Chenopodiaceae	40	
Spinach	*Spinacia oleracea*	Chenopodiaceae	99[b]	20
Beetroot	*Beta vulgaris*	Chenopodiaceae	40	
Marrow	*Cucurbita pepo*	Cucurbitaceae	99[b]	21

Cucumber	*Cucumis sativus*	Cucurbitaceae	34	
Shepherd's purse	*Capsella bursa-pastoris*	Cruciferae	99[b]	22
Spring cabbage	*Brassica oleracea*	Cruciferae	40	
Cauliflower	*Brassica oleracea*	Cruciferae	40	
Tomato	*Lycopersicon esculentum*	Solanaceae	36	
Potato	*Solanum tuberosum*	Solanaceae	99[b]	23
Tobacco	*Nicotiana tabacum*	Solanaceae	40	
Solanum	*Solanum crispum*	Solanaceae	40	
Snapdragon	*Antirrhinum majus*	Scrophulariaceae	36	
Foxglove	*Digitalis purpurea*	Scrophulariaceae	40	
Broom	*Cytisis ballendieri*	Leguminosae	42	
Lupin	*Lupinus* sp.	Leguminosae	40	
Robinia	*Robinia pseudoacacia*	Leguminosae	40	
Daviesia	*Daviesia latifolia*	Leguminosae	40	
Pea	*Pisum sativa*	Leguminosae	36	
Clover	*Trifolium medium*	Leguminosae	40	
Broad bean	*Vicia faba*	Leguminosae	99[b]	24
Mung bean	*Phaseolus aureus*	Leguminosae	35	
French bean	*Phaseolus vulgaris*	Leguminosae	99[b]	25
Comfrey	*Symphytum* X *uplandicum*	Boraginaceae	96	
Echium	*Echium plantadinium*	Boraginaceae	37	
Japanese cherry	*Prunus serrulata splendens*	Rosaceae	40	
Hawthorn	*Crataegus monogyna*	Rosaceae	40	
Magnolia	*Magnolia* X *soulangeana*	Magnoliaceae	40	
Tulip tree	*Liriodendron tulipifera*	Magnoliaceae	40	
Hawkweed	*Hieracium* sp.	Compositae	40	
Dandelion	*Taraxacum officinalis*	Compositae	40	
Lettuce	*Lactuca sativa*	Compositae	99[b]	26
Salsify	*Tragopogon porrifolius*	Compositae	40	

continued

Table 1. *continued*

Common English name	Botanical name	Family	No. of residues sequenced	Reference
Angiosperms				
Thistle	*Cirsium vulgare*	Compositae	40	
Knapweed	*Centaurea niger*	Compositae	37	
Groundsel	*Senecio vulgaris*	Compositae	40	
Sunflower	*Helianthus annuus*	Compositae	40	
Niger	*Guizotia abyssinica*	Compositae	40	
Tansy	*Chrysanthemum vulgare*	Compositae	40	
Bindweed	*Convolvulus* sp.	Convolvulaceae	32	
Larkspur	*Delphinium ajacis*	Ranunculaceae	40	
Ribes	*Ribes sanguinium*	Saxifragaceae	40	
Chickweed	*Stellaria medica*	Caryophylaceae	40	
Tree mallow	*Lavatera arborea*	Malvaceae	37	
Ruta	*Ruta graveolans*	Rutaceae	40	
Nettle	*Urtica dioica*	Urticaceae	90	
Dog's mercury	*Mercurialis perenis*	Euphorbiaceae	99[b]	22
Willowherb	*Epilobium angustifolium*	Onagraceae	40	
Balsam	*Impatiens glandulifera*	Balsaminaceae	38	

[a]Where no reference is given, the plastocyanin was purified and the sequence data obtained at Durham.
[b]Complete sequence.

The crude extract is fractionated and the plastocyanin concentrated (Step 2) by precipitation with either cold acetone (30, 33, 34) or ammonium sulfate (27, 31, 32). The precipitated crude plastocyanin is then dialyzed and adsorbed onto a column of DEAE-cellulose. A variety of different buffers, e.g., phosphate, acetate, or Tris-HCl, has been used successfully for equilibration (10, 30, 32, 35). This column is then eluted with NaCl to separate the plastocyanin from ferredoxin and polyphenolic materials, which remain bound to the column (30, 33, 34). Elution may be directly with 0.2 M NaCl or, when large amounts of polyphenolic material are present with a stepwise elution pattern. This step may be followed by another fractionation with ammonium sulfate (27, 30, 34) or, alternatively, purification is achieved by means of an additional chromatographic step (28, 32, 33).

The final purification stage originally consisted of repeated chromatography on DEAE-cellulose columns (27), but the final yield of plastocyanin with this method was low. More recently, final purification has been achieved by gel filtration (30, 32–35), followed in some cases by a further ion exchange separation on DEAE-Sephadex (32). However, with plastocyanins from blue-green algae, CM-cellulose or CM-Sephadex may have to be used because these proteins may not necessarily be acidic like other plastocyanins (16, 36).

A procedure involving gel filtration rather than repeated ion exchange chromatography gives higher yields of purified plastocyanin, i.e. 10 mg/kg leaf material, fresh weight (30, 34) as compared to 3 mg/kg obtained by Katoh et al. (27). However, this method, unlike ion exchange methods, does not remove deamidation products or any apoplastocyanin which is present.

The total amount of plastocyanin in the chloroplast is about 30 μg of plastocyanin/mg of chlorophyll (1 mol of plastocyanin/300 mol of chlorophyll), and this ratio is obtained whether it is determined after isolation of the components (10) or by using a sensitive enzymic assay (37) or by EPR studies on intact chloroplasts (12).

The main impurities present in plastocyanin preparations are polyphenolic compounds with similar behavior to plastocyanin on both ion exchange chromatography and gel filtration (30, 34, 35), and these can only be satisfactorily separated from plastocyanin by repeated use of these methods. Another contaminant, which is often difficult to remove, is bound ferrocyanide (38), which originates from the potassium ferricyanide added in the early stages of purification to oxidize the protein. Its presence causes an increase in absorbance at 278 nm similar to that due to bound polyphenolic compounds. Apoprotein and deamidation products may also be present to variable extents depending on the initial extraction and concentration methods used.

Criteria of Purity

Because plastocyanin has no conveniently measured enzymic activity, the usual way of following the progress of the purification is to measure the absorbance ratio $A_{278} : A_{597}$ (27). The lowest value of this ratio so far reported for a pure plastocyanin is 0.8 for spinach (27), although typically values reported for other sources lie in the range 1.2–1.4 (Table 2). However, Katoh et al. (27) reported that samples with an 0.8 purity ratio were difficult to handle and that concentration of the sample by means of a small ion exchange column was sufficient to decrease the apparent purity to a 1.0 ratio. Various factors can influence the values obtained, and it is possible for a preparation to have an absorbance ratio as high as 2.0 and yet still appear perfectly homogeneous by other methods such as electrophoresis (32, 34, 38) and copper content (39).

High values may be caused not only by the presence of bound polyphenolic compounds or ferrocyanide but also by variations in the amount of aromatic amino acids present in the plastocyanin obtained from different sources (32, 34, 35). So far, most higher plant plastocyanins examined have been found to contain 3 residues of tyrosine and no tryptophan, but some species, e.g., nettle, contain an additional tyrosine residue. Tryptophan has been found in the plastocyanin isolated from *Chlorella fusca* (17). By carefully controlled chromatography on columns of DEAE-Sephadex, it is possible to remove any apoprotein which is present, and plastocyanin from a wide variety of species can be reproducibly purified to give a final ratio of 1.1.

PROPERTIES

Oxidation and Reduction

Plastocyanin can exist in the oxidized blue form and the reduced, colorless form; it is readily reduced, e.g., by ascorbic acid or sodium dithionite, and can be reoxidized by, for example, potassium ferricyanide (27, 28, 34, 46). It has also been noted that plastocyanin preparations can be reduced by exposure to light (47). The reduced form is not readily auto-oxidizable, and the protein does not bind oxygen or carbon monoxide (27, 34).

Katoh and Takamiya (48) showed, on the basis of its reaction with cupric and cuprous-specific chelating agents, that oxidized plastocyanin contains cupric copper, while reduced plastocyanin contains cuprous copper. Although this type of experiment is open to the criticism that the oxidation state of the metal may change during the reaction of the protein with the chelating agent, the result was subsequently confirmed by EPR spectroscopy (45). The change in valence state of the copper atom and hence of the charge on the protein during oxidation and reduction explains the observation of Wells (47) that the oxidized and reduced forms of plastocyanin could be separated by starch gel electrophoresis.

The oxidation-reduction potential of purified plastocyanin, measured spectrophotometrically, is +390 mV (*C. ellipsoidea*) (46); +370 mV (*Chlamydomonas reinhardtii*) (28); +370 mV (spinach) (27); and +350 mV (marrow) (34). The oxidation-reduction potential of membrane-bound plastocyanin in spinach chloroplasts determined by EPR measurements was +340 mV (49). This same method gave a value of +370 mV for purified spinach plastocyanin, and, although the determined values of the oxidation-reduction potential vary somewhat, it seems likely that in the membrane-bound state in the chloroplast, plastocyanin has a lower potential, thereby making its interaction with cytochrome *f* more favorable (49). Nernst plots have shown that the oxidation and reduction of plastocyanin involves a single electron.

The oxidation-reduction potentials of plastocyanin and the other type I copper proteins are all considerably higher than that of the hydrated ions Cu^{2+}/Cu^{+} (e.g., 50), and it has been suggested that this is due to a type of copper bonding which stabilizes the cuprous form (51, 52).

Kinetic studies of the reduction of plastocyanin by chromous ion (53) gave values for the rate constant k (25°C) between $1\text{–}3.3 \times 10^4$ M^{-1} s^{-1} and for $\Delta H\ddagger$ of about 1 kcal/mol at pH 4.2. Kinetic studies on the reduction of various blue copper proteins by $Fe(EDTA)^{2-}$ suggest that a normal outer sphere mechanism appears to operate in the reduction of plastocyanin (54). A nuclear magnetic resonance (NMR) investigation of electron transfer in bean plastocyanin has shown that the self-exchange between Cu(II) and Cu(I)-plastocyanin is slow on the NMR time scale, $k_{ex} \ll 2 \cdot 10^4$ M^{-1} s^{-1} at 50°C, but that electron transfer in the presence of ferricyanide is rapid, $k \gg 1 \cdot 10^5$ M^{-1} s^{-1} (55).

The effect of the pH value on the oxidation-reduction potential has been examined with the use of both spinach and marrow plastocyanins (27, 34); in each case, the potential remained constant over the pH range 6.5–9.9 (see, for example, Figure 1). At pH values below about 6, the potential increases at a limiting rate of 57 mV/pH unit, indicating that a single ionizable group with a pK of about 5.4 is closely involved in the copper binding site. The shape of the pH dependence curve is very similar to that found for copper imidazole complexes (51) and may indicate that histidine is a copper ligand in plastocyanins, especially as the pK of the change is within the range of pK values found for histidine in proteins.

Spectral Properties

The ultraviolet and visible absorbance spectrum of plastocyanin shows a protein absorbance peak centered at 278 nm, as well as characteristic peak at 597 nm and minor maxima at 775 nm and 460 nm. The peak in the ultraviolet absorbance region shows superimposed fine structure at 269, 265, 259, and 253 nm, probably due to phenylalanine, and at 284 nm due to tyrosine. These phenylalanine fine structure bands are absent or much reduced in magnitude in the spectra of plastocyanin from the green algae *C. reinhardtii* (28), *C. ellipsoidea* (46), and the blue-green alga *Anabaena variabilis* (36), although they are present in all higher plant plastocyanins. The absence of the fine structure bands may be due to the presence of tryptophan in the green algal proteins, although this amino acid has only been demonstrated in the plastocyanin from *C. fusca* (17). The presence of tryptophan would account for the fact that, in the green algal plastocyanins, the absorbance ratio ($A_{278} : A_{597}$) is higher than that of the

Table 2. Molecular properties of plastocyanins

Source	Molecular weight	Cu content		$A_{278}:A_{597}$	Reference
		%	Atom/molecule		
Chlamydomonas reinhardtii	13,000[a]			1.9	28
Chlorella fusca	10,410[b]			2.0	17
Anabaena variabilis	11,220[b]			1.57	16, 36
Spinacia oleracea (spinach)	21,000[c] }	0.58	2	0.8	
	10,800[d] }				27
	9,000[c] }				20
	10,480[b] }				
	40,000[e]				40
Sambucus nigra (elder)	15,900[d]	0.4	1	1.1	32
	11,300[a]				19
	10,800[e]				
	10,520[b]				
Cucurbita pepo (marrow)	10,600[c]	0.56	1	1.15	34
	11,400[d]				21
	12,200[a]				
	10,580[b]				
Phaseolus vulgaris (French bean)	10,780[c]	0.51	1	1.1	.35
	12,500[d]				25
	10,800–14,100[a]				
	10,590[b]				
Urtica dioica (nettle)	14,300[a]	0.44	1	1.2	32
	11,300[a]				
	10,600[e]				

Mercurialis perennis (dog's mercury)	11,900[d]	0.53	1	1.3	32
	11,300[a]				
	10,500[e]				
Galium aparine (goosegrass)	11,300[a]			1.7	32
	9,200[e]				
Symphytum officinale (comfrey)	13,400[d]	0.47	1	1.3	32
	11,300[a]				
	10,600[e]				
Lactuca sativa (lettuce)	11,300[a]			1.2	32
	10,300[e]				
Zea sp. (maize)	14,000[d]	0.45	1	2.0	41
	23,000[a]				
Petroselinum sp. (parsley)	7,200–12,900[c]		1	2.1	38
	10,100[d]				
Triticum sp. (wheat)	27,600[d]	0.23			42
	13,800[d]	0.46	1		33, 43
	15,000[a]				
Cucumis sativus (cucumber)	11,000[d]	0.57	1	1.6	44
	13,000–16,000[a]				
Chenopodium album (fat hen)	11,500[f]	1.1	2		45

[a]Molecular weight determined from gel filtration.
[b]Molecular weight determined by amino acid sequence.
[c]Molecular weight determined by ultracentrifugation.
[d]Molecular weight determined by copper content.
[e]Molecular weight determined by SDS-gel electrophoresis.
[f]Method not given.

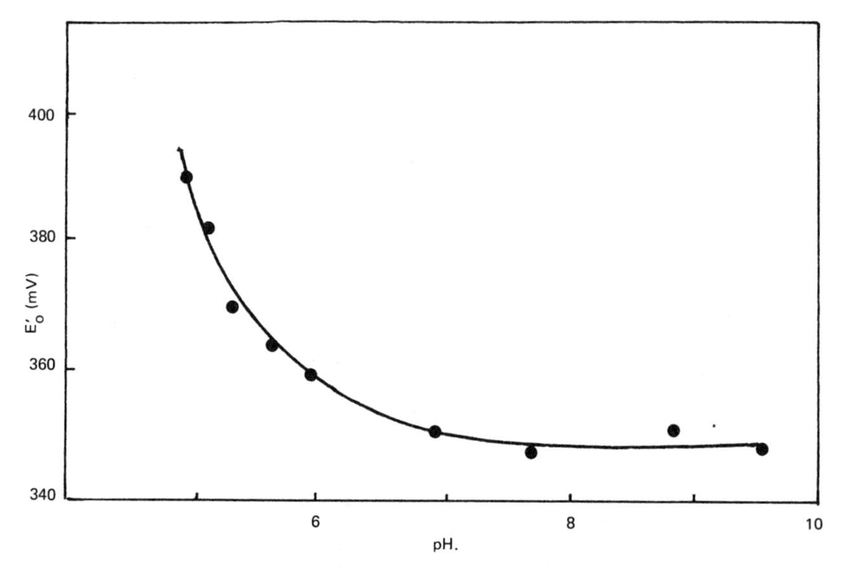

Figure 1. Effect of pH on the oxidation-reduction potential of marrow plastocyanin. Data from Scawen et al. (34) and M. D. Scawen, unpublished observations.

higher plant plastocyanins (Table 2). However, for the blue-green alga *Anabaena,* sequence studies have demonstrated the absence of tryptophan (16), which suggests that other reasons for these differences may also exist.

A peak at 597 nm is present in all oxidized plastocyanins, and in every case the extinction coefficient is 4.5×10^3–4.8×10^3 liter mol^{-1} cm^{-1} (Table 3). Similarly, the minor maxima at 785 and 460 nm are present in all the plastocyanins studied. All three visible absorbance peaks are lost on reduction of the protein (27), as well as on treatment with 6 M guanidinium chloride (34, 35), pyridine, diethylamine (56), *p*-chloromercuribenzoate, and mercuric acetate (34, 48). The fluorescence maximum of plastocyanin at 315 nm when excited at 275 nm is probably caused by the phenolic side chain of some or all of the tyrosine residues. The excitation spectrum is structureless and shows a maximum at 275 nm (38).

The visible optical rotatory dispersion spectrum of plastocyanin from *Chenopodium album* (45) consists of three major Cotton effects centered at each of the three optical absorbance maxima, showing that the copper coordination center is asymmetric. No optical rotatory dispersion or circular dichroic spectra in the ultraviolet region have been obtained for any plastocyanin, although such studies have been carried out on other "blue" copper proteins, e.g., umecyanin (66) and azurin (67).

Electron paramagnetic resonance spectra have been determined for plastocyanin from *C. album,* marrow, spinach, parsley, and pea (34, 38, 45, 57, 68). In each case, the $g_{\parallel}$, $g_{\perp}$, and $A_{\parallel}$ values are similar (Table 3) and suggest that the

Table 3. The optical and EPR properties of plastocyanin and some other type I copper proteins

Protein	Source	λ max	$10^{-3} \times \epsilon$ at λ_{max}	$A\|\|(cm^{-1})$	$g\|\|$	$g\perp$	Reference
Plastocyanin	*Spinacia oleracea* (spinach)	597	4.9	0.005	2.23	2.05	27
	Chenopodium album (fat hen)	597	4.7	0.006	2.23	2.05	45
	Cucumis sativus (cucumber)	597	4.1	0.006	2.226	2.060	44
	Cucurbita pepo (marrow)	597	4.7	0.006	2.23	2.05	34
	Petroselinum sp. (parsley)	597	5.0	0.007	2.23	2.06	38
	Pisum sp. (pea)	597		0.006	2.211	2.045	57
Azurin	*Pseudonomas aeruginosa*	625	3.5	0.006	2.260	2.052	58
	Pseudomonas fluorescens	625	3.5	0.006	2.261	2.052	58
	Bordetella bronchiseptica	625	3.5	0.006	2.273	2.05	59, 60
Stellacyanin	*Rhus vernicifera*	604	4.08	0.0035	2.287	2.025[a] 2.077[b]	61
Umecyanin	*Armaracia lapathifolia*	610	3.4	0.0035	2.317	2.05	62, 63
Blue protein	*Oryza sativa* (rice bran)	600	4.4	0.002		2.09	64
Plantacyanin	*Cucumis sativus* (cucumber)	592	0.9	0.005	2.232	2.114	44
	Spinacia oleracea (spinach)	593	0.8				65

[a]Value for g_x.
[b]Value for g_y.

copper ion may be in a distorted tetragonal binding site with an out-of-plane distortion of around 12.5° (69). The low values for $g\|$ compared to purely ionic Cu^{2+} complexes may indicate that the copper bonding in plastocyanin is partly covalent. The hyperfine splitting constant ($A\|$) is unusually low compared with typical copper complexes with low molecular weight chelating agents, and this may also be due to some covalency in the copper bonding. In the native protein, there is no super-hyperfine structure.

The optical and magnetic properties of plastocyanin are typical of those found in a number of other copper proteins which are also intensely blue in color. These proteins have been classified as type I copper proteins (70, 71), and they are characterized by an intense absorbance maximum at about 600 nm, with extinction coefficients in the range of 3,400–5,700/g atom of type I copper and very low values ($<10 \times 10^{-3}$ cm^{-1}) for the hyperfine splitting constant ($A\|$). The values of $g\|$ for type I copper proteins, although more variable, are all lower than the values found for copper complexes with EDTA and other chelating agents (72, 73). However, cusacyanin, while having an $A\|$ value typical of type I copper proteins, has a much lower extinction coefficient of 900 (44).

The other major group of copper proteins is called type II copper proteins. These proteins, e.g., monoamine oxidase and superoxide dismutase (74), are colorless or nearly so and have little absorbance at about 600 nm, similar to low molecular weight copper complexes which typically have extinction coefficients of about 100 (75). The hyperfine splitting constant for this type of protein copper is greater than 14×10^{-3} cm^{-1}, more like the values found for low molecular weight copper complexes (72, 73). Some copper proteins like ascorbate oxidase (76), ceruloplasmin (77), and laccase (61) contain both types of copper, in which case their blue color is due entirely to the type I copper content (71, 77).

Stability, Effects of Denaturing Agents, and Surface Properties

Plastocyanin is a fairly stable protein, and there is little deterioration even after several months when stored frozen at −20°C in the oxidized form (30, 37). However, repeated freezing and thawing of the solution does lead to deterioration (39), probably through irreversible copper loss. At room temperature and at 40°C, parsley plastocyanin is less stable, particularly when in the reduced state (38). Spinach plastocyanin appears stable to heat; 5-min incubations at temperatures up to 45°C do not lead to changes in the absorption spectrum, and increasing the temperature to 60°C did not lead to irreversible loss of blue color or to denaturation (48).

Plastocyanin is irreversibly denatured by 6 M guanidinium chloride (34, 35); the denaturation is accompanied by the loss of the phenylalanine fine structure bands in the ultraviolet spectrum, although there is no change in extinction coefficient. This effect is similar to that of exposure of the protein to high pH values (>12) and may be caused by aromatic residues, which normally exist in a hydrophobic environment, being exposed to solvent; this denaturation process leads to the complete loss of the absorbance at 597 nm.

Urea will also denature plastocyanin, although it is not as effective a denaturant as guanidinium chloride (35). However, 4 M urea caused a parallel loss of absorbance at 597 nm and of the EPR signal with pea plastocyanin (56). This was apparently due to reduction of the copper atom because the addition of potassium ferricyanide immediately restored both the absorbance at 597 nm and the EPR signal, although after 12–16 hr of treatment the changes were irreversible (56).

Treatment with alkali to give a pH value of 11 caused a slow decolorization of the protein and gave rise to minor changes in the shape and intensity of the EPR spectrum (38, 48). Again, these changes could be reversed by the addition of potassium ferricyanide, at least for a short time after the adjustment in pH (38).

At pH values >11.5 or in 30% pyridine or in diethylamine, the EPR spectrum of the protein was similar to that of the copper-biuret complex, with well resolved nitrogen super-hyperfine lines (56). Other solvents such as dimethylsulfoxide, formamide, and dimethylformamide had no effect on either the visible or EPR spectra of plastocyanin (56).

Spinach plastocyanin is very sensitive to lowered pH values and is rapidly and irreversibly decolorized and precipitated at pH 4.6 and below (48). Higher plant plastocyanins have a low isoelectric point (about pH 4 (32)), which precludes the use of isoelectric focusing as a means of purification because the protein is unstable at its isoelectric point, particularly in a medium of low ionic strength. Plastocyanin from the blue-green alga *A. variabilis,* however, has a much higher pI value, because CM-cellulose must be used to purify the protein (36).

After acid denaturation of plastocyanin, the copper atom is still in an axial environment, although the hyperfine splitting constant of wheat plastocyanin was increased from 6×10^{-3} cm^{-1} to 14×10^{-3} cm^{-1} (56).

Brody (78, 79) has studied the properties of plastocyanin in monomolecular films. The area/molecule (A) and surface potential (ΔV) of reduced plastocyanin was measured as a function of pH. At pH 7.5, a maximum was observed for A and a minimum observed for ΔV. At a surface pressure of 5 dyn/cm, the maximum A_5 = 378 Å^2 and minimum ΔV_5 = 335 mV. Oxidized plastocyanin and chlorophyll interacted in a mixed film possibly forming a complex. Upon irradiation of the film in air, the surface potential and area of the films decreased significantly, indicating a photoreaction between the chlorophyll and oxidized plastocyanin. Mixed films of reduced plastocyanin and chlorophyll also showed an interaction between these components, and, upon irradiation of the film in nitrogen, the area of the film decreased, whereas the surface potential increased, a result consistent with the conversion of reduced plastocyanin to the oxidized form.

Preparation of Apoplastocyanin

Apoplastocyanin was first prepared by Katoh and Takamiya (48), by precipitation of spinach plastocyanin in 40% saturation ammonium sulfate solution at pH 2. The precipitate was free of copper, but when it was redissolved at pH 7 it reacted with added Cu^{2+} to form the holoprotein. Graziani et al. (38) prepared apoplastocyanin by dialyzing the reduced protein against sodium cyanide at neutral pH. These workers found that the apoprotein was very unstable and was particularly sensitive to oxidation, presumably because of the presence of an unprotected thiol group; this was also found by Milne and Wells (35). The apoprotein had ultraviolet absorbance and fluorescence spectra similar to that of the native protein; it reacted slowly with added Cu^{2+} to form the holoprotein, although an excess of copper caused the protein to precipitate.

Scawen et al. (34) prepared apoplastocyanin by treating the native protein with mercuric acetate, which presumably displaced the copper from the thiol group; the mercury atom was then removed by treatment with reduced glutathione. The apoprotein was stable in the presence of the thiol reagent and reacted with added copper (Cu^{2+}) to form the holoprotein, which could be oxidized by potassium ferricyanide, after removal of the reduced glutathione. The ease with which the apoprotein can be prepared is of interest because it has proved possible with both stellacyanin, azurin, and plastocyanin to replace the copper atom by cobalt (80, 81), thus giving information about the possible nature of the metal ligands.

STRUCTURE

General Characteristics

Studies have been made on the molecular weight and copper content of plastocyanins purified from a wide range of species (Table 2). The first plastocyanin to be examined in detail was that from spinach (27); the molecular weight of this plastocyanin was determined as 21,000, and because the copper content was 0.58%, this suggested 2 copper atoms per molecule. Analyses of plastocyanins from different sources, however, have resulted in a great variety of values for the determined copper content and molecular weight. For example, Blumberg and Peisach (45) reported that plastocyanin from *C. album,* a plant from the same family as spinach, had a molecular weight of 11,500 and contained 2 copper atoms. Gorman and Levine (28), on the other hand, working with plastocyanin from the green alga *C. reinhardtii,* reported that this protein had a molecular weight of 13,000 ± 2,000 and contained 1 copper atom per molecule. Plastocyanin from wheat was originally reported to have a copper content of 0.23%, which is equivalent to a molecular weight of about 27,000 (42). A later publication by the same group, which used a sample that was probably much purer than that used originally, gave a copper content for wheat plastocyanin of 0.46%; this is equivalent to a molecular weight of 13,700 per copper atom (43).

This value is more consistent with the value of 15,000, which these workers found using gel filtration chromatography (33). Plastocyanin from corn, another monocotyledonous plant like wheat, is reported to have a copper content of 0.45% (41); this would give a molecular weight of 14,000 per copper atom. However, when this protein was examined by chromatography on Sephadex G-100 there were two plastocyanin bands, which eluted in volumes corresponding to molecular weights of 45,000 and 23,000 (41). Recently, the plastocyanins from French bean, parsley, and marrow have been studied in great detail (34, 38, 82). For all of these proteins, a molecular weight of about 10,500 was established by a variety of methods, and in each case the copper content was 1 atom per molecule. After further studies on several other plastocyanins from plants of broad taxonomic distribution, Ramshaw et al. (32) concluded that all higher plant plastocyanins probably had a molecular weight of about 10,500 and contained 1 copper atom per molecule. Various methods were used in this study, including sodium dodecylsulfate (SDS)-gel electrophoresis, and in many cases the molecular weight value has subsequently been confirmed by amino acid sequence studies (18–21, 23, 24).

In the light of more recent information, some of the apparently anomalous molecular weight values may be explained. The amino acid sequence of spinach has been completed and gives a molecular weight of 10,415 (20). This value was checked independently by sedimentation equilibrium analysis, which gave a molecular weight of 9,000. The discrepancy between this value and that originally reported by Katoh et al. (27) probably arises from the value obtained by Katoh et al. (27) for the diffusion coefficient, $D_{20,w}$. Katoh et al. (27) determined this coefficient as 6.6×10^{-7} cm^2 s^{-1}, which is typical of a globular protein of about 70,000 molecular weight (83). In contrast, the $D_{20,w}$ value for marrow plastocyanin was reported to be 14.6×10^{-7} cm^2 s^{-1} (34, 82), whereas that for parsley plastocyanin varied between 6 and 11×10^{-7} cm^2 s^{-1} in different experiments (38), the variations being attributed to protein instability. The sedimentation coefficients, $S_{20,w}$, for these proteins, on the other hand, are all very similar; 1.72 S for spinach, 1.4 S for parsley, and 1.69 S for marrow (27, 34, 38, 82).

An alternative explanation of the high molecular weight value obtained by Katoh et al. (27) for spinach plastocyanin is that dimerization had occurred. Milne and Wells (35) considered the possibility of a monomer-dimer interacting system for French bean plastocyanin, but found no evidence of it. Siegelman et al. (40), however, have obtained a value of 40,000 for the molecular weights of spinach and *Scenedesmus* plastocyanins by using SDS-gel electrophoresis under nonreducing conditions. However, when heat denaturation was used in addition, bands of 20,000 and 10,000 molecular weights were also observed. The apoprotein obtained by removal of Cu at pH 2 (48) or reduction with dithiothreitol under strong denaturing conditions had a molecular weight of 10,000. Attempts to reconstitute the plastocyanin by addition of Cu^{2+} to the apoprotein gave a 40,000 molecular weight tetramer which was active in a biological assay. Although this tetramer went blue on oxidation, the EPR spectrum gave a value for $A\|$ of 157 G (0.015 cm^{-1}), which is more typical of nonblue type II copper proteins (see above). It has yet to be demonstrated that native, rather than denatured or reconstituted, plastocyanin is present or active in a tetrameric form. From the amino acid sequence data now available for spinach (see below), it is clear that a tetrameric molecule would contain 4 copper atoms and 4 polypeptide chains with essentially identical amino acid sequences.

In some cases, the apparently anomalous values for the molecular weight may be explained in terms of the experimental methods used. The molecular weight values reported for wheat (33), corn (41), and *Chlamydomonas* (28) plastocyanins were all determined from elution volumes in gel filtration experiments. When these values are compared with those of other species, they are all high. Milne and Wells (35) have shown that the nature of the buffer used in gel filtration experiments can dramatically alter the elution volume of French bean plastocyanin, apparently without detectable protein association. When they employed the buffers used for the wheat and *Chlamydomonas* plastocyanin determinations with French bean plastocyanin, both buffers gave larger molecular weight values than other systems. This may also be the explanation of the high molecular weight obtained for corn plastocyanin (41), although it does not explain the presence of two corn plastocyanin bands, which may indicate that a stable dimer does exist for this species.

All the plastocyanins examined show 1 copper atom per molecular weight of about 10,000 (Table 2), except that from *Chenopodium album* (45); the different values obtained with spinach have been discussed earlier. The anomalous value for *Chenopodium album* plastocyanin may arise from an analytical error, because the first 40 residues are clearly homologous with other higher plant sequences (Boulter et al., unpublished observations), but no details of either the preparation or the analysis were given. If the protein does contain 2 copper atoms, then its extinction coefficient is considerably less than for other plastocyanins.

There is little evidence on whether plastocyanin has any carbohydrate as part of its structure, although some reports have mentioned the presence of small quantities of associated carbohydrate. Katoh et al. (27) found that with spinach plastocyanin, hydrolysis by 1 N HCl at 100°C for 4 hr, followed by analysis of the hydrolysate using paper chromatography, revealed the presence of four carbohydrate components. Two of these were identified as glucose and arabinose, another was possibly hexosamine, while the fourth component was unidentified. Using similar qualitative analytical methods, Mutuskin et al. (33) found the same four components associated with wheat plastocyanin. Katoh et al. (27) suggested that the carbohydrate content in spinach was 5.4% (w/w), but marrow plastocyanin contained only 1.0% anthrone-reactive carbohydrate, and there was no detectable hexosamine present (34). It is also possible that any carbohydrate present is labile, because Kelly and Ambler (84) reported that during sequence analysis some unidentified labile group, which could be carbohydrate, was indicated in the plastocyanin of *C. fusca.*

The calculated minimum carbohydrate content of spinach and wheat plastocyanins is 6.4%, based on the presence of four different sugar moieties and a 99-residue polypeptide chain (see below).

Only Katoh et al. (27) have given a quantitative value for carbohydrate content, quoting 5.4% (w/w) for spinach. This value was not derived from quantitative sugar analyses but apparently from the incomplete recovery of material on amino acid analysis; this in itself cannot be regarded as a very accurate or reliable method. When the amino acid analysis results which they used were compared to the known sequence (20), several differences are found; notable are the low recoveries of the more labile amino acids, for which no corrections for destruction during hydrolysis were made. This suggests that the recovery value given by Katoh et al. (27) for the amino acid analysis may be too low and, therefore, the suggested sugar content too high and, therefore, inconsistent with the calculated minimum required by the qualitative studies.

Clearly, the carbohydrate content of plastocyanin needs further investigation. Present evidence that there is more than zero or 1 residue of carbohydrate per molecule is not strong. If, as suggested by Kelly and Ambler (84), a labile group, which may be carbohydrate, is associated with the protein, then even under mild purification conditions it may often be lost.

Amino Acid Sequence

Amino Acid Composition Amino acid compositions for several plastocyanins were published (27, 33, 35, 38, 82) prior to any complete amino acid sequences being reported, but now sequence data are also available (see Figure 2).

All the higher plant plastocyanins so far examined, with the possible exception of wheat, lack both arginine and tryptophan residues. Arginine is also absent from the algal proteins, but some algal plastocyanins contain tryptophan. In all cases, there is a predominance of acidic over basic residues, which accounts for the low, acidic pI observed (see above). Similar amounts of glycine and of proline occur in all plastocyanins, and all have a fairly large content of hydrophobic residues, notably valine. All show the presence of a minimum of 2 histidine residues and all have a single cysteine residue. In none of the analyses was there any evidence of glucosamine or of any unusual amino acids being present.

Amino Acid Sequence Studies The initial work on the primary structure of French bean plastocyanin (35) indicated that it consisted of a single polypeptide chain of either 99 or 100 residues. Kelly and Ambler (84) established unequivocally the presence of a unique amino acid sequence of 98 residues, indicating a single polypeptide chain with the plastocyanin of *C. fusca.* Subsequently, completion of the French bean (25) and the determination of other higher plant plastocyanin sequences (18–21, 23, 24) has shown that plastocyanin from higher

```
                -2  -1   1                                    10                                                20                                                30
Anabaena:       Glu-Thr-Tyr-Thr-Val-Lys-Leu-Gly-Ser-Asp-Lys-Gly-Leu-Leu-Val-Phe-Glu-Pro-Ala-Lys-Leu-Thr-Ile-Lys-Pro-Gly-Asp-Thr-Val-Glu-Phe-Leu-Asn-Asn-Lys-Val-
Chlorella:          Asp-Val-Thr-Val-Lys-Leu-Gly-Ala-Asp-Ser-Gly-Ala-Leu-Val-Phe-Glu-Pro-Ser-Ser-Val-Thr-Ile-Lys-Ala-Gly-Glu-Thr-Val-Thr-Trp Val-Asn-Asn-Ala-Gly-
Rumex:                  Ile-Glu-Ile-Lys-Leu-Gly-Gly-Asp-Asp-Gly-Ala-Leu-Ala-Phe-Val-Pro-Gly-Ser-Phe-Thr-Val-Ala-Ala-Gly-Glu-Lys-Ile-Val-Phe-Lys-Asn-Asn-Ala-Gly-
Sambucus:               Val-Glu-Ile-Leu-Leu-Gly-Gly-Glu-Asp-Gly-Ser-Leu-Ala-Phe-Ile/Val-Pro-Ser/Gly-Asn-Phe-Ser-Val-Pro-Ser-Gly-Glu-Lys-Ile-Thr-Phe-Lys-Asn-Asn-Ala-Gly-
Spinacia:               Val-Glu-Val-Leu-Leu-Gly-Gly-Gly-Asp-Gly-Ser-Leu-Ala-Phe-Leu-Pro-Gly-Asp-Phe-Ser-Val-Ala-Ser-Gly-Glu-Glu-Ile-Val-Phe-Lys-Asn-Asn-Ala-Gly-
Cucurbita:              Ile-Glu-Val-Leu-Leu-Gly-Gly-Asp-Asp-Gly-Ser-Leu-Ala-Phe-Ile-Pro-Asn-Asp-Phe-Ser-Val-Ala-Ala-Gly-Glu-Lys-Ile-Val-Phe-Lys-Asn-Asn-Ala-Gly-
Capsella:               Ile-Glu-Val-Leu-Leu-Gly-Gly-Gly-Asp-Gly-Ser-Leu-Ala-Phe-Val-Pro-Asn-Asp-Phe-Ser-Ile-Ala-Lys-Gly-Glu-Lys-Ile-Val-Phe-Lys-Asn-Asn-Ala-Gly-
Solanum:                Leu-Asp-Val-Leu-Leu-Gly-Gly-Asp-Asp-Gly-Ser-Leu-Ala-Phe-Ile-Pro-Gly-Asn-Phe-Ser-Val-Ser-Ala-Gly-Glu-Lys-Ile-Thr-Phe-Lys-Asn-Asn-Ala-Gly-
Vicia:                  Val-Glu-Val-Leu-Leu-Gly-Ala-Ser-Asp-Gly-Gly-Leu-Ala-Phe-Val-Pro-Asn-Ser-Phe-Glu-Val-Ser-Ala-Gly-Asp-Thr-Ile-Val-Phe-Lys-Asn-Asn-Ala-Gly-
Phaseolus:              Leu-Glu-Val-Leu-Leu-Gly-Ser-Gly-Asp-Gly-Ser-Leu-Val-Phe-Val-Pro-Ser-Glu-Phe-Ser-Val-Pro-Ser-Gly-Glu-Lys-Ile-Val-Phe-Lys-Asn-Asn-Ala-Gly-
Mercurialis:            Leu-Asp-Val-Leu-Leu-Gly-Ser-Asp-Asp-Gly-Glu-Leu-Ala-Phe-Val-Pro-Asn-Asn-Phe-Ser-Val-Pro-Ser-Gly-Glu-Lys-Ile-Thr-Phe-Lys-Asn-Asn-Ala-Gly-
Lactuca:                Ala-Glu-Val-Leu-Leu-Gly-Ser-Ser-Asp-Gly-Gly-Leu-Val-Phe-Glu-Pro-Ser-Thr-Phe-Ser-Val-Ala-Ser-Gly-Glu-Lys-Ile-Val-Phe-Lys-Asn-Asn-Ala-Gly-

                    40                                      50                                              60                                              70
Pro-Pro-His-Asn-Val-Val-Phe-Asp-Ala-Ala-Leu-Asn-Pro-Ala-Lys-Ser-Ala-Asp-Leu-Ala-Lys-Ser-Leu-Ser-His-Lys-Gln-Leu-Leu-Met-Ser-Pro-Gly-Gln-Ser-Thr-Ser-Thr-Thr-Phe+
Phe-Pro-His-Asn-Ile-Val-Phe-Asp-Glu-Asp-Glu-Val-Pro-Ser-Gly-Ala-Asn-Ala-Glu-Ala-Leu-Ser- * - * -His-Glu-Asp-Tyr-Leu-Asn-Ala-Pro-Gly-Glu-Ser-Tyr-Ser-Ala-Lys-Phe-
Phe-Pro-His-Asn-Ile-Val-Phe-Asp-Glu-Asp-Glu-Val-Pro-Ala-Gly-Val-Asp-Ala-Ser-Lys-Ile-Ser-Met-Ser-Glu-Glu-Asp-Leu-Leu-Asn-Ala-Pro-Gly-Glu-Thr-Tyr-Ala-Val-Thr-Leu-
Phe-Pro-His-Asn-Val-Val-Phe-Asp-Glu-Asp-Glu-Val-Pro-Ser-Gly-Val-Asp-Ser-Ala-Lys-Ile-Ser-Met-Ser-Glu-Asp-Asp-Leu-Leu-Asn-Ala-Pro-Gly-Glu-Thr-Tyr-Ser-Val-Thr-Leu-
Phe-Pro-His-Asn-Val-Val-Phe-Asp-Glu-Asp-Glu-Ile-Pro-Ser-Gly-Val-Asp-Ala-Ala-Lys-Ile-Ser-Met-Ser-Glu-Glu-Asp-Leu-Leu-Asn-Ala-Pro-Gly-Glu-Thr-Tyr-Lys-Val-Thr-Leu-
Phe-Pro-His-Asn-Val-Val-Phe-Asp-Glu-Asp-Glu-Ile-Pro-Ser-Gly-Val-Asp-Ala-Gly-Lys-Ile-Ser-Met-Asn-Glu-Glu-Asp-Leu-Leu-Asn-Ala-Pro-Gly-Glu-Val-Tyr-Lys-Val-Asn-Leu-
Phe-Pro-His-Asn-Val-Val-Phe-Asp-Glu-Asp-Glu-Ile-Pro-Ser-Gly-Val-Asp-Ala-Ser-Lys-Ile-Ser-Met-Asp-Glu-Asn-Asp-Leu-Leu-Asn-Ala-Ala-Gly-Glu-Thr-Tyr-Glu-Val-Ala-Leu-
Phe-Pro-His-Asn-Val-Val-Phe-Asp-Glu-Asp-Glu-Ile-Pro-Ala-Gly-Val-Asp-Ala-Ser-Lys-Ile-Ser-Met-Ala-Glu-Glu-Asp-Leu-Leu-Asn-Ala-Ala-Gly-Glu-Thr-Tyr-Ser-Val-Thr-Leu-
Phe-Pro-His-Asn-Val-Val-Phe-Asp-Glu-Asp-Glu-Ile-Pro-Ser-Gly-Val-Asp-Ala-Ala-Lys-Ile-Ser-Met-Pro-Glu-Glu-Asp-Leu-Leu-Asn-Ala-Pro-Gly-Glu-Thr-Tyr-Ser-Val-Lys-Leu-
Phe-Pro-His-Asn-Val-Val-Phe-Asp-Glu-Asp-Glu-Ile-Pro-Ala-Gly-Val-Asp-Ala-Val-Lys-Ile-Ser-Met-Pro-Glu-Glu-Glu-Leu-Leu-Asn-Ala-Pro-Gly-Glu-Thr-Tyr-Val-Val-Thr-Leu-
Phe-Pro-His-Asn-Val-Val-Phe-Asp-Glu-Asp-Glu-Ile-Pro-Ser-Gly-Val-Asp-Ala-Ser-Lys-Ile-Ser-Met-Asp-Glu-Ala-Asp-Leu-Leu-Asn-Ala-Pro-Gly-Glu-Thr-Tyr-Ala-Val-Thr-Leu-
Phe-Pro-His-Asn-Val-Val-Phe-Asp-Glu-Asp-Glu-Ile-Pro-Ala-Gly-Val-Asp-Ala-Ser-Lys-Ile-Ser-Met-Ser-Glu-Glu-Asp-Leu-Leu-Asn-Ala-Pro-Gly-Glu-Thr-Tyr-Ala-Val-Thr-Leu-

                    80                                      90                                              100
Asp-Ala‡-Ala-Gly-Glu-Tyr-Thr-Phe-Tyr-Cys-Glu-Pro-His-Arg-Gly-Ala-Gly-Met-Val-Gly-Lys-Ile-Thr-Val-Ala-Gly
Asp-Thr-Ala-Gly-Thr-Tyr-Gly-Tyr-Phe-Cys-Glu-Pro-His-Gln-Gly-Ala-Gly-Met-Lys-Gly-Thr-Ile-Thr-Val-Gln
Ser-Glu-Lys-Gly-Thr-Tyr-Ser-Phe-Tyr-Cys-Ser-Pro-His-Gln-Gly-Ala-Gly-Leu-Val-Gly-Lys-Val-Thr-Val-Gln
Thr-Glu-Ser-Gly-Thr-Tyr-Lys-Phe-Tyr-Cys-Ser-Pro-His-Gln-Gly-Ala-Gly-Met-Val-Gly-Lys-Val-Thr-Val-Asn
Thr-Glu-Lys-Gly-Thr-Tyr-Lys-Phe-Tyr-Cys-Ser-Pro-His-Gln-Gly-Ala-Gly-Met-Val-Gly-Lys-Val-Thr-Val-Asn
Thr-Glu-Lys-Gly-Ser-Tyr-Ser-Phe-Tyr-Cys-Ser-Pro-His-Gln-Gly-Ala-Gly-Met-Val-Gly-Lys-Val-Thr-Val-Asn
Thr-Glu-Ala-Gly-Thr-Tyr-Ser-Phe-Tyr-Cys-Ala-Pro-His-Gln-Gly-Ala-Gly-Met-Val-Gly-Lys-Val-Thr-Val-Asn
Ser-Glu-Lys-Gly-Thr-Tyr-Thr-Phe-Tyr-Cys-Ala-Pro-His-Gln-Gly-Ala-Gly-Met-Val-Gly-Lys-Val-Thr-Val-Asn
Asp-Ala-Lys-Gly-Thr-Tyr-Lys-Phe-Tyr-Cys-Ser-Pro-His-Gln-Gly-Ala-Gly-Met-Val-Gly-Gln-Val-Thr-Val-Asn
Asp-Thr-Lys-Gly-Thr-Tyr-Ser-Phe-Tyr-Cys-Ser-Pro-His-Gln-Gly-Ala-Gly-Met-Val-Gly-Lys-Val-Thr-Val-Asn
Thr-Glu-Lys-Gly-Ser-Tyr-Ser-Phe-Tyr-Cys-Ser-Pro-His-Gln-Gly-Ala-Gly-Met-Val-Gly-Lys-Val-Thr-Val-Asn
Thr-Glu-Lys-Gly-Thr-Tyr-Ser-Phe-Tyr-Cys-Ala-Pro-His-Gln-Gly-Ala-Gly-Met-Val-Gly-Lys-Val-Thr-Val-Asn
```

Figure 2. Complete sequences of plastocyanin. Numbering is with higher plant NH_2-terminus as 1. See Table 1 for references. *, no residue; +, insert –Pro–Ala– here (see Aitken (16)); ‡, insert –Pro– here (see Aitken (16)).

plant sources is a single polypeptide chain of 99 residues, whereas that from the blue-green algae, *A. variabilis,* has 105 residues (16).

The sequence of plastocyanin from potato, which can be considered typical of higher plant plastocyanins, is given in Figure 3. Because there is no NH_2-terminal blocking group, automatic sequencing methods (85) are particularly useful in studying this protein. With as little as 300 nmol of protein, the NH_2-terminal 40 residues of the sequence can be readily established.

Milne and Wells (35) showed that French bean plastocyanin contained 2 methionine residues and that specific cleavage of the polypeptide at these residues by cyanogen bromide (86) gave three fragments which were readily separated by gel filtration. The majority of higher plant plastocyanins also contain 2 methionine residues (Figure 2), an exception being dock plastocyanin (18). Thus, a similar strategy provides a good starting point for sequencing this protein. As shown by the potato plastocyanin sequence (Figure 3), the methionine residues occupy positions 57 and 92; therefore, fragments of 57, 35, and 7 residue length are obtained after cyanogen bromide cleavage. This range in size readily explains their ease of purification by gel filtration. NH_2-terminal and COOH-terminal analyses establish the order of these fragments in the sequence. The sequences of the purified fragments were established by conventional enzymic cleavage followed by dansylphenylisothiocyanate analysis of purified peptides. The enzyme specificities observed during the examination of the potato plastocyanin are shown in Figure 3. Examination of the largest fragment, much of whose sequence is known from an automatic analysis of the intact protein, is most usefully accomplished by the use of thermolysin and papain digestions. Although trypsin cleaves at lysine residues 26, 30, and 54, the fragments are difficult to purify; similarly, the peptides from chymotryptic digestion of this fragment were also difficult to purify and, in many cases, were present in low yield (24).

The second fragment can be digested to give useful peptides by several enzymes, and thermolysin, trypsin, and chymotrypsin have all been used successfully. This fragment can also be partially sequenced automatically, by use of a "peptide" program (e.g., Beckman Prog. No. 071872, as supplied with the automatic sequencer). The smallest fragment can be sequenced manually. The complete sequences of plastocyanin from several different sources have now been determined and, in most cases, the sequencing strategy outlined above has been used (19–21, 23, 24). These sequences have been aligned in Figure 2, from which it can be seen that a large amount of similarity exists, despite the wide range of sources. The sequences from algae, however, both differ from the higher plant plastocyanins. The *Chlorella* sequence shows an additional residue at the NH_2 terminus, and there is a 2-residue deletion around residues 56 and 57 in the sequence (17). The *Anabaena* sequence has 2 additional residues at its NH_2 terminus, 3 additional residues around residue 74, and 1 extra residue at the COOH terminus (see Figure 2) (16). However, the high degree of similarity

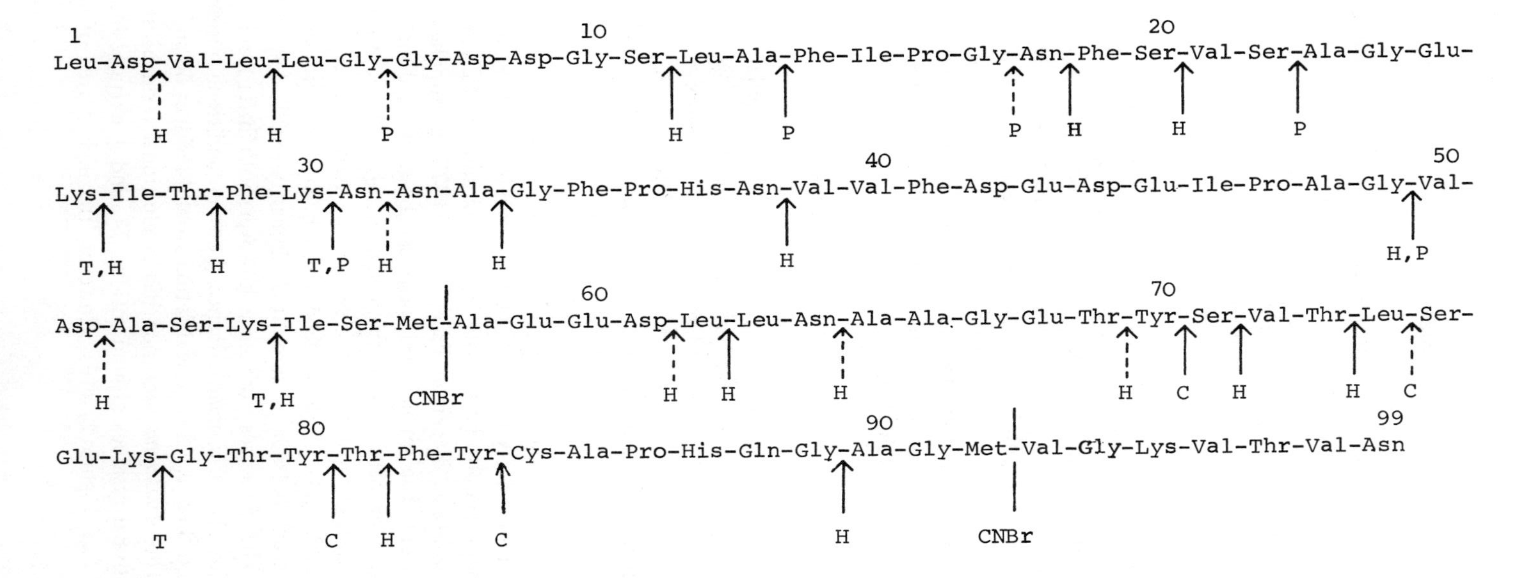

Figure 3. Amino acid sequence of potato plastocyanin showing observed enzymic cleavages of the cyanogen bromide fragments (23). CNBr, cyanogen bromide cleavage; T, trypsin; H, thermolysin; P, papain; C, chymotrypsin; ↑, major; ⇡, minor.

between the sequences from such a wide variety of sources suggests that it is unlikely that the wide variety of structures suggested by the early analyses exists.

In addition to the data from the complete sequences, incomplete sequences, mostly obtained from studies of the NH_2-terminal regions by automated sequencing, are also available. The sources and the amount of sequence data from each are listed in Table 1. Figure 4 shows the listing of every residue found at each position in the sequence. It can be seen from the data in this figure that there are 27 residues which are invariant for all species. If only the higher plant sequences are considered, then this value is 50.

The distribution of variability over the sequence is not random, but care must be taken in attributing significance to the variability because the number of different sequences and partial sequences used is not the same for every residue position. Bearing this limitation in mind, two regions of highly conserved sequence appear to exist; these are residues 31–44 and 84–91. These areas of sequence include both of the invariant histidine residues and the single invariant cysteine residue. Of the other invariant residues, 7 are glycine and 3 are proline, which suggests the steric and structural importance of these residues. Several of the hydrophobic residues remain invariant, and when changes occur they tend to be conservative. Where aromatic residues occur in the sequence, they are either invariant of only very conservatively substituted in almost all cases, and this may reflect the importance of the aromatic ring in promoting electron transfer in the molecule. Most of the variability observed is in positions occupied by hydrophilic residues, which is to be expected, because these residues are probably exposed on the surface of the molecule.

Secondary and Tertiary Structure

The determination of both the secondary and tertiary structure of plastocyanin must await the completion of an x-ray crystallographic study. Plastocyanin has been crystallized from *C. album* (45), marrow (34), and French bean (personal communication from H. C. Freeman, cited by Milne et al. (25)), but no x-ray data have as yet been reported.

Currently available methods for establishing tertiary structure do not allow unambiguous predictions to be made. However, when methods for predicting helical content (87–90) are applied to the plastocyanin sequence data, they suggest that little or no α- helix is present in the molecule (J. A. M. Ramshaw, unpublished observations). Because these methods tend to give overestimates of helical content, this prediction may represent a close approximation to the actual situation. As yet, there is no estimate of helical content in plastocyanin based on circular dichroic or other studies.

Copper Binding Site

The geometry of the copper binding site and the ligands involved in binding the copper in plastocyanin are not known and await the completion of an x-ray crystallographic study. In the meantime, a variety of less powerful methods have

-2	-1	1								10										20						
-	-	Ala	Asp	Ile	Glu	Leu	Gly	Ala	Asp	Asp	Gly	Ala	Leu	Ala	Phe	Glu	Pro	Ala	Asp	Phe	Glu	Ile	Ala	Ala	Gly	Asp
Glu	Ala	Phe	Glu	Leu	Lys	Met		Asp	Glu	Glu		Glu		Val		Ile		Asp	Glu	Ile	Asn	Leu	Asp	Lys		Glu
	Asp	Ile	Leu	Val	Leu			Gly	Gly	Gly		Gly				Leu		Gly	Lys	Leu	Gln	Val	Lys	Pro		
	Thr	Leu	Thr		Asn			Ser	Asn	Lys		Leu				Ser		Lys	Asn	Val	Ser		Asn	Ser		
		Met							Ser	Asn		Asn				Val		Asn	Gln		Thr		Pro	Val		
		Gln								Ser		Ser						Gln	Ser				Ser			
		Val										Thr						Ser	Thr				Thr			
		Tyr										Val														

				30										40										50		
Glu	Ile	Glu	Phe	Ile	Asn	Asn	Ala	Ala	Phe	Pro	His	Asn	Ile	Ile	Phe	Asp	Ala	Ala	Ala	Ile	Pro	Ala	Gly	Ala	Ala	Ala
Lys	Val	Ile	Trp	Lys			Lys	Gly	Pro				Val	Val			Glu	Asp	Glu	Asn		Ser	Lys	Ser	Asp	Asp
Ser		Thr		Leu			Ser	Val	Tyr										Leu	Val				Val	Asn	Ser
Thr		Val		Val					Lys																	

							60										70									
Ala	Ala	Ile	Ser	-	-	Glu	Ala	Asp	Leu	Leu	Met	Ala	Ala	Gly	Glu	Ser	Thr	Ala	Ala	Ala	Phe +	Asp	Ala ‡	Ala	Gly	Glu
Asp	Lys	Lys		Leu	Ala	His	Asp	Glu	Tyr		Asn	Ser	Pro		Gln	Thr	Tyr	Glu	Thr	Lys	Leu	Ser	Glu	Lys		Ser
Glu		Leu		Met	Asp		Glu	Leu								Val		Lys	Val	Asn		Thr	Thr	Ser		Thr
Gly					Asn		Lys	Gln										Ser		Thr						
Leu					Pro		Asn											Val								
Ser					Ser																					
Val																										

80										90										100
Tyr	Gly	Phe	Phe	Cys	Ala	Pro	His	Ala	Gly	Ala	Gly	Leu	Lys	-	Lys	Ile	Thr	Val	Ala	-
	Lys	Val	Tyr		Glu			Gln				Met	Val	Gly	Gln	Val			Asn	Gly
	Ser	Tyr			Ser			Arg							Thr				Gln	
	Thr																			

Figure 4. Variability of each residue position for plastocyanin data set. Underlined residues are unique to the algal sequences. Numbering is done as for higher plant sequences. +, insert –Pro–Ala– here (see Aitken (16)); ‡, insert –Pro– here, in the *Anabaena* sequence (see Aitken (16)).

been used which suggest S, N, and/or O as potential ligand atoms in a wide variety of possible geometries.

Chemical evidence of the reaction of plastocyanin with heavy metals (27) indicated that a free sulfhydryl group may be an essential ligand. Brill et al. (51) suggested on theoretical considerations that the observed oxidation-reduction value implied the presence of a sulfur or unsaturated ligand(s) and that a sulfur ligand would be needed to stabilize the Cu(I) state. Hemmerich (52) has suggested that only cysteine and histidine would bind strongly enough to copper in both its oxidation states to ensure stability. Recently, resonance Raman spectroscopy using a tunable laser at about 600 nm as the exciting light source has been used to study plastocyanin and azurin (91). The results of this study were interpreted to show that the ligand geometry approximated to a trigonal bipyramid with 2 nitrogen atoms and 1 sulfur atom in the equatorial plane and with less strongly bound nitrogen or oxygen atoms in the axial positions. This geometry could potentially stabilize the reduced Cu^+ form in accordance with the entatic state concept proposed by Vallee and Williams (92). Studies on plastocyanin which used electron spectroscopy for chemical analysis (ESCA) technique purport to demonstrate the presence of a sulfur ligand by examination of the sulfur 2*p* binding energies (93). However, the results of these authors are not yet conclusive, because oxidized sulfur has a peak of binding energy close to that ascribed by Solomon et al. (93) to the effect of delocalization due to a Cu-S ligand. Furthermore, the reported shift in the binding energy peak is greater than that found for Cu-Samino acid model compounds (94). A sulfhydryl has been widely postulated as an essential ligand in many other type I copper proteins (80, 81, 91, 95).

The relationship between oxidation-reduction potential and pH value (27, 34) indicates the probable involvement of histidine in the oxidation-reduction change. The NMR spectra of French bean plastocyanin also show notable changes in the histidine proton resonances on change of oxidation state (55). On the other hand, the resonance Raman studies on plastocyanin suggested that histidine was not a ligand. However, the band assignments were not definitive and were based on aqueous solution model studies (91). Although the EPR spectrum gives no information regarding the nature of the copper ligands in plastocyanin, super-hyperfine structure due to nitrogen ligands has been observed in the spectra of azurin and stellacyanin at pH 12. It is questionable, however, whether the copper ligands in the native state are the same because the optical, circular dichroic and EPR spectra indicate that at this pH considerable changes have occurred in the conformation of the protein and probably, also, in the coordination geometry. Electron nuclear double resonance spectroscopy has shown that, in stellacyanin, at least 1 nitrogen atom is a copper ligand (96). Overall, the evidence suggests the importance of histidine, either as a Cu ligand or in the oxidation-reduction process. In a "nonblue" copper protein, superoxide dismutase, the three-dimensional structure has now been determined to 3.0 Å resolution (97) and shows that the copper is coordinated by the imidazole side chains of 4 histidine residues. Both difference spectra and chemical reactivities suggest that a tyrosine residue is part of the copper "binding" site of plastocyanin, or at least is closely associated with it (98). Fluorescence studies also suggest that a tyrosine residue is closely associated with copper bindings (38).

Although the sequence data cannot positively identify the copper ligands, they can be used to limit the possibilities, because residues which contribute the copper ligands should be invariant in the data set. Even though the number of invariant residues in the present data set is still large, the number of side chains which could potentially act as copper ligands is limited because many of the invariant residues, e.g., the 7 glycine residues and the 3 leucine residues, do not have side chains that could act as ligands. It is also possible that the polypeptide backbone could provide one or more of the copper ligands.

Most plastocyanins contain 2 methionine and 1 cysteine residues and, of these, only cysteine-84 is invariant (see Figure 4), i.e., could supply a sulfur ligand. Several side chains could provide nitrogen ligands but few are invariant. Tryptophan is found only in certain algal plastocyanins, and arginine is totally absent. Lysine residues are distributed in a very variable manner throughout the chain, and none is invariant. The α amino group at the NH_2-terminal of the protein is also unlikely to be involved, because in certain plastocyanins the NH_2-terminal region is longer than in others. Plastocyanin contains 2 invariant histidine residues, histidine-37 and histidine-87, which could, therefore, supply ligands; neither residue can be reacted chemically without irreversible loss of the blue color (98).

There are several side chains of invariant residues which could provide oxygen ligands—for example, tyrosine-80, serine-56, threonine-97, and aspartic acid-42. Studies with model copper complexes (99), however, indicate that a strong oxygen ligand is unlikely. Miskowski et al. (91) suggested that oxygen ligands could occupy weak axial positions in a trigonal bipyrimidal geometry. The most interesting of the potential oxygen ligands is tyrosine-80.

Finally, carbohydrate has been postulated as being of importance in type I copper binding because many other type I copper proteins (e.g., stellacyanin (100) and rice blue protein (64)) contain as much as 20% carbohydrate. In the case of plastocyanin, there is little, if any, carbohydrate present (see under "Amino Acid Sequence"), and since other type I copper proteins (e.g., azurin) exist which contain no carbohydrate, it is very unlikely that this is involved.

The location of the copper atom within the molecule is not known, although measurements of proton relaxation rates (45) of water proton spin-lattice relaxation times (101) and the slow rate of reaction with PCMB (*p*-chloromercuri-benzoate) (48), have been interpreted as evidence that the copper atom is completely buried within the protein.

The number of ligands required for the copper coordination is not known. Currently, models favor either a distorted tetrahedral or a five-coordinate site (80, 81, 91), in which a sulfur atom is present. Such geometries would be intermediate between those preferred in stable Cu(I) and Cu(II) oxidation states.

The sulfur atom is proposed to allow a ligand-to-metal charge transfer transition in the Cu(II)-S-(Cys) unit, which could account for the intense blue color. The large numbers of bands observed in the visible region of the circular dichroic spectra of other type I blue proteins indicate that some form of charge transfer and possibly also ligand transitions must be considered in the discussion of the electronic structure of the type I copper site. In conclusion then, sequence data taken in conjunction with the chemical and physicochemical evidence strongly suggest that cysteine-84, histidines-37 and -87, and tyrosine-80 may be involved in the type I copper binding site, or very closely associated with it. In addition, the polypeptide backbone may play a role.

Structural Homologies with Other Type I Copper Proteins

Of the other type I copper proteins, sequence data exist only for azurin, in which the amino acid sequences of the protein from nine different sources have now been determined (102). When the sequence data of plastocyanin and azurin are compared by the method of Fitch (103), the results suggest that slight homology may exist between the two proteins (J. A. M. Ramshaw and M. D. Scawen, unpublished observations); similarities between the COOH-terminal regions have been noted (23, 84) (Figure 5). In plastocyanin, this region contains the invariant cysteine residue, which is a possible copper ligand and also one of the 2 invariant histidine residues, and the invariant tyrosine residue, for all of which apparently homologous residues exist in the azurin structure. This may

	33		35		37			40	41
PLASTOCYANIN:	Ala	Gly	Phe	Pro	His	Asn	Ile	Ile	Phe
	Lys	Val	Pro				Val	Val	
	Ser	Ala	Tyr						
			Lys						

	42			45	46				50
AZURIN:	Ala	Ala	Met	Gly	His	Asn	Leu	Val	Ile
	Asn	Val					Trp		Leu
									Val

	78		80				84			87			90
PLASTOCYANIN:	Gly	Glu	Tyr	Gly	Phe	Phe	Cys	Ala	Pro	His	Ala	Gly	Ala
		Ser		Lys	Val	Tyr		Glu			Gln		
		Thr		Ser	Tyr			Ser			Arg		
				Thr									

	106				110		112			115		117			120
AZURIN:	Asp	Ala	Tyr	Ala	Phe	Phe	Cys	Ser	Phe	Pro	Gly	His	Phe	Ala	Ile
	Glu	Asp		Glu	Tyr			Thr					Gly	Ser	Leu
	Gly	Lys		Gly									Ile		Met
		Gln		Met									Asn		
		Ser		Thr									Ser		
													Trp		

Figure 5. Possible regions of homology between the plastocyanin and azurin sequences.

suggest that these residues are part of the copper binding site or at least those closely associated with it.

It is possible that the invariant histidine-46 in azurin is homologous to histidine-37 in plastocyanin, because in both molecules it is followed by an invariant asparagine residue and conservatively substituted hydrophobic positions (see Figure 5).

In the absence of any detailed structural information on any of the other type I copper proteins, it is difficult to establish whether any evolutionary homologies exist. However, the circular dichroic spectra of azurin and umecyanin are almost identical (66), while being very different from the circular dichroic spectra of other blue proteins (104). Also, it is of interest that the amino acid composition of umecyanin shows the presence of a free sulfhydryl group (105). The only other type I copper proteins that have been studied in detail are stellacyanin (61,100) and cusacyanin (44, 65). Both these proteins show EPR spectra with rhombic shape as opposed to plastocyanin, which shows an axial shape (45, 68). The extinction coefficient for cusacyanin is considerably lower than those for other blue proteins. It would seem, therefore, that apart from azurin and possibly umecyanin the diversity of properties of the blue type I copper proteins suggests little evolutionary homology and that the structure of the copper site may vary between the different proteins. The unusual intense blue color and the low hyperfine splitting constant $A_{\|}$ shared by all may be due to convergent evolution and only reflect the presence of a sulfhydryl group as one of the copper ligands in a distorted geometry.

BIOLOGICAL ROLE

Localization of Plastocyanin

The chloroplast contains two types of lamellae, the stroma lamellae and those adpressed in grana, grana lamellae (106). Various fractionation methods have been used to produce photochemically active particles which consist predominantly of one or other of the two types (see review of Park and Sane (107)). For example, incubation of chloroplasts with the nonionic detergent digitonin (108) produced two particles, the heavier of which showed photosystem II and photosystem I activity, while the lighter fraction showed only photosystem I activity. Similar particles were obtained from chloroplasts by use of the French-press (109), and it has been suggested that the heavy particles consist of grana and the light particles of stroma lamellae (110). However, while plastocyanin has been shown to be bound to the chloroplast lamellae as such (10), it is still not certain whether one or both types of lamellae are involved, although most of the evidence suggests that they both are. Plastocyanin was found in both lamellae by Sane and Hauska (111), although the stroma fractions contained more plastocyanin than did the grana fraction, as was also reported by Baszynski et al. (112). Arntzen et al. (113) also found that both grana and stroma lamellae contained plastocyanin but that the distribution of the protein between the two fractions was approximately equal. These results are in contrast to those of Arnon et al. (114), who found only minimum amounts of plastocyanin in stroma lamellae preparations.

Plastocyanin appears to be situated within the membrane rather than on its surface. Hauska et al. (115) have shown that plastocyanin-dependent electron transport is not affected by the presence of antibody prepared against the copper protein, whereas sonication of the chloroplast fragments in the presence of the antibody leads to a marked inhibition. These results suggest that plastocyanin is located on the inner surface of the chloroplast membrane in equilibrium with plastocyanin in the thylakoid space (115).

Further support for this suggestion is given by the results of Katoh (116), who found that the effectiveness of added electron donors in the photosystem I reactions of the chloroplast or chloroplast fragments depends on the structure of the thylakoid membrane. Ascorbate or ascorbate plus plastocyanin were inefficient as electron donors for photosystem I in intact chloroplasts, but their efficiency was enhanced by mechanical- or detergent-mediated disruption of the membranes.

Role in Photosynthetic Electron Transport

Katoh and Takamiya (117) observed a light-dependent reduction of purified plastocyanin in the presence of spinach chloroplasts. Such reduction was inhibited by dichlorophenyldimethylurea and phenanthroline, but stimulated by the addition of NADP. Photoreduction of plastocyanin was also greater under conditions in which phosphorylation could take place or when electron transport was uncoupled from phosphorylation. If digitonin was added, the reducing activity of the chloroplasts was suppressed and photo-oxidation of reduced plastocyanin was observed (117). This reaction was unaffected by the addition of dichlorophenyldimethylurea or heating the chloroplasts to 50°C. Because both factors inhibit photosystem II activity (118), it was proposed that plastocyanin was reduced by photosystem II and oxidized by photosystem I (see Figure 6).

Because the oxidation-reduction potential of plastocyanin is +370 mV (27), the position of the protein in the electron transport chain must be very close to photosystem I. However, some uncertainty exists as to the exact position of plastocyanin because cytochrome *f* has a nearly identical oxidation-reduction potential of 365 mV (120), so that the order of the two components cannot be established on this basis. A series of experiments has been carried out with the use of chloroplasts or chloroplast fragments depleted of plastocyanin by sonication; this system allows measurement of the photo-oxidation of cytochrome *f* and NADP reduction with DPIP (diphenolindophenol) ascorbate, in the presence and absence of added plastocyanin. Most experiments of this type suggest that cytochrome *f* photo-oxidation is inhibited by the removal of plastocyanin, the inhibition being reversed on addition of plastocyanin (121–123). Also, NADP photoreduction by chloroplast fragments required the presence of plastocyanin

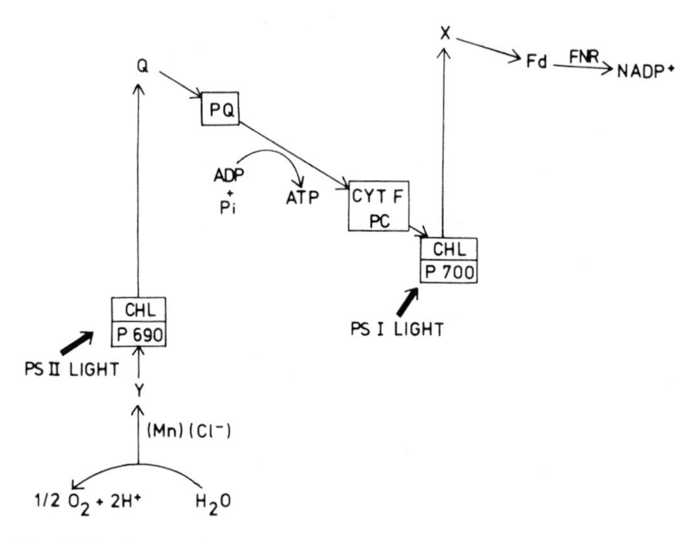

Figure 6. The "Z" scheme of photosynthetic electron transport derived from Hill and Bendall (119). PC, plastocyanin; PQ, plastoquinone; CHL, chlorophyll; CYT F, cytochrome *f*; Fd, ferredoxin; FNR, ferredoxin-NADP reductase. X and Q are the unknown primary electron acceptors of photosystems I and II, respectively.

(124, 125). Gorman and Levine (126) found that chloroplast fragments from a plastocyanin-deficient mutant of *C. reinhardtii* required the addition of purified plastocyanin for photo-oxidation of cytochrome *f*. The above results suggest the following order:

$$\text{Cytochrome } f \rightarrow \text{plastocyanin} \rightarrow \text{P700 (photosystem I)}$$

Studies with plastocyanin inhibitors such as polylysine and mercury have shown that these compounds inhibit electron flow between cytochrome *f* and photosystem I in chloroplasts, suggesting that plastocyanin is the immediate electron donor for photosystem I. However, the specificity of these inhibitors has been questioned (see Trebst (1) for references). Additional evidence for this reaction pathway has been given by Wood (127) and Wood and Bendall (128), who showed that electrons are readily transferred from reduced cytochrome *f* to plastocyanin in vitro, and that reduced plastocyanin is a much more efficient electron donor to P700 than is cytochrome *f* in chloroplasts treated with Triton X-100 or digitonin.

However, Fork and Murata (129) found that photosystem I particles prepared from spinach chloroplasts by the use of the French-press showed the same level of cytochrome *f* photo-oxidation as control chloroplasts, even though the plastocyanin content of the former was only 15% of that of the control chloroplasts. Knaff and Arnon (130) demonstrated that sonicated chloroplasts could photo-oxidize cytochrome *f* and that the photo-oxidation was unaffected by the addition of purified plastocyanin. In order to check that the sonication removed all plastocyanin from the chloroplasts, Knaff (131) repeated the experiments measuring the level of plastocyanin remaining in sonicated particles by the size of the EPR signal produced by the oxidized copper. Again, although less than 5% of the plastocyanin remained, cytochrome *f* photo-oxidation was unaffected. However, cytochrome *f* photoreduction was inhibited in these particles and this activity could be restored on addition of plastocyanin. These results suggested that plastocyanin is on the reducing side of cytochrome *f*:

$$\text{Plastocyanin} \rightarrow \text{cytochrome } f \rightarrow \text{P700 (photosystem I)}$$

Haehnal (132) measured the kinetics of electron donation from the primary donor (PD) to chlorophyll a_1, the reaction center of photosystem I. High time resolution flash photometry revealed the existence of two reactions with half-times of 10 μs and 200 μs, respectively. It was concluded that these two reactions involved plastocyanin and cytochrome *f*, which could act as alternative electron donors to chlorophyll a_1. Kok et al. (133) also concluded that plastocyanin and cytochrome *f* can act as alternative electron donors on the basis of kinetic measurements in spinach chloroplasts which had been sonicated in the presence of Tween 20. Lightbody and Krogmann (36) found that cytochrome *f* could partially replace plastocyanin in the photoreduction of NADP by DPIP with the use of photosynthetic particles prepared from *Anabaena,* although Wood and Bendall (128) suggested that higher plant cytochrome *f* may not be analogous with the soluble algal cytochromes *f*. Recently, Arnon et al. (134) and Park and Sane (107) have presented electron transport schemes involving three light reactions in which two photosystem II light reactions are connected in series and contain plastocyanin, whereas the photosystem I reaction is unconnected to photosystem II and contains cytochrome *f*.

The difficulty in interpreting the results developed from disruptive methods raises the question as to whether they represent the situation in the intact chloroplast and underline the advantage of using mutants.

BIOSYNTHESIS

Plastocyanin has been found in etiolated leaves of barley (37, 123, 135) and French bean (136). In barley leaves, its content remained constant during the greening process if results were expressed on a fresh weight basis (123, 135), but, in greening bean leaves, the level of plastocyanin per leaf increased after an initial lag period, up to a constant chlorophyll to plastocyanin ratio of 200 (136); the final ratio was independent of the light intensity during greening. The synthesis of plastocyanin in etiolated bean leaves was stimulated by red light, whereas red light treatment followed immediately by far-red produced no such stimulation (136), suggesting that phytochrome may be involved in the control of synthesis of plastocyanin.

The synthesis of plastocyanin is thought to occur on 80 S cytoplasmic ribosomes, because inhibitors such as D-threochloramphenicol and lincomycin, inhibitors of protein synthesis by chloroplast ribosomes, had no significant effect on the synthesis of plastocyanin in bean leaves (136). In addition, Surzicki et al. (137) found that the level of plastocyanin in mutant cells of *C. reinhardtii* was comparable to that of the wild type, even though the mutant strain had only 6% of the number of chloroplast ribosomes.

The gene responsible for plastocyanin synthesis appears to be located in the nucleus rather than in the chloroplast. Genetic experiments with a mutant strain of *C. reinhardtii* lacking plastocyanin have shown that the mutation is due to a single gene which is inherited in a Mendelian manner (138).

MOLECULAR EVOLUTION

At the present time, 12 complete amino acid sequences of plastocyanin are known (Figure 2), and a further 57 partial sequences have been determined by the Durham group (see Table 1). Because the partial sequences are as yet unpublished and, therefore, their experimental validation has not been given, they are not presented in this review. However, the mass of data is in the preliminary stage of analysis, and a number of general features with respect to the evolutionary implications are apparent and are discussed.

Evolutionary Rate

The higher plant complete sequence data set is sufficiently large to allow comparisons to be made between the number of differences in the cytochromes of several pairs of taxa and the number of differences in the plastocyanins of the same pairs of taxa or closely related taxa. These comparisons indicate that plastocyanin is evolving about 1.5 times as fast as cytochrome *c*.

Relative Usefulness of Partial, Compared with Complete, Sequences

A statistical model of the evolutionary change in proteins (139) suggests that the precision in measuring the evolutionary distance between the proteins is inversely proportional to the square root of the number of variable residues in the sequence data set; using the full sequence of plastocyanin (about 100 residues) will be on average 1.6 times as accurate as using the first 40 residues only. However, by using automated methods and determining the first 40 NH_2-terminal residues, the rate of data collection is an order of magnitude greater than if complete sequences are established. Because the amount of sequence data available is the limiting factor in constructing dendrograms (sensu Mayer et al. (140)), the greater amount of information gained in a given period of time from several partial sequences more than compensates for the greater precision of using complete sequences.

Haslett and Boulter (141) have discussed the suitability of the criteria to be used with automatic sequence methods and especially the problems arising from the relative lack of redundant information as compared with manual sequencing methods. Automatic methods, which give relatively little redundant information, are particularly suited to evolutionary studies, because many different sequences of the same protein are compared.

For those taxa for which a full sequence is available, it is possible to compare the number of differences in the first 40 residues between any two taxa of the data set with those of the remaining residues (41 to the COOH terminus) between each pair. This comparison by means of the 12 sequences of Figure 2 shows a correlation of +0.72, which is significant at the 0.1% level (66 comparisons).

When the dendrogram obtained from the complete sequences by using the ancestral sequence method of Dayhoff and Eck (142) is compared to that obtained from their first 40 residues, the two dendrograms show no real incompatibilities within the limits of error expected with this method and this protein (see Peacock and Boulter (143)).

General Phylogenetic Implications

The results in Figure 7 show that the number of differences between the prokaryote algal plastocyanin sequence is less compared to the eukaryotic green

	Anabaena	Chlorella	Rumex	Sambucus	Spinacia	Cucurbita	Capsella	Solanum	Vicia	Phaseolus	Mercurialis	Lactuca
Anabaena:	0											
Chlorella:	50	0										
Rumex:	59	43	0									
Sambucus:	60	47	20	0								
Spinacia:	60	48	19	12	0							
Cucurbita:	61	48	19	19	12	0						
Capsella:	61	49	22	20	15	14	0					
Solanum:	58	47	19	17	17	17	18	0				
Vicia:	58	42	24	21	17	20	23	22	0			
Phaseolus:	57	47	22	19	16	19	20	19	20	0		
Mercurialis:	59	49	21	16	17	15	17	15	22	16	0	
Lactuca:	57	47	18	20	14	17	17	17	21	13	15	0

Figure 7. Matrix of amino acid differences. Data are from Figure 2.

algal sequence than compared to the higher plant sequences. However, it would be premature to use such data for phylogenetic inference, because we do not know the extent of the variation of plastocyanin sequences from either the blue-green or green algal groups. In any case, even assuming that the two sequences presented here are highly representative of the groups as a whole, the differences between them are so large as to make the confidence limits with the use of present data-handling procedures so great that their application to the data at this point is unjustified. However, it is of interest to note the similarity in the amino acid sequences between the plastocyanins of these different pro- and eukaryotic groups. The extent of this similarity is so great as to suggest that it is unlikely to be due to convergent evolution (16). Future work with plastocyanin will, undoubtedly, be directed to answering the question as to whether or not the whole organism or the chloroplast of photosynthetic eukaryotes has descended from blue-green algal ancestors. Only from higher plants are sufficient data available to discern any cladistic relationships, and, even then, there are severe limitations on the interpretation of the data.

Higher Plants

For 12 plant families, partial plastocyanin sequences from more than one genus are available (see Table 1). These sequences have been subjected to analysis by using the ancestral sequence program of Dayhoff and Eck (142), and the results show that, with three exceptions, the sequences fall into groups equivalent to the taxonomic families. An example of how the sequences are arranged within a family is given in Figure 8 for the Compositae; an examination of the sequence data set shows that the Compositae which are considered by most to form a distinct natural group (144) have three positions where there are amino acid substitutions unique to the family. The three exceptions noted above are: *Beta vulgaris,* relative to the Chenopodiaceae, and the two *Phaseolus* species relative to the other members of the Leguminosae. The sequence from *B. vulgaris* is

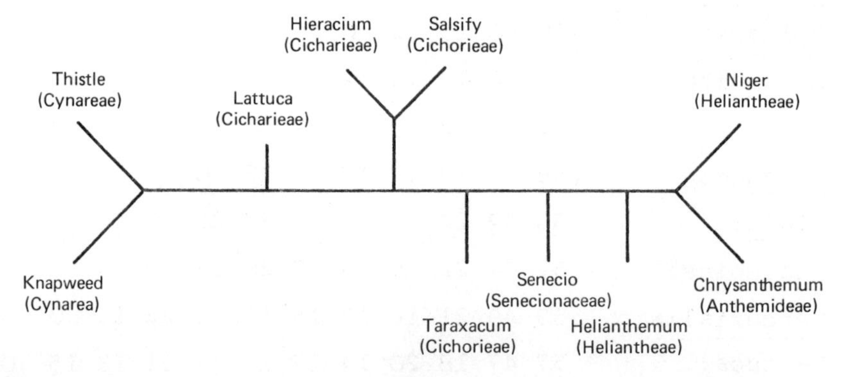

Figure 8. Molecular cladogram of plastocyanin sequences from the Compositae. Constructed with the use of the Dayhoff and Eck (142) method.

exceptional because it contains six unique amino acid substitutions in comparison with the rest of the flowering plants: glutamate-8, glycine-9, valine-11, glutamine-17, asparagine-20, and tyrosine-35, although two of these also occur in yew, which is a gymnosperm. The amino acid sequences of *Phaseolus aureus* and *P. vulgaris* contain no unique amino acid substitutions and are unexceptional, apart from their low familial affinity, having somewhat greater affinities toward members of the Chenopodiaceae and Compositae than toward other members of the Leguminosae.

While the sequences, with the above exceptions, form groups according to the taxonomic families, several different interfamilial trees are generated by the Dayhoff and Eck procedure.

Polymorphism

Fifteen sequences have one, two, or three positions that have definitely shown polymorphism; many of these polymorphic positions are in the first 26 residues. However, more sequence data are available in this region of the sequence, and also, the use of quantitative results obtained from the automatic sequencer analysis allow more definitive identification of polymorphic changes than do the qualitative results obtained by dansyl-Edman analysis. Consequently, polymorphism in the COOH-terminal part of the sequence may escape notice unless the substitution is large or radical. The polymorphism which has been observed has usually been chemically conservative, involving alternatives such as Ile–Leu or Glu–Asp; however, non-conservative polymorphism such as Glu–Val has been observed. Although plastocyanin has usually been obtained from a population of plants, in the case of Malva it was obtained from a single plant bush, indicating the presence of two different plastocyanins in the same plant.

Serological Comparisons

In spite of the relative ease with which sequence data can now be obtained, it still requires considerable time to examine a large number of species by this method. Because plastocyanin antibodies can be prepared (115, 145), it may be possible to use immunological comparisons as a supplementary method since, in certain groups of proteins such as cytochrome *c* and avian lysozymes, there is a correlation between immunological cross-reactivity and sequence difference (146, 147). Wallace and Boulter (145, 148), using about 20 plastocyanins from various higher plant families, have shown that the immunological comparisons using microcomplement fixation do give approximate measures of the plastocyanin sequence differences. Once again, because of the small number of comparisons presented, it would be premature to discuss in detail possible phylogenetic implications.

Conclusions

The faster evolutionary rate of plastocyanin compared to that of cytochrome *c* may prove particularly useful in providing more detailed phylogenetic insights, especially within plant families. Indeed some genera such as *Phaseolus* and *Solanum* are sufficiently variable in their plastocyanins to show evolution within the genus, because the two *Phaseolus* sequences differ by 4 and the two *Solanum* sequences by 6 residues, respectively. Again, in the case of marrow plastocyanin, it has been shown that the sequences of the proteins prepared from several varieties differ in at least one position (21, 34).

The sequence dendrograms so far constructed are in good agreement with the established familial groupings and also show the relationship between the genera within families such as the Compositae. The variability, however, found within the families is almost as large as the interfamilial variability, indicating the need to have sufficient sequences from one family to ensure true representation. With the present data set, several equally probable interfamilial dendrograms are generated, due to the relatively fast evolutionary rate of plastocyanin and its associated high proportion of convergent amino acid substitutions, estimated at 40–50% (143). Furthermore, the presence of polymorphic positions also complicates the analysis because these positions are excluded with the use of the Dayhoff and Eck method (142). With the number of partial sequences now available (about 70), computing time is made unreasonably long, and future strategy will be to determine partial sequences from several additional proteins from plant species already examined for plastocyanin.

ACKNOWLEDGMENTS

We wish to thank Mrs. M. Creighton and Mr. J. S. Redhead for help in preparing the manuscript.

REFERENCES

1. Trebst, A. (1974). Annu. Rev. Plant Physiol. 25:423.
2. Somner, A. L. (1939). Plant Physiol. (Lancaster) 6:339.
3. Green, L. F., McCarthy, J. F., and King, C. G. (1939). J. Biol. Chem. 128:447.
4. Neish, A. C. (1939). Biochem. J. 33:300.
5. Whatley, F. R., Ordin, L., and Arnon, D. I. (1951). Plant Physiol. (Lancaster) 26:414.
6. Arnon, D. I. (1949). Plant Physiol. (Lancaster) 24:1.
7. Katoh, S., and Takamiya, A. (1961). Nature 189:665.
8. Nieman, R. H., and Vennesland, B. (1959). Plant Physiol. (Lancaster) 34: 255.
9. Nieman, R. H., Nakamura, H., and Vennesland, B. (1959). Plant Physiol. (Lancaster) 34:262.
10. Katoh, A., Suga, I., Shiratori, I., and Takamiya, A. (1961). Arch. Biochem. Biophys. 94:136.
11. De Kouchkovsky, Y., and Fork, D. C. (1964). Proc. Natl. Acad. Sci. U.S.A. 52:2329.
12. Malkin, R., and Bearden, A. J. (1973). Biochim. Biophys. Acta 292:169.
13. Visser, J. W. M., Amesz, J., and Van Gelder, B. F. (1974). Biochim. Biophys. Acta 333:279.
14. Brown, J., Bril, C., and Urback, W. (1965). Plant Physiol. 40:1086.
15. Kunert, K. -J., and Böger, P. (1975). Z. Naturforsch. 30c:190.
16. Aitken, A. (1975). Biochem. J. 149:675.
17. Kelly, J., and Ambler, R. P. (1974). Biochem. J. 143:681.
18. Haslett, B., Bailey, C. J., Ramshaw, J. A. M., Scawen, M. D., and Boulter, D. (1974). Biochem. Soc. Trans. 2:1329.

19. Scawen, M. D., Ramshaw, J. A. M., Brown, R. H., and Boulter, D. (1974). Eur. J. Biochem. 44:299.
20. Scawen, M. D., Ramshaw, J. A. M., and Boulter, D. (1975). Biochem. J. 147:343.
21. Scawen, M. D., and Boulter, D. (1974). Biochem. J. 143:257.
22. Scawen, M. D., Ramshaw, J. A. M., Brown, R. H., and Boulter, D. Phytochemistry, manuscript in preparation.
23. Ramshaw, J. A. M., Scawen, M. D., Bailey, C. J., and Boulter, D. (1974). Biochem. J. 139:583.
24. Ramshaw, J. A. M., Scawen, M. D., and Boulter, D. (1974). Biochem. J. 141:835.
25. Milne, P. R., Wells, J. R. E., and Ambler, R. P. (1974). Biochem. J. 143:691.
26. Ramshaw, J. A. M., Scawen, M. D., Jones, E. A., Brown, R. H., and Boulter, D. (1976). Phytochemistry 15:1199.
27. Katoh, S., Shiratori, I., and Takamiya, A. (1962). J. Biochem. (Tokyo) 51:32.
28. Gorman, D. S., and Levine, R. P. (1966). Plant Physiol. 41:1637.
29. Hewitt, E. J., Hucklesby, D. P., and Betts, G. F. (1968). *In* E. J. Hewitt and C. V. Cutting (eds.), Recent Aspects of Nitrogen Metabolism in Plants, pp. 47–81. Academic Press, London.
30. Borchert, M. T., and Wessels, J. S. C. (1970). Biochim. Biophys. Acta 197:78.
31. Wells, J. R. E. (1965). Biochem. J. 97:228.
32. Ramshaw, J. A. M., Brown, R. H., Scawen, M. D., and Boulter, D. (1973). Biochim. Biophys. Acta 303:269.
33. Mutuskin, A. A., Pshenova, K. V., Alekhina, S. K., and Kolesnikov, P. A. (1971). Biokhimiia 36:236.
34. Scawen, M. D., Hewitt, E. J., and James, D. M. (1975). Phytochemistry 14:1225.
35. Milne, P. R., and Wells, J. R. E. (1970). J. Biol. Chem. 245:1566.
36. Lightbody, J. J., and Krogmann, D. W. (1967). Biochim. Biophys. Acta 131:508.
37. Plesnicar, M., and Bendall, D. S. (1970). Biochim. Biophys. Acta 216:192.
38. Graziani, M. T., Agro, A. F., Rotilio, G., Barra, D., and Mondovi, B. (1974). Biochemistry 13:804.
39. Anderson, M. M., and McCarty, R. E. (1969). Biochim. Biophys. Acta 189:193.
40. Siegelman, M. H., Rasched, I., and Böger, P. (1975). Biochem. Biophys. Res. Commun. 65:1456.
41. Gamayunova, M. S., Lichadeev, G. I., and Grigora, M. Y. (1972). Fiziol. Biokhim. Kul't. Rast. 4:482.
42. Pshenova, K. V., Mutuskin, A. A., and Kilesnikov. P. A. (1966). Dokl. Acad. Nauk SSSR 170:457.
43. Mutuskin, A. A., and Pshenova, K. V. (1967). V. V. Koval'skii (ed.), Biologicheskaya Rol Medi Simpozium, pp. 52–55. Nauka, Moscow.
44. Markossian, K. A., Aikazyan, V. T., and Nalbandyan, R. M. (1974). Biochim. Biophys. Acta 359:47.
45. Blumberg, W. E., and Peisach, J. (1966). Biochim. Biophys. Acta 126:269.
46. Katoh, S. (1960). Nature 186:533.
47. Wells, J. R. E. (1966). Nature 212:896.
48. Katoh, S., and Takamiya, A. (1964). J. Biochem. (Tokyo) 55:378.
49. Malkin, R., Knaff, D. B., and Bearden, A. J. (1973). Biochim. Biophys. Acta 305:675.
50. Hughes, M. N. (1972). The Inorganic Chemistry of Biological Processes. John Wiley & Sons, London.
51. Brill, A. S., Martin, R. B., and Williams, R. J. P. (1964). *In* B. Pullman (ed.), Electronic Aspects of Biochemistry, pp. 519–557. Academic Press, London.
52. Hemmerich, P. (1966). *In* P. Eisen, W. E.. Blumberg, and J. Peisach (eds.), The Biochemistry of Copper, pp. 16–34. Academic Press, New York.
53. Dawson, J. W., Gray, H. B., Holwerda, R. A., and Westhead, E. W. (1972). Proc. Natl. Acad. Sci. U.S.A. 69:30.
54. Wherland, S., Holwerda, R. A., Rosenberg, R. C., and Gray, H. B. (1975). J. Am. Chem. Soc. 97:5260.
55. Beattie, J. K., Fensom, D. J., Freeman, H. C., Woodcock, E., Hill, H. A. O., and Stokes, A. M. (1975). Biochim. Biophys. Acta 405:109.
56. Aikazyan, V. T., Mutuskin, A. A., Pshenova, K. V., and Nalbandyan, R. M. (1973). FEBS Lett. 34:103.

57. Nalbandyan, P. M., Mutuskin, A. A., and Pshenova, K. V. (1971). Dokl. Akad. Nauk SSSR 201:1396.
58. Brill, A. S., Bryce, G. F., and Maria, H. J. (1968). Biochim. Biophys. Acta 154:342.
59. Broman, L., Malmström, B. G., Aasa, R., and Vanngard, T. (1963). Biochim. Biophys. Acta 75:365.
60. Sutherland, I. W., and Wilkinson, J. F. (1963). J. Gen. Microbiol. 30:105.
61. Malmström, B. G., Reinhammar, B., and Vanngard, T. (1970). Biochim. Biophys. Acta 205:48.
62. Paul, K. G., and Stigbrand, T. (1970). Biochim. Biophys. Acta 221:255.
63. Stigbrand, T., Malmström, B. G., and Vanngard, T. (1971). FEBS Lett. 12:260.
64. Morita, Y., Wadano, A., and Ida, S. (1971). Agr. Biol. Chem. 35:255.
65. Aikazyan, V. T., and Nalbandyan, R. M. (1975). FEBS Lett. 55:272.
66. Stigbrand, T., and Sjöholm, I. (1972). Biochim. Biophys. Acta 263:244.
67. Tang, S. P., Coleman, J., and Myer, Y. (1968). J. Biol. Chem. 243:4286.
68. Malkin, R. (1973). *In* G. L. Eichhorn (ed.), Inorganic Biochemistry, Vol. 2, pp. 689–709. Elsevier, New York.
69. Brill, A. S., and Bryce, G. F. (1968). J. Chem. Physiol. 48:4398.
70. Malmström, B. G., Reinhammar, B., and Vanngard, T. (1968). Biochim. Biophys. Acta 156:67.
71. Malkin, R., Malmström, B. G., and Vanngard, T. (1969). Eur. J. Biochem. 7:253.
72. Malkin, R., and Malmström, B. G. (1970). Adv. Enzymol. 33:177.
73. Vanngard, T. (1972). *In* H. M. Swartz, J. R. Bolton, and D. C. Borg (eds.), Biological Application of Electron Spin Resonance, pp. 411–477. John Wiley & Sons, London.
74. Malmström, B. G., and Vanngard, T. (1960). J. Mol. Biol. 2:118.
75. Cotton, F. A., and Wilkinson, G. (1966). Comprehensive Inorganic Chemistry. John Wiley & Sons, London.
76. Nakamura, T., Makino, M., and Ogura, Y. (1968). J. Biochem. (Tokyo) 64:189.
77. Andreasson, L. E., and Vanngard, T. (1970). Biochim. Biophys. Acta 200:247.
78. Brody, S. S. (1973). Z. Naturforsch. 28c:397.
79. Brody, S. S. (1975). Z. Naturforsch. 30c:318.
80. McMillin, D. R., Holwerda, R. A., and Gray, H. B. (1974). Proc. Natl. Acad. Sci. U.S.A. 71:1339.
81. McMillin, D. R., Rosenberg, R. C., and Gray, H. B. (1974). Proc. Natl. Acad. Sci. U.S.A. 71:470.
82. Scawen, M. D., and Hewitt, E. J. (1971). Biochem. J. 124:32P.
83. Tanford, C. (1961). Physical Chemistry of Macromolecules. John Wiley & Sons, New York.
84. Kelly, J., and Ambler, R. P. (1973). Biochem. Soc. Trans. 1:164.
85. Edman, P., and Begg, G. (1967). Eur. J. Biochem. 1:80.
86. Steers, E., Craven, G. T., Anfinsen, C. B., and Bethune, J. L. (1965). J. Biol. Chem. 240:2478.
87. Prothero, J. W. (1966). Biophys. J. 6:367.
88. Kotelchuck, D., and Scheraga, H. A. (1969). Proc. Natl. Acad. Sci. U.S.A. 62:14.
89. Lim, V. I. (1974). J. Mol. Biol. 88:873.
90. Kabat, E. A., and Wu, T. T. (1973). Biopolymers 12:751.
91. Miskowski, V., Tang, S.-P. W., Spiro, T. G., Shapiro, E., and Moss, T. H. (1975). Biochemistry 14:1244.
92. Vallee, B. L., and Williams, R. J. P. (1968). Proc. Natl. Acad. Sci. U.S.A. 59:498.
93. Solomon, E. I., Clendening, P. J., Gray, H. B., and Grunthaner, F. J. (1975). J. Am. Chem. Soc. 97:3878.
94. Grunthaner, F. J. (1974). Ph.D. thesis, California Institute of Technology.
95. Siiman, O., Young, N. M., and Carey, P. R. (1974). J. Am. Chem. Soc. 96:5583.
96. Rist, G. H., Hyde, J. S., and Vanngard, T. (1970). Proc. Natl. Acad. Sci. U.S.A. 67:79.
97. Richardson, J. S., Thomas, K. A., Rubin, B. H., and Richardson, D. C. (1975). Proc. Natl. Acad. Sci. U.S.A. 72:1349.
98. Searle, G. F. W. (1969). Ph.D. thesis, University of Warwick.

99. Freeman, H. C. (1973). *In* G. L. Eichhorn (ed.), Inorganic Biochemistry, Vol. 1, pp. 121–166. Elsevier, New York.
100. Peisach, J., Levine, W. G., and Blumberg, W. E. (1967). J. Biol. Chem. 242:2847.
101. Boden, N., Holmes, M. C., and Knowles, P. F. (1974). Biochem. Biophys. Res. Commun. 57:845.
102. Ambler, R. P. (1971). Recent Developments in the Chemical Study of Protein Structures, pp. 289–305. INSERM, Paris. Proceedings of Colloque Montpellier, organized by A. Previero, J. F. Pechere, and M. A. Coletti-Previero.
103. Fitch, W. M. (1966). J. Mol. Biol. 16:9.
104. Falk, K. E., and Reinhammar, B. (1972). Biochim. Biophys. Acta 285:84.
105. Stigbrand, T. (1971). Biochim. Biophys. Acta 236:246.
106. Steinmann, E., and Sjöstrand, F. S. (1955). Exp. Cell Res. 8:15.
107. Park, R. B., and Sane, P. V. (1971). Annu. Rev. Plant Physiol. 22:395.
108. Anderson, J. M., and Boardman, N. K. (1966). Biochim. Biophys. Acta 112:403.
109. Michel, J. M., and Michel-Wolwertz, M. R. (1967) Carnegie Institution Year Book 67: pp. 508–514.
110. Sane, P. V. Goodchild, D. J., and Park, R. B. (1970. Biochim. Biophys. Acta 216:162.
111. Sane, P. V., and Hauska, G. A. (1972). Z. Naturforsch. 27:932.
112. Baszynski, T., Brand, J., Krogmann, D. W., and Crane, F. L. (1971). Biochim. Biophys. Acta 234:537.
113. Arntzen, C. J., Dilley, R. A., Peters, G. A., and Shaw, E. R. (1972). Biochim. Biophys. Acta 256:85.
114. Arnon, D. I., Chain, R. K., McSwain, B. D., Tsujimoto, H., and Knaff, D. B. (1970). Proc. Natl. Acad. Sci. U.S.A. 67:1404.
115. Hauska, G., McCarty, R. E., Berzborn, R. J., and Racker, E. (1971). J. Biol. Chem. 246:3524.
116. Katoh, S. (1972). Biochim. Biophys. Acta 283:293.
117. Katoh, S., and Takamiya, A. (1963). 4:335.
118. Katoh, S., and San Peitro, A. (1967). Arch. Biochem. Biophys. 122:144.
119. Hill, R., and Bendall, F. (1960). Nature 186:136.
120. Davenport, H. E., and Hill, R. (1952). Proc. R. Soc. Lond. (Biol.). 139:327.
121. Avron, M., and Shneyour, A. (1971). Biochim. Biophys. Acta 226:498.
122. Hind, G. (1968). Biochim. Biophys. Acta 153:235.
123. Plesnicar, M., and Bendall, D. S. (1973). Eur. J. Biochem. 34:483.
124. Vernon, L. P. Shaw, E. R., and Ke, B. (1966). J. Biol. Chem. 241:4101.
125. Wessels, J. S. C. (1966). Biochim. Biophys. Acta 126:581.
126. Gorman, D. S., and Levine, R. P. (1966). Plant Physiol. 41:1648.
127. Wood, P. M. (1974). Biochim. Biophys. Acta 357:370.
128. Wood, P. M., and Bendall, D. S. (1975). Biochim. Biophys. Acta 387:115.
129. Fork, D. C., and Murata, N. (1971). Photochem. Photobiol. 13:33.
130. Knaff, D. B., and Arnon, D. I. (1970). Biochim. Biophys. Acta 223:201.
131. Knaff, D. B. (1973). Biochim. Biophys. Acta 292:186.
132. Haehnel, W. (1973). Biochim. Biophys. Acta 305:618.
133. Kok, B., Rurainski, H. J., and Harmon, E. A. (1964). Plant Physiol. 39:513.
134. Arnon, D. I., Knaff, D. B., McSwain, B. D., Chain, R. K., and Tsujimoto, H. Y. (1971). Photochem. Photobiol. 14:397.
135. Plesnicar, M., and Bendall, D. S. (1972). Proceedings of the Second International Congress Photosynthesis Research, Stresa 3:2367.
136. Haslett, B. G., and Cammack, R. (1974). Biochem. J. 144:567.
137. Surzicki, S. J., Goodenough, U. W., Levine, R. P., and Armstrong, J. J. (1970). S. E. B. Symp. 24:13.
138. Levine, R. P. (1968). Science 162:768.
139. Dayhoff, M. O. (1972). Atlas of Protein Sequence and Structure, Vol. 5. National Biomedical Research Foundation, Silver Spring, Maryland.
140. Mayer, E., Linsley, E. G., and Usinger, R. L. (1953). Methods and Principles of Systematic Zoology. McGraw-Hill, New York.
141. Haslett, B. G., and Boulter, D. (1976). Biochem. J. 153:33.

142. Dayhoff, M. O., and Eck, R. V. (1966). Atlas of Protein Sequence and Structure, Vol. 2. National Biomedical Research Foundation, Silver Spring, Maryland.
143. Peacock, D., and Boulter, D. (1975). J. Mol. Biol. 95:513.
144. Cronquist, A. (1955). Phylogeny and taxonomy of the Compositae. Am. Midland Naturalist 53:478.
145. Wallace, D. G., and Boulter, D. (1974). Carnegie Institution Year Book 73:1069.
146. Prager, E. M., and Wilson, A. C. (1971). J. Biol. Chem. 246:5978.
147. Prager, E. M., and Wilson, A. C. (1971). J. Biol. Chem. 246:7010.
148. Wallace, D. G., and Boulter, D. (1976). Phytochemistry 15:137.

International Review of Biochemistry
Plant Biochemistry II, Volume 13
Edited by D. H. Northcote

2 Electron and Proton Transfer in Chloroplasts

D. S. BENDALL
Department of Biochemistry, University of Cambridge, Cambridge, England

---

REACTION CENTERS IN PURPLE NONSULFUR BACTERIA 43

PHOTOSYSTEM I 45
- Transmembrane Model for Photosystem I 47
- Evidence Against the Transmembrane Model for Photosystem I 48

PHOTOSYSTEM II 49
- Transmembrane Model for Photosystem II 51
- Evidence Against the Transmembrane Model for Photosystem II 52

ELECTRON TRANSFER BETWEEN PHOTOSYSTEMS 55
- Structural Organization of Chain 55
- H^+ Flow Coupled to Electron Transfer Across Membrane 56
- Mechanism of Plastoquinol Oxidation 58
- Electron Transfer from Pool to P700 60
- Electron Transfer from Q to Pool 62

CONTROL OF ELECTRON TRANSPORT 67

CONCLUSION 70

An overall view of the electron transport system of the chloroplast depends on three major developments. The first of these is the Z scheme, formulated 15 years ago (1) and now firmly established as an outline description of electron transport. In essence, two photochemical reactions, distinguished by the oxidation-reduction potential range which they span, are considered to be linked in series by a "chain" of electron or hydrogen carriers. One reaction is more closely concerned with reduction of NADP, the other with oxidation of water to molecular oxygen. The major argument about the validity of the Z scheme centered upon attempts to observe the Emerson enhancement effect with isolated chloroplasts. It is now realized that while the enhancement phenomenon was valuable evidence it was not a necessary condition for two light reactions. Hill's emphasis on oxidation-reduction potentials seems the more significant. Nevertheless, it should be noted that the complete failure to observe enhancement in certain types of experiment (2) has not yet been adequately explained.

The second development concerns structure. Although the importance of structure in the mechanism of photosynthesis has long been appreciated, only recently has it become possible to discuss membrane structure in a coherent fashion based on determinable properties of proteins and lipids. Anderson (3) has discussed the application of the fluid-mosaic model of membrane structure to chloroplasts. Here we must content ourselves with noting the features of this model that are significant for photosynthetic electron transport: the fluid lipid phase, the distinction between intrinsic and extrinsic proteins, the possibility that some proteins span the membrane, the relatively free lateral diffusion but highly restricted transmembrane diffusion ("flip-flop") of most of the proteins and lipids, and the probability that the two sides of the membrane are not identical in composition.

The third development is the chemiosmotic hypothesis of Mitchell (5–7), according to which proton-translocating electron transport and ATP synthase systems may be linked to form a phosphorylating proton circuit. Mitchell suggested (6, 7) that in chloroplasts the primary photochemical electron transfers would provide the electrogenic arms of the loops, whereas proton release into the thylakoid interior could result from the oxidation of water and plastoquinol. This vectorial Z scheme (Figure 1) has derived a great deal of experimental support from two schools of research in particular, those of Witt (8) and of Trebst (9). Together with the proton-translocating reversible ATPase, established by the classical experiments of Jagendorf, Hind, and colleagues (10–12), it provided a simple and satisfying chemiosmotic system.

The aim of this review is to discuss the molecular mechanisms of electron and hydrogen transfer in the proton-translocating electron transport system of the chloroplast. An important question is the strength of the evidence that electron and proton transfer are directly linked. A strong case can be made for such a linkage in the cyclic electron flow of purple photosynthetic bacteria (13), but is beyond the scope of this review. Emphasis will be placed on electron transfer between the two light reactions, a topic which has been relatively neglected in recent reviews. The general standpoint adopted here is similar to that taken by Trebst (9) so that in the present review particular attention is paid to recent developments and to questions which are still controversial.

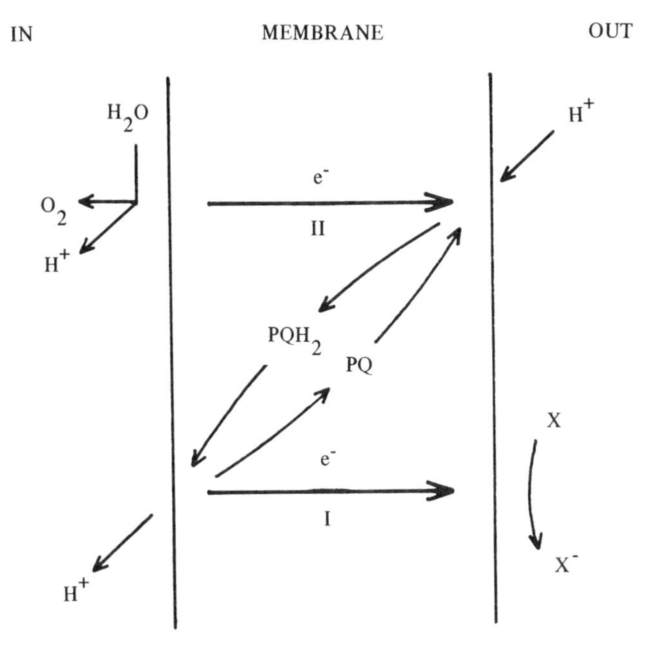

Figure 1. Vectorial Z scheme of electron and proton transfer in chloroplasts. Stoichiometry has not been considered.

It is convenient to start with a brief account of recent work on the reaction center of purple nonsulfur bacteria, because a clearer picture has been obtained with this system than has yet proved possible with photosystems of chloroplasts.

REACTION CENTERS IN PURPLE NONSULFUR BACTERIA

Primary processes of bacterial photosynthesis have been reviewed in detail (14, 15). For the present purposes, it is sufficient to draw attention to molecular properties of reaction center preparations that lead to a view of the primary process as an electron transfer organized within a membrane-spanning protein.

Fractionation of bacterial chromatophore preparations with detergent can separate the bulk of the pigments from a small bacteriochlorophyll-protein complex that retains photochemical activity. The best preparations have been obtained with the nonionic detergent lauryldimethylamine oxide, and the most thoroughly studied material has come from the carotenoidless mutant R-26 of *Rhodopseudomonas spheroides* (16, 17), although similar preparations have also been obtained from the wild type and some other purple nonsulfur bacteria (18–20). The preparation from *R. spheroides* R-26 is a protein with a minimum molecular weight of 73,000 (21). It contains three different subunits of molecular weights 28,000, 24,000, and 21,000, which occur in the molar proportions 1:1:1. The protein is very hydrophobic, as would be expected of an intrinsic membrane protein, with 68% nonpolar residues (71% in the 21,000 subunit) (22). It is now clear that each particle of 73,000 contains 4 molecules of bacteriochlorophyll and 2 of bacteriophaeophytin (23). Circular dichroism spectra suggest that the bacteriochlorophyll molecules form a complex closely linked by exciton coupling (24, 25) (see Sauer (18, 26) for discussions of the application and interpretation of circular dichroism to photosynthetic materials), which is broken when the complex undergoes a 1-electron photo-oxidation. Oxidation causes a bleaching at 865 nm which is usually attributed to a pigment designated P870. The circular dichroism spectra suggest that the absorption band at 865 nm is a property of the whole complex, but there is now strong evidence from the application of electron spin resonance (ESR) (27, 28) and electron nuclear double resonance (29) techniques that the bleached form, $P870^+$, is a dimer of bacteriochlorophyll which has lost 1 electron and, therefore, behaves as a free radical.

P870 is the primary electron donor of the photosystem. The fact that reaction center preparations show photochemical bleaching of P870 indicates that they also contain the primary acceptor. Analyses have shown them to contain 1 nonheme iron atom per P870 (17) and some tightly bound ubiquinone (30). Irradiation of the preparation with a single flash seems to cause reduction of the iron atom, as judged by the appearance of a complex ESR signal in liquid helium (31), whereas no free radical signal appears that might be attributed to ubisemiquinone. On the other hand, photochemical activity is lost reversibly if the ubiquinone is extracted (32), but removal of iron (33, 34) or its partial substitution by manganese (33) has no effect. After removal of iron, reduction of the primary acceptor gives a free radical ESR signal (34). Optical absorption changes characteristic of the reduction of ubiquinone to its semiquinone (35) have been observed in iron-containing preparations (36, 37). The natural conclusion to draw from these observations is that the primary acceptor is a molecule of ubiquinone bound to an iron atom. Mössbauer studies (33) have indicated that the iron is Fe^{2+} and does not undergo oxidation and reduction. Binding of the ubiquinone to the reaction center protein could explain the much greater stability of the semiquinone than is the case in solution at neutral pH. It is not clear why reduction to the fully reduced form should not occur with a second photochemical act. Two anomalous observations require explanation. Wraight et al. (38) reported a good correlation between the rate of decay of the 450 nm absorbance of ubisemiquinone and the recovery of the ability to photo-oxidize P870 a second time, but when secondary electron acceptors were added, the agreement broke down, the semiquinone decay being slightly retarded and the recovery of activity greatly accelerated. A second difficulty is that while most oxidation-reduction titrations purporting to determine the potential of the primary acceptor have yielded values in the range −15 to −160 mV (15) (c.f. the value of +65 mV reported by Urban and Klingenberg (39) for the midpoint potential of ubiquinone in submitochondrial particles), claims that the true potential is very much lower have been made recently (40–42). Recent work on the acceptor complex of photosystem II might be relevant to some of these problems (see below).

There is little direct evidence yet available about the molecular structure of the reaction centers, but there is significant indirect evidence about the arrangement of P870 and the primary acceptor. There are now compelling reasons for considering that the primary charge separation takes place across the chromatophore membrane. The arguments will not be discussed here, but in essence they are similar to many of those referred to below in relation to the chloroplast photosystems. They have led to the suggestion that the protein complex of the reaction centers spans the membrane, with the P870 bacteriochlorophyll complex accessible to electron donors (cytochrome c_2) at the inner surface and to the ubiquinone acceptor complex near the outer surface. Charge separation across the lipid phase of the membrane would be expected to hinder the spontaneous recombination of charges. Studies of the rate of the back reaction (which can occur when secondary transfer is blocked by *O*-phenanthroline or low temperature) have provided information on the distance across which the primary charge separation occurs. The decay kinetics of $P870^+$ after illumination has been followed by optical or ESR measurements under conditions in which the reduced primary acceptor is the only plausible donor (43–45) (although simultaneous decay of the primary acceptor has not yet been demonstrated). The reaction occurs at a significant rate ($\tau \approx 30$ ms) at temperatures near to absolute zero, and the rate does not vary between 1.6 and 100°K. Above 100°K, the rate *declines* as the temperature rises. These observations have been explained by invoking quantum mechanical tunneling as a mechanism for the nonthermal return of the electron. Calculations suggest that tunneling occurs through an energy barrier about 30 Å wide. The decreasing rate at higher temperatures could be explained by slight increases in the barrier width.

The primary charge separation thus seems to occur across an appreciable distance, of the same order as the width of the chromatophore membrane. The pathway by which electrons from excited P870 can cross this distance in a few nanoseconds or less is an intriguing question. The other pigment molecules of the reaction center complex or aromatic amino acid residues of the protein might be involved. Junge (46) points out that the mobility of electrons photoinjected into pure hydrocarbon phases is very high.

PHOTOSYSTEM I

Photosystem I of chloroplasts differs in important respects from the bacterial photosystem. Nevertheless, although there is less precise information available about its organization because of the lack of a genuine reaction center preparation, there is now substantial evidence for an underlying similarity in that, in both types of photosystem, the primary photochemical event is electron transfer over an appreciable distance across the membrane. Recent reviews dealing with photosystem I are those of Trebst (9) and Bearden and Malkin (47).

Attempts to purify the reaction center have so far always yielded preparations which retained substantial amounts of light-harvesting chlorophyll. The best preparations are substantially free of chlorophyll *b* and secondary electron carriers such as cytochromes. For example, the chlorophyll-protein complex I isolated by Thornber (48) by fractionation with the detergents sodium dodecylsulfate or Triton X-100 has a molecular weight of 110,000, 14 molecules of chlorophyll *a* per molecule of complex, and a ratio of chlorophyll to P700 which equals 40–45. These figures show that only about one in three of the particles contains an active center; it is not known whether the heterogeneity of pigment is reflected in heterogeneity of protein. There is conflicting evidence in the literature concerning the polypeptide composition of chlorophyll-protein complex I. Anderson and Levine (49) found it ran as a single band in polyacrylamide gels in the presence of sodium dodecylsulfate and was resolved into two polypeptides of 50–60,000 molecular weight by removal of lipid and pigment. Machold (50) reported a single lipid-free polypeptide of 69,000. Nelson and Bengis (51, 52) obtained a particle which gave a polypeptide with a molecular weight of 70,000 in polyacrylamide gels (lipid does not seem to have been removed); the ratio of chlorophyll to P700 was similar to that of Thornber's preparations.

Although the purest preparations show photobleaching of P700, the shift in its peak position to wave lengths below 700 nm suggests that it has been modified and photochemical activity with added electron donors or acceptors is either absent or different in character from that of the intact system. A preparation obtained by Malkin (53) by use of lauryl dimethylamine oxide showed a very high rate of light-induced oxygen uptake with ascorbate as donor, whereas chloroplasts disrupted by gentler procedures do not show this kind of activity to a significant extent. Pure chlorophyll *a* dispersed in micelles of Triton X-100 can show high rates of light-induced electron transfer. There is thus a need for a preparation which shows normal kinetics and specificity in its photochemical reactions. Nelson and Bengis (51, 52) found that chromatography of a Triton extract of chloroplasts on DEAE-cellulose yielded a preparation which was active in NADP reduction in the presence of plastocyanin, ferredoxin, and ferredoxin-NADP reductase; this preparation contained polypeptides of molecular weights 70,000, 25,000, 18,000, and 16,000. Purification of the 70,000 fragment by treatment with sodium dodecylsulfate and sucrose density gradient centrifugation led to loss of ability to reduce NADP and a shift in the peak position of P700 to shorter wave lengths.

As with the bacterial P870, there is good evidence from studies involving ESR (27, 54), electron nuclear double resonance (55), and circular dichroism (56) that P700 is a chlorophyll dimer. The oxidation-reduction potential of P700 seems to be significantly higher than originally estimated by Kok (57) (E_0' = +430 mV, pH 7). Ruuge and Izawa (58) have determined a midpoint potential of +500 mV in intact chloroplasts and reported that acetone treatment, as used by Kok, lowers the value. Knaff and Malkin (59) obtained a value of +520 mV in digitonin-photosystem I particles.

Transmembrane Model for Photosystem I

Trebst (9) has critically assessed the evidence for the view that the primary process of photosystem I may be regarded as charge separation, positive inwards, across the thylakoid membrane. Briefly, the argument rests upon two types of experiment. The first attempts to localize individual components associated with the donor or acceptor side by labeling either with specific antibodies or with nonpenetrant chemical labels. The second involves observations on the light-induced absorbance increase in the region 515–520 nm, which Witt (8) has interpreted as an effect of the membrane potential (electrochromic effect), generated by the primary charge separation, on the spectrum of pigments in the membrane. Strichartz and Chance (60) showed that a similar absorbance change could be caused in the dark by salt addition to impose a diffusion potential (cf. the experiments of Jackson and Crofts (61) with bacterial chromatophores). Crofts et al. (62) have recently expressed serious doubts about the interpretation of the related phenomenon in bacterial chromatophores as an electrochromic effect, but this need not invalidate the general argument.

Some simple but ingenious experiments recently carried out by Fowler and Kok (63) provide direct evidence that the primary reaction generates an electric field. Illumination of a chloroplast suspension with a nonsaturating flash was shown to give rise to a weak transient electric field in the direction of the flash. Attenuation of the flash by pigment absorption would give progressively fewer charge separations and hence a net electric dipole as the beam passed through the suspension. The fact that the electrode nearest to the light source became negatively charged determined the polarity of the primary charge separation according to this interpretation: the surface of the thylakoid in contact with the medium gained a negative charge. The signal would decay rapidly by compensating diffusion of ions in the suspending medium.

A further source of information about the photochemical charge separation is the spontaneous back reaction. Simultaneous reduction of P700 and oxidation of an acceptor molecule should be observed. At room temperature, the back reaction can be seen when secondary donors and acceptors are absent (64, 65). Hiyama and Ke (64) observed a reaction with $t_{1/2} = 45$ ms between P700 and a pigment P430 identified by its optical absorption band at 430 nm in the oxidized state. P430 was later shown (66) to be accounted for by a bound iron sulfur protein of the ferredoxin type, discovered by Malkin and Bearden (67), who observed its photoreduction by photosystem I at 77°K by ESR spectroscopy. Oxidation-reduction titration revealed more than one species of bound ferredoxin. Evans et al. (68) analyzed their results in terms of two components, one with g values at 2.05, 1.94, and 1.86, with $E_0' = -553$ mV, the other with g values 2.05, 1.92, and 1.89, with $E_0' = -594$ mV. A species with $g = 1.94$ has been isolated by Malkin et al. (69) and has been shown to be a protein of molecular weight 8,000, containing 4 atoms each of iron and inorganic sulfide per molecule.

Photo-oxidation of P700 at low temperature has frequently been found to be irreversible, although partial reversibility has been reported (70, 71). Both Visser et al. (72) and Ke et al. (73) have reported a back reaction between a reduced iron sulfur protein and $P700^+$ at temperatures below about 200°K. The proportion of centers reacting in this way decreased as the temperature was lowered. Both groups reported a biphasic reaction with rate constants which were independent of temperature below 150–200°K, although the actual values of the constants obtained differed markedly. In contrast to these results, McIntosh et al. (74) and Evans and Cammack (75) found a dark reduction of P700 at low temperature that was not associated with reoxidation of iron-sulfur proteins of the ferredoxin type. McIntosh et al. found a new ESR signal with *g* values at 1.75 and 2.07 that was complementary in its behavior to that of P700. This component might represent a different class of iron-sulfur protein. Evans and Cammack observed a completely reversible photo-oxidation of P700 at 20°K when the bound ferredoxins were chemically reduced with dithionite before freezing. This type of observation suggests that the bound ferredoxins are secondary acceptors.

Although the results described above seem somewhat contradictory, they all have in common the observation of a temperature-independent back reaction which could most readily be explained by quantum mechanical tunneling of the electron through a barrier of appreciable thickness. Warden et al. (76) calculated a barrier width of about 50 Å and values around 40 Å were derived by Visser et al. (72). These distances are of the same order as the thickness of the thylakoid membrane and therefore lend support to the transmembrane model of the primary reaction.

Evidence Against the Transmembrane Model for Photosystem I

Some observations that seem to run counter to this model concern the location of plastocyanin within the chloroplast. Evidence that plastocyanin donates directly to P700 is now strong (see below). If P700 is internal, as discussed above, plastocyanin should be too; a transmembrane reaction between plastocyanin by a tunneling mechanism would be too slow and plastocyanin cannot be photo-oxidized at 77°K (77). Experiments supporting the internal location of plastocyanin have been reported (78, 79), but contrary evidence has been obtained by the use of nonpenetrating reagents that react with plastocyanin. The first type of reagent of this kind was polylysine and similar polycations, which are assumed to be too large to penetrate the thylakoid membrane. In solution, polylysine will form a complex with plastocyanin, which is an acidic molecule (in higher plants), and polylysine appears to inhibit photosystem I activities in the vicinity of plastocyanin (80, 81). Direct evidence for reaction between plastocyanin in the chloroplast and externally added polylysine is lacking, and

Trebst (9) suggested that polycations would combine with negative charges on the surface of the thylakoids and might induce a conformational change that would indirectly interfere with the normal functioning of plastocyanin. More direct evidence for an external location was provided by the labeling experiments of Selman et al. (82), who obtained an appreciable level of labeling of plastocyanin by [^{35}S] diazoniumbenzene sulfonate, which was considered to be a nonpenetrant protein label, even with apparently intact thylakoid preparations. Disruption of the chloroplasts increased the incorporation of reagent into plastocyanin by a factor of about three. It is not clear whether these experiments should be interpreted as indicating that all the plastocyanin is external but partially protected or alternatively whether two-thirds is completely inaccessible and one-third completely accessible. One-third seems too large a fraction of the total to be accounted for simply by the proportion of broken thylakoids in the preparation. Further support for the accessibility of at least some of the plastocyanin comes from observations that an antiserum to purified plastocyanin can cause agglutination of chloroplasts and a partial inhibition of electron transport (83). The fact that plastocyanin added to intact preparations is not photo-oxidized (79) suggests that any external plastocyanin is inactive. The remaining internal plastocyanin would probably be enough to maintain high rates of electron transport, because the total amount of plastocyanin in chloroplasts is 3 times the amount of P700 (77, 84, 85).

PHOTOSYSTEM II

Photosystem II is more labile and complex than the photosystems discussed so far, and precise chemical information has been difficult to obtain. The fundamental problem is how the primary process, which transfers one equivalent, donates four equivalents to water to liberate a molecule of O_2. The solution to the problem has come from the study of the yield of O_2 from each flash of a series (86–89), which was made possible by the development of sensitive O_2 electrodes by Joliot (90); in principle, the flashes are short and intense enough to bring about single, synchronous turnovers of all the reaction centers. The most satisfactory general explanation of the pattern of damped oscillations of period four is the linear four-step model of Kok et al. (87), which postulates that an appropriate carrier (presumably a manganese protein) has to accumulate four successive oxidizing equivalents from the same reaction center before a molecule of O_2 can be evolved. (Recent reviews concerning photosystem II are found in references 47, 91, and 92.)

Membrane fragments with photosystem II activity can be separated (93–96) from photosystem I and from light-harvesting chlorophyll which may probably be identified with the light-harvesting chlorophyll-protein complex ("Complex II") of Thornber (48). The Triton X-100 particles, TSF-2a of Vernon et al. (93, 94), and the digitonin particles, FII, of Wessels et al. (95) have been referred to as reaction center preparations, but, in fact both are rather complex particles which retain a considerable amount of bulk chlorophyll. Their overall composition is similar, with a ratio of chlorophyll *a* to *b* ≈ 25, a ratio of β-carotene to xanthophyll higher than in chloroplasts, and the presence of some oxidation-reduction carriers, i.e., the pigment C550, cytochrome *b*-559, and a trace of plastoquinone. FII contains six or more different polypeptides (97), but is notably lacking in polypeptides II*a* and II*b* of Levine et al. (98), which have a molecular weight of about 25,000 and have been associated with the light-harvesting chlorophyll-protein complex by Anderson and Levine (49). Both types of particle have lost the ability to photo-oxidize water, but have retained the ability to bring about reactions between artificial donors and acceptors which are sensitive to 3-(3,4-dichlorophenyl)-1,1-dimethylurea (DCMU) and characteristic of photosystem II.

The optical and ESR spectra of P680, the primary electron donor of system II, are closely similar to those of P700 (99, 100), and, therefore, P680 may also be a special dimer of chlorophyll *a*. The most obvious difference between P700 and P680 lies in the relatively high oxidation-reduction potential of the latter. Malkin and Bearden (99) observed the photo-oxidation of P680 in the presence of ferricyanide and concluded that its midpoint potential must be greater than +600 mV, and clearly it should be high enough to bring about decomposition of water (+820 mV, or more). Bendall and Sofrová (101) pointed out that, considering the potentials of the one-equivalent steps of the oxygen-evolving system in the model of Kok et al. (87), the couple S_0/S_1 probably has a potential considerably lower than +800 mV and S_3/S_4 considerably higher. P680 would have to have a potential at least as high as S_3/S_4. Kok et al. (102) have recently been able to estimate the midpoint potentials for each step relative to that of P680 and found for S_0/S_1, $\Delta E = -320$ mV, for S_1/S_2, $\Delta E = -260$ mV, and for S_2/S_3, $\Delta E = -160$ mV. From these data, one can calculate a minimum potential for P680 (see 101). If 5.5 is taken as the pH of the thylakoid interior, $E'_0 \geqslant +910$ mV (the potential of the couple H_2O/O_2 at this pH) can be written for the four-equivalent couple S_0/S_4. This leads to $E'_0 \geqslant +1.1$ V for P680; the theoretical maximum is about +1.3 V (103).

Reduction and oxidation of the primary acceptor of photosystem II have been detected mainly by the variable fluorescence quenching ("Q" of Duysens and Sweers (104)) or by absorbance changes. Surprisingly, ESR signals attributable to the primary acceptor have not been observed. The component C550 has many properties one would expect of the primary acceptor, e.g., apparent photoreduction at low temperature (101, 105, 106), oxidation-reduction potential (107), and correlation with fluorescence oscillations in a flash sequence at −40°C (108). Nevertheless, photochemistry does not depend on the presence of C550 (109, 110), and it seems more likely that it acts as an indicator of the primary reaction, as suggested by Butler and Okayama (111). The chemical nature of C550 has not been established, but there are indications that it may be a β-carotene protein (110, 112).

A second type of absorbance change with properties qualifying it to be a possible primary acceptor was discovered by Stiehl and Witt (113) and called X-320. X-320 shows very rapid turnover on flash illumination and can be photoreduced at low temperatures (114). The suggestion of Stiehl and Witt that it represents the reduction of plastoquinone to its semiquinone has received support from the spectra of plastosemiquinones in solution obtained by Bensasson and Land (35). A detailed spectrum measured in photosystem II particles prepared with deoxycholate has been reported by Van Gorkom (115), and it corresponds well with the difference spectrum for reduction of plastoquinone to its semiquinone anion, except for a shift of about 12 nm toward longer wave lengths, possibly under the influence of the detergent. These results indicate that the primary acceptor is likely to be a special bound form of plastoquinone in which the semiquinone anion is stabilized. Pulles et al. (116) have shown that with chloroplasts at −40°C subjected to a series of flashes the acceptor plastoquinone and C550 behave in a similar manner which is dependent on the oxidation-reduction state of the oxygen-evolving system; these results are consistent with the view that C550 indicates the oxidation-reduction state of a primary plastoquinone acceptor molecule. However, the possibility exists that two different types of primary acceptor are involved (108, 117, 118).

The fact that the fluorescence quencher Q is fully reduced by a single flash indicates that the primary acceptor behaves as a 1-electron carrier. The failure of the bound $PQ^{\overline{\cdot}}$ to be reduced to PQH_2 needs explanation. Dismutation of $PQ^{\overline{\cdot}}$ would be prevented by tight binding of a single molecule to each reaction center.

Transmembrane Model for Photosystem II

As with photosystem I, there are compelling reasons for the view that the primary process of system II is charge separation across the thylakoid membrane. Perhaps the simplest evidence is provided by the experiments of Fowler and Kok (63) discussed above. These authors found that the transient electric dipole generated by flash illumination of a chloroplast suspension received equal contributions from each of the two photosystems. For additive contributions, the two systems must be similarly oriented in the membrane, with the donor inside and the acceptor outside.

The above conclusion is consistent with the observations of Schliephake et al. (119) and Joliot and Delosme (120) that the electric field indicated by the absorbance change at 515–520 nm receives additive contributions from the two photosystems. Similarly, delayed fluorescence, which is thought to arise from a spontaneous back reaction between the oxidized primary donor and reduced primary acceptor of photosystem II, shows a contribution from the membrane potential and provides evidence of orientation (103, 121).

At room temperature, the back reaction is quantitatively insignificant, but at low temperatures it may become more important, owing to the slowing down of electron transfer from Q to secondary acceptors (122, 123). A single, saturating flash at the temperature of liquid nitrogen does not cause a stable reduction of C550 (71, 124, 125) or of the fluorescence quencher Q (125, 126). This has been shown to be the result of a rapid back reaction between the primary acceptor (as indicated by C550) (127) and $P680^+$ (99, 128), which occurs at the same rate at 35°K and 77°K (99). The temperature insensitivity may reflect the return of the electron by a tunneling mechanism, but $t_{1/2}$ is much shorter (5 ms) than has been found with photosystem I. The barrier width is, therefore, likely to be smaller, possibly reflecting a buried position of Q (see below).

The transmembrane model for photosystem II suggests that water oxidation takes place at the inner surface of the thylakoid membrane. The same argument applies here as with the generation of a membrane potential by the primary charge separation in the two photosystems; proton pumping by two proton-translocating sites implies that both sites have the same orientation. In this case, because protons are pumped inwards, liberation of H^+ by water oxidation must occur inside the thylakoid. Direct evidence for this was recently provided by Fowler and Kok (129) who measured proton exchanges during a series of short, intense flashes with a rapidly responding glass electrode. A progressive proton uptake was observed with each flash. In the presence of methylamine, proton uptake gave way to a proton release, which was interpreted as coming from the dissociation of protonated amine molecules as the free base diffused inwards to take up internally liberated protons. The liberation of internal protons showed oscillations of period four with the maximum release on the third flash. This pattern characterized the protons as originating in the photo-oxidation of water and demonstrates not only that this occurs in the thylakoid interior but also that all the protons from water are liberated in synchrony with the O_2 molecule.

Fowler and Kok's experiments do not, of course, provide any direct evidence that O_2 itself is liberated to the thylakoid interior. That this is, in fact, the case is indicated by the experiments of Blankenship and Sauer (130) on the fate on endogenous manganese in Tris-washed chloroplasts. The process of washing with a high concentration of Tris buffer reversibly inactivates oxygen evolution and also causes the reversible appearance of the ESR signal of free Mn^{2+} The free Mn^{2+} is not liberated into the medium and so presumably must be released into the interior space of the thylakoid.

Evidence Against the Transmembrane Model for Photosystem II

A basic difficulty in trying to understand the topology of photosystem II has been the accessibility of the system toward artificial electron donors and acceptors. For example, Q has seemed to be buried in the membrane, judging by its inability under normal circumstances to donate directly to water-soluble acceptors such as ferricyanide and dichlorophenolindophenol. However, the same is true of lipid-soluble acceptors (quinones, phenylenediamines), because their reduction is inhibited by DCMU even though they might be expected to penetrate into the lipid interior of the membrane. Recently it has been shown

(131, 132) that silicotungstate and silicomolybdate are able to act as system II acceptors even in the presence of DCMU and so can presumably interact directly with Q. At the same time, ferricyanide reduction becomes insensitive to DCMU. These observations suggest that Q may not be so much buried in the membrane as masked by protein removable by silicotungstate, which is known to dissociate the chloroplast coupling factor CF_1 (133). Ausländer and Junge (134) have suggested that at the external surface of the thylakoid membrane there is a layer of hydrophobic material that acts as a barrier to free diffusion of protons, and presumably of other ions too.

Articial donors such as I^-, Mn^{2+}, and $Fe(CN_6)^{4-}$ also present a problem; the donor side of photosystem II is supposedly inside, and these hydrophilic ions are unlikely to be able to penetrate the lipid barrier of the membrane. A distinction between the response of photosystem II to nonpenetrant hydrophilic donors and penetrant lipophilic donors was made by Babcock and Sauer (135), who studied a rapidly responding ESR signal of type II which they referred to as II_f. Signal II_f behaved as if it were caused by the oxidized form of a secondary donor (Y) to P680, if one considers electron transfer in photosystem II to take the following form: $H_2O \rightarrow Z \rightarrow Y \rightarrow P680 \rightarrow Q$; in this scheme, Z indicates the component which stores four successive oxidizing equivalents. The light-induced signal II_f was stabilized in Tris-washed chloroplasts which were no longer capable of oxygen evolution, and the decay was restored by lipophilic, but not by hydrophilic, donors. On the other hand, Mn^{2+}, a hydrophilic donor, inhibited the formation of the signal, presumably by competing with Y for oxidizing equivalents from P680. These observations suggested that P680 is external and Y internal. A difficulty for this model is that it would require the light-induced charge separation which is sensed by the absorption change at 515 nm to involve the secondary donor Y. The rise time for the 515 effect is less than 20 ns (136), whereas $t_{1/2}$ for electron transfer from Y to P680 is probably 35 μs (137), based on the decay time of delayed light (138) and the turnover time of P680 (139). An alternative viewpoint, which allows the continued belief in an internal P680, is that hydrophilic donors donate via an externally located endogenous carrier. The natural function of such an endogenous carrier might be to prevent overoxidation in the reaction center by providing a pathway for electrons from ascorbate available in the stroma. The best known alternative donor is cytochrome b-559_{HP}, which is thought to be externally located (140, 141); other less well characterized donors have been postulated (124, 142), but their position in the membrane has not been established. Cytochrome b-559_{HP} is oxidized by photosystem II when water oxidation is impaired. Cramer and Horton (143) argued from the external location of the cytochrome that P680 is external also. However, the temperature insensitivity of cytochrome oxidation (144) suggests electron transfer by a tunneling mechanism, possibly across the membrane. The half-time for cytochrome oxidation at low temperature (4.6 ms) (144) is the same as for the back reaction between Q (or C550) and P680, but this time is admittedly very much shorter than that required for the back reaction in photosystem I or the reaction center of purple bacteria. Cytochrome *b*-559$_{HP}$ might well provide a reasonable pathway for oxidation of ferrocyanide and iodide, although the possibility exists, especially for iodide, that at the high concentrations required they are able to diffuse through the membrane (145). Oxidation via the cytochrome is much less likely for Mn^{2+} because of the high oxidation-reduction potential of the Mn^{2+}/Mn^{3+} couple, but evidence that Mn^{2+} can be oxidized at a high rate by preparations with undamaged membranes is lacking.

Specific inhibition of oxidation of water and not of artificial donors by nonpenetrant reagents requires explanation if O_2 liberation occurs internally. Selman and Bannister showed that exposure of thylakoid membranes to trypsin inactivated the photo-oxidation of water more rapidly than that of diphenylcarbazide (146), even though it does not cause gross damage to the membrane integrity (147). Similarly, Giaquinta et al. (148) found that O_2 evolution is sensitive to diazoniumbenzene sulfonate and that electron transport could be restored by diphenylcarbazide. Observations of this type might be explained in terms of an attack on the outer surface of a membrane-spanning protein complex.

Dilley and coworkers have developed the idea that illumination results in a major conformational or even topological change in photosystem II (132, 148–150). The effect of light was attributed specifically to electron transport in the Q to plastoquinone region of the chain (149), and the conformational change was detected either by an increased binding of ^{35}S-labeled diazoniumbenzene sulfonate or by measurements on proton uptake or phosphorylation. Light-induced electron transport with methylviologen or dichlorophenolindophenol as acceptor brought about increased labeling which was sensitive to DCMU. However, as mentioned above, electron transport to silicomolybdate continued in the presence of DCMU, and the additional labeling no longer occurred (132). Neither proton uptake nor photophosphorylation could be detected when silicomolybdate was the acceptor, and it was postulated that in this system the electrons from water were liberated to the exterior rather than to the interior of the thylakoid. In most experiments, silicomolybdate behaved as an uncoupler when present in the oxidized form, but not in the reduced form. Giaquinta and Dilley (132) rejected this relatively trivial explanation, and it seems unlikely that a large polyanion could act as a proton ionophore. An effect on the membrane that would make it leaky to protons seems more plausible, but the failure of the reduced form to act similarly is difficult to explain. Recent work of Lockau and Selman (151) has shown that experiments on labeling with diazoniumbenzene sulfonate must be interpreted with caution in view of their finding that the reagent can be photoreduced by chloroplasts and that increased labeling in the light seems to result from the reactivity of some unstable intermediate of reduction.

ELECTRON TRANSFER BETWEEN PHOTOSYSTEMS

Structural Organization of Chain

The structural relation between photosystems I and II and, consequently, the mechanism by which they interact, are still very much an open question; the term "electron transfer chain" is convenient, but has implications that are too precise. The interaction occurs at two levels, that of energy transfer between light-harvesting chlorophyll molecules and that of electron transfer. Here, only the latter can be discussed, but the two are obviously interrelated.

The reaction centers of photosystems I and II seem to be present in approximately equal numbers. In grana-containing chloroplasts, system I centers slightly exceed system II centers (102, 152), but as much as 25% of the system I centers may reside in stroma lamellae which lack system II, so that, in the noncyclic electron transport system of the grana, system I centers are slightly fewer than those of system II (152). Each photosystem is organized into a protein complex, as discussed above, but whether or not these are associated in the membrane is unknown. Electron microscopic studies by the freeze-etch technique reveal at least two types of particle embedded in the membrane. Large particles occur mainly in grana on the inner fracture face of the thylakoid membrane and tend to be correlated with photosystem II; small particles occur on the outer fracture face of both grana and stroma membranes and are correlated with photosystem I (3). Ojakian and Satir (153) have shown from studies of the redistribution of particles during destacking of thylakoids in vitro that the two types of particle are able to move independently by lateral diffusion within the membrane. If the particles are in fact physical manifestations of the two photosystems, the observations of Ojakian and Satir indicate the need for a mobile electron carrier. However, the freeze-etch particles are primarily manifestations of chlorophyll-protein complexes, the major intrinsic proteins of thylakoids, rather than of photosystems per se, and so they are a poor guide to the physical arrangement of the photosystems (3). Thus, rather little can be said about the limitations imposed by structure on the way the electron transport system is organized, except that the transmembrane orientation of the photosystems discussed above requires a mechanism for "dark" electron transfer through the lipid barrier of the membrane.

When dark-equilibrated chloroplasts are illuminated, the fluorescence yield rises and the rate of oxygen evolution at first increases to a peak value and then declines more slowly to a relatively low steady state value which depends on light intensity (154). Both types of experiment provide evidence for a pool of some electron acceptor which is present in considerably larger quantity than the system II centers. The yields of fluorescence and of oxygen will depend on the oxidation-reduction state of Q, which in turn depends on the oxidation-reduction state of the pool. Estimates of the size of the pool vary, but lie in the range

10–20 equivalents per system II center (155). The only carrier known to be present on this scale is plastoquinone, and optical measurements at 265 nm demonstrate reduction of plastoquinone by system II and its reoxidation by system I (156). The size of the functional plastoquinone pool has been estimated from absorbance changes to be 15 electron equivalents per P700 (157) or 14 equivalents per P680 (158). Chemical estimates of the total plastoquinone content of chloroplasts are usually larger than this; much of the discrepancy can be accounted for by the presence of plastoglobuli which provide an inactive reservoir of plastoquinone.

The pool as defined by the fluorescence induction and oxygen gush curves is not necessarily composed of a carrier that is mobile. Evidence that plastoquinone does behave as a diffusible carrier was provided by Siggel et al. (159), who showed by partial inhibition with DCMU that any one photosystem II center could interact with a much larger quantity of plastoquinone than the average 14 equivalents mentioned above. Such "interchain" electron transfers might also be mediated by plastocyanin, which behaves as an extrinsic protein only loosely bound to the membrane; this notion would not be inconsistent with the results of Siggel et al. Some indirect evidence for a distributive function of plastocyanin was reported by Bouges-Bocquet (160, 161), and further support comes from observations that extreme shade plants contain considerably less than one cytochrome *f* molecule per P700 (162, 163).

The plastoquinone pool seems to perform three distinct functions. First, there is the distributive function mentioned above. Second, it provides a pool of oxidizing equivalents to allow the proper operation of the ratchet mechanism for oxygen evolution. Third, it provides a simple mechanism for transfer not only of electrons but also of protons from one side of the membrane to the other. Proton transfer would arise from the fact that the neutral quinone and quinol molecules could probably diffuse across fairly readily, whereas the dissociated form of the quinol (PQ^{2-}) would remain anchored to one side by its charge. However, direct evidence that plastoquinone does behave in this way is lacking. The evidence that proton transfer occurs in this region of the chain is discussed in the following section.

H^+ Flow Coupled to Electron Transfer Across Membrane

Most measurements of the ratio $H^+:e^-$ for overall noncyclic electron transport have yielded values between 1 and 2 (see Dilley and Giaquinta (164) for a recent review). Many of the technical difficulties have been overcome by measuring changes in proton concentration in a chloroplast suspension in response to single turnovers of the photosystems. Junge and Ausländer (165) used the absorption changes of cresol red to indicate pH changes in the medium and a repetitive flash technique with signal averaging to increase the signal-to-noise ratio. When benzylviologen or ferricyanide at a low concentration (0.3 mM) were present as electron acceptors, electron flow through both photosystems was possible. Under these conditions, each flash caused the uptake of 1 proton with ferricyanide and 2 with benzylviologen (1 of these would be required for oxygen reduction by the reduced dye). Similar measurements in the presence of the proton ionophore carbonylcyanide-*p*-trifluoromethoxyphenylhydrazone (FCCP) showed that 2 protons were liberated to the thylakoid interior on each flash. On the other hand, 10 mM ferricyanide provided conditions under which only photosystem II was operative (no P700 turnover was detected), and then there was no proton uptake from the medium and 1 proton was released to the interior. These experiments provided evidence for a site of proton translocation between the photosystems with $H^+:e^- = 1$, and a site of proton release to the interior (presumably from water oxidation) associated with photosystem II.

A surprising feature of Junge and Ausländer's experiments was that proton uptake from the medium was considerably slower ($t_{1/2}$ = 60 ms) than electron transfer (for reduction of plastoquinone $t_{1/2}$ = 0.6 ms (166)). Ausländer and Junge (134) showed that proton uptake could be accelerated by grinding the chloroplasts with sand or by exposure to digitonin or uncouplers, and they suggested that the site of plastoquinone reduction is shielded from the external medium by a removable barrier (presumably proteinaceous) to free diffusion of protons. This explanation would require the presence of a reservoir of protons beneath the barrier. Under steady state conditions, when the rates of proton uptake and electron transfer must be equal, the rate of proton diffusion through the barrier would be increased by depletion of the reservoir.

Dilley and Giaquinta (164) concluded that most measurements of the steady state rates of proton uptake and overall electron transport, when made near pH 6, gave $H^+:e^- \approx 2$, as expected from the results discussed above and other experiments, whereas at pH 8 a value near 1 was usually found. They discussed several possible explanations for low values and suggested that at pH 8 protonation of PQ^{2-} is slowed down to such an extent that a significant electrophoresis of PQ^{2-} across the membrane occurs under the influence of the membrane potential. This explanation would be consistent with the idea of a barrier to free diffusion of H^+ at the outer surface of the thylakoid and with the observation that valinomycin in the presence of potassium ions tends to raise the value of $H^+:e^-$ toward 2.

Measurements of the $H^+:e^-$ ratio have thus provided strong evidence for a site of proton translocation (and no more than one site) associated with the oxidizing side of photosystem I or with the intersystem chain. Energy considerations suggest that this site is located between plastoquinol and cytochrome *f*, and this is supported by evidence for a phosphorylation site in this region of the chain (167, 168) from observation of the effect of ADP on the oxidation-reduction state of the carriers. The transmembrane model discussed above would require plastoquinone to be reduced externally and cytochrome *f* to be internal. Evidence for the internal location of cytochrome *f* has been presented (169, 170). Contrary to earlier suggestions, this site of proton translocation is probably not used by artificial donor systems such as ascorbate and dichlorophenolindophenol or diaminodurene. These donor systems do not reduce the pool and are probably oxidized mainly through plastocyanin. Moreover, their oxidation is not inhibited by 2,5-dibromomethylisopropyl-*p*-benzoquinone (DBMIB) (171), which is regarded as a specific inhibitor of plastoquinol oxidation. Leucoindophenol and diaminodurene probably liberate a proton when they are oxidized at the inner thylakoid surface and so act as artificial proton pumps (9, 172–174).

Mechanism of Plastoquinol Oxidation

The reaction in which plastoquinol is oxidized is likely to be a proton-translocating step and tends to be rate-limiting for the whole electron transport system (158, 175). It is also chemically complex, in that plastoquinol is a two-equivalent carrier, whereas cytochrome *f*, or any other known acceptor which might be considered, is a one-equivalent carrier.

Stiehl and Witt (158) made a detailed study of the kinetics of oxidation and reduction of plastoquinone in chloroplasts by spectrophotometric measurements of the absorbance changes at 265 nm. They concluded that the rate of reoxidation of plastoquinol depends on the concentration of the quinone (PQ) as well as that of the quinol (PQH_2):

$$\frac{d(PQH_2)}{dt} = -k(PQ)(PQH_2)$$

They suggested, therefore, that the reduction of cytochrome *f* is by the semiquinone formed by back dismutation. Semiquinones react rapidly with cytochromes, whereas back dismutation tends to be slow so that the rate-limiting step was thought to be the reaction:

$$PQ + PQH_2 \rightarrow 2\ PQH\cdot .$$

This mechanism is similar to that proposed by Yamazaki and Ohnishi (176) for the reduction of mammalian cytochrome *c* by hydroquinone, a reaction which is slow in the absence of added quinone and is autocatalytic. A major disadvantage of Stiehl and Witt's mechanism is that it predicts an inhibition of overall electron transport at high light intensities as the concentration of PQ falls; such an inhibition has not been reported. A further weakness is that it fails to explain why electron transfer from plastoquinol is very much faster than from simpler quinols (177).

The problem has been approached in a different way by Wood and Bendall (178), who used a model system in which disrupted chloroplasts catalyzed a dark electron transfer between plastoquinol-1 (a synthetic analogue of the natural plastoquinol-9 with a side chain containing only 1 isoprenoid unit) and plastocyanin (or, more conveniently, *Pseudomonas* cytochrome *c*-551). The reaction was catalyzed by a particulate fraction prepared from chloroplasts by treatment with digitonin, followed by density gradient centrifugation as described by Wessels (179, 180). These particles contain *f* and some type *b* cytochromes but are almost devoid of chlorophyll. The reaction was shown to be dependent upon the membrane-bound cytochrome *f*, and purified preparations of the cytochrome were inactive. Plastoquinol-1 was a successful donor in this system, presumably because it has sufficient water solubility to gain access to the system; plastoquinol-9 merely formed inert micelles when an ethanolic solution of it was added to an aqueous chloroplast suspension. Plastoquinol-1 could replace the endogenous plastoquinol-9, because prior extraction of the latter had no effect on the rate. On the other hand, the isoprenoid side chain seems to be important, because trimethylquinol was a very much poorer donor, in agreement with the results of Trebst and Eck (177). Plastoquinol-1 oxidation was inhibited by DBMIB, as would be expected if the reaction is analogous to plastoquinol-9 oxidation in the physiological system.

These experiments provide evidence for the view that the reduction of plastocyanin by plastoquinol in vivo can only proceed via a membrane-bound cytochrome *f*. The type *b* cytochromes present in the digitonin particles did not seem to be directly involved in electron transfer. The experiments also suggest that membrane-bound cytochrome *f* can react with oxidation-reduction reagents by two types of mechanism. Lipophilic reagents of suitable potential react by a DBMIB-sensitive process, provided an isoprenoid side chain is present. Plastoquinol-1 shows a maximum rate (V) as defined by:

$$\frac{d(\text{plastocyanin})}{dt} = -V(\text{cytochrome } f)$$

with $V \approx 70\text{s}^{-1}$. This is very similar to the maximum rate of plastoquinol oxidation in chloroplasts (158). On the other hand, hydrophilic reagents apparently react with both membrane-bound (178) and soluble (181) cytochrome *f* by a simple bimolecular collision process which does not show a saturating rate. Thus, cytochrome *f* might be described as a two-faced protein. The activity of the lipophilic face depends on as yet unidentified factors.

Thermodynamic considerations suggest that plastoquinol oxidation occurs by a hydrogen atom transfer process (178), perhaps directly to cytochrome *f*. The oxidation of plastoquinol-1 does not occur by back dismutation, as suggested by Stiehl and Witt (158), because the rate is unaffected by addition of the oxidized form. Direct electron transfer from PQH_2 is improbable, because it would involve either the dissociated forms PQH^- or PQ^{2-}, which are present in extremely low proportions at neutral pH, or PQH_2 itself, which would give rise to $PQH_2^{\dagger}$. The oxidation-reduction potential of the couple $PQH_2/PQH_2^{\dagger}$ would probably be greater than +900 mV, compared with +365 mV for cytochrome *f*, so the rate of simple electron transfer would again be vanishingly small. This argument is probably valid even in the face of drastic displacement of the equilibrium by rapid deprotonation of $PQH_2^{\dagger}$, because the absolute rate would be expected to depend on relative oxidation-reduction potentials (cf. Marcus theory for outer sphere electron transfer reactions of transition metal complexes (182)).

Although cytochrome *f* is overall an electron carrier (its midpoint potential is independent of pH below pH 8 (183)), mechanistically it could behave as a hydrogen carrier provided loss of a proton were coupled to uptake of a hydrogen atom (184). If these two processes were to involve different groups in the protein, it is possible to begin to see how the oxidation of plastoquinol by cytochrome *f* might be coupled to proton flux across the membrane. On the other hand, if plastoquinol were oxidized by an electron transfer mechanism, the protons would tend to electrophorese outwards again unless the oxidation reaction took place firmly in the inner hydrophilic region.

The immediate product of hydrogen atom transfer would be the semiquinone PQH·. Further oxidation of the semiquinone is unlikely to be rate-limiting. Alternatively, two molecules of PQH· might react together and dismute. In any event, the proton needs to be conserved and deposited specifically within the thylakoid. The danger is that PQH· ($pK \approx 5{-}6$ in solution) might dissociate and allow the proton to escape to the exterior before oxidation of the semiquinone has taken place. Possibly the failure to achieve this accounts for the tendency for the ratio $H^+:e^-$ to fall at higher pH values.

Electron Transfer from Pool to P700

The classic experiments of Amesz, Duysens, and coworkers (157, 185) implied that cytochrome *f* is an essential link in the noncylic chain and connects plastoquinone with P700. The series scheme plastoquinol → cytochrome *f* → plastocyanin → P700 has an appealing simplicity. The function of plastocyanin might be primarily to act as a relatively mobile carrier only loosely attached to the inner surface of the thylakoid membrane (79); this would allow a cytochrome *f* molecule the possibility of reacting with more than one P700. However, considerable controversy surrounds this scheme, and the roles of both cytochrome *f* and plastocyanin have been called in question. Most of the claims that cytochrome *f* oxidation can proceed rapidly in the absence of plastocyanin can be explained by incomplete removal of plastocyanin, especially when one considers that higher plant chloroplasts contain a 3-fold excess of plastocyanin (on a copper basis) over P700 and that considerably less than one plastocyanin may be adequate to maintain high rates of cytochrome *f* oxidation. Knaff's experiments (185) are a possible exception to this. The series scheme is further supported by the effects of penetrant inhibitors, i.e., $HgCl_2$ (187) and KCN (188, 189).

The most serious difficulties that remain arise from kinetic observations on the intact system. These observations tend to confirm the participation of plastocyanin, but throw in doubt the role of cytochrome *f*. Before discussing these problems, two further observations may be made that favor the series scheme. First, it is difficult to see how a small hydrophilic protein such as plastocyanin could interact directly with plastoquinol, and the same applies to the readily soluble low molecular weight forms of cytochrome *f* that occur in some algae. In one or two algae, a bound form of cytochrome *f* has been found in addition to the soluble form (190, 191). Second, the interactions suggested in the series scheme have now been given a sound kinetic basis by studies with soluble systems. On the one hand, Wood (181) has determined a high value for the bimolecular rate constant ($k = 3.6 \times 10^7$ M^{-1} s^{-1}) for electron transfer from cytochrome *f* to plastocyanin (both purified from parsley), whereas values at least 30 times lower were obtained for the reduction of plastocyanin by red algal cytochrome *c*, *Pseudomonas* cytochrome *c*, or mammalian cytochrome *c*. On the other hand, Wood and Bendall (192) have shown a very high rate of reaction between soluble plastocyanin and P700 in disrupted chloroplasts ($k = 8 \times 10^7$ $\text{M}^{-1}\ \text{s}^{-1}$). Soluble cytochrome *f* from the red alga *Plocamium coccineum* reacted at about one-fourth of this rate, but the rate with parsley cytochrome *f* was at least 1,000 times slower. These studies indicate the chemical propensities of cytochrome *f*, plastocyanin, and P700, but their behavior might be significantly modified in vivo by binding to the membrane. The much slower rate ($k = 2 \times 10^5$ $\text{M}^{-1}\ \text{s}^{-1}$) reported by Warden and Bolton (65) for the reaction between plastocyanin and P700 in photosystem I particles (TSF-1) was presumably the result of inhibition by detergent.

The controversy centers around the apparent equilibrium constants between cytochrome *f*, plastocyanin, and P700 and the electron capacity of the carriers lying between the plastoquinone pool and P700. Apparent inconsistencies in the behavior of cytochrome *f* in light and dark led Marsho and Kok (193) and Haehnel (194) to suggest a light-induced change in the kinetic properties of the system, but this has not been substantiated. Although no model is likely to reconcile all the conflicting data, most authors agree upon the existence of a secondary donor which cannot be identified with cytochrome *f* and is likely to be plastocyanin.

Marsho and Kok (193) monitored the oxidation-reduction states of cytochrome *f* and P700 during oxidation by weak far-red light and during their dark reduction. They considered the secondary donors to represent a "pool" of two electron equivalents per P700, one of these being identified with cytochrome *f* and the other, tentatively, with plastocyanin. Similar figures were given by Stiehl and Witt (158) and by Haehnel (195). Recently, Bouges-Bocquet (161) has studied the turnover of photosystem I by measuring the quantity of reduced methylviologen formed by a second flash at various time intervals after the first; the secondary donors and the plastoquinone pool were initially in the reduced state following illumination by strong white light and a brief period of darkness. These experiments suggested that only one intermediate is involved between plastoquinone and P700. However, the flashes caused numerous "double hits," and consequently the calibration of electron equivalents was uncertain. The most careful estimation seems to be in the more recent work of Haehnel (194), who determined that the relative functional amounts of P700:plastocyanin:cytochrome *f* are 1:2:0.45 (plastocyanin is treated here as a protein with 1 copper atom per molecule (196)). These results are in closer agreement with the chemical analyses which show an excess of plastocyanin over cytochrome *f* and P700 (77, 84, 85).

In the experiments of Marsho and Kok (193), the oxidation-reduction state of plastocyanin was deduced indirectly. Plastocyanin seemed to behave as if it were in equilibrium with P700 ($K \approx 20$ or $\Delta E'_0 \approx 80$ mV), whereas cytochrome

f did not equilibrate (the apparent K varied according to conditions). These experiments were extended recently by Haehnel (194, 195), who was able to follow oxidation and reduction of plastocyanin by its absorbance changes at 584 nm, an isosbestic point for P700; the absorption spectrum of the changes between 580 and 650 nm, when corrected for the contribution of P700, agreed well with that of soluble plastocyanin. Chloroplasts were first illuminated with weak far-red light to oxidize all components between the light reactions and then given a variable number of brief, intense white flashes which injected a known number of electrons into the plastoquinone pool. The absorption changes of plastoquinone, cytochrome f, plastocyanin, and P700 were followed in the dark period after the last flash. Again, the apparent equilibrium constant between cytochrome f and P700 appeared to vary with conditions, having a value of 50–100 after one or two flashes and lower values after more flashes. In particular, after 10 flashes cytochrome f reduction proceeded rapidly without any detectable lag; such a time course seemed to imply $K \leqslant 5$. Similar results were obtained by Cox (197) by means of a different technique in which the dark reductions of cytochrome f and P700 were followed in chloroplasts suspended in fluid media at temperatures less than 0°C. All measurements of oxidation-reduction potentials of P700 and its donors imply a significantly larger value for the equilibrium constant, and both authors conclude that the pathway for electrons between plastoquinol and P700 cannot be only through cytochrome f. Haehnel suggested that plastocyanin is the sole intermediate and that reduction of cytochrome f occurs in a side reaction; oxidation of cytochrome f by P700 might occur in the light, but not in the dark.

Recent determinations (198, 199) have shown that the midpoint potential of plastocyanin is significantly lower when bound to the membrane than when free in solution. The potential of P700 has also been revised upwards. For example, with small digitonin particles (D-144), Knaff and Malkin (199) have reported E'_0 values of +375, +320, and +520 mV for cytochrome f, plastocyanin, and P700, respectively. These values correspond to equilibrium constants of 260 for the reaction cytochrome $f \rightleftharpoons$ P700 and 2150 for plastocyanin $\rightleftharpoons$ P700. With such a high equilibrium constant, the back reaction is negligably slow, and plastocyanin and P700 cannot react together as if they were near equilibrium throughout the time course. On the other hand, in Haehnel's experiments, cytochrome f and plastocyanin do behave as if they were in equilibrium with each other and, hence, were reacting rapidly together throughout the time course, with a gap of about 30 mV between their midpoint potentials, cytochrome f being more positive. Thus, Haehnel's kinetic experiments can be reconciled with the series scheme if one accepts that the potential of plastocyanin is shifted to a value slightly negative compared with cytochrome f by binding to the membrane.

Electron Transfer from Q to Pool

The problems connected with transfer from Q to the pool are in many ways the inverse of those of transfer from the pool to P700. In particular, there is the question of how Q, which behaves as a 1-equivalent carrier, reduces plastoquinone, which behaves as a 2-equivalent carrier, according to the spectral evidence.

There are a number of indications that this process is not a simple one. For example, the pool, as defined by experiments on the oxygen gush or fluorescence induction, is complex and can be divided into at least two components of unequal size (155). Moreover, the apparent equilibrium constant between Q and the pool varies according to conditions: a value of about 1 can be observed in the light, and in the dark higher values, perhaps as high as 100 (122, 200–204).

Despite these indications of complexity, the only chemical component positively identified with this region of the chain is plastoquinone. Although suggestions for the involvement of a cytochrome *b*-559 have been made (205, 206), the experiments of Amesz et al. (157) clearly show that the only cytochrome that may be considered as a normal carrier between the light reactions is cytochrome *f*.

The key to the problem has been provided by a series of ingenious experiments (207–210). Bouges-Bocquet (207, 208) determined the number of electrons flowing through photosystem I and the number remaining in the pool with each flash of a series by polarographic measurement of the quantities of methylviologen reduced. From these results, she was able to calculate that the number of electrons entering the pool from photosystem II in response to each flash exhibited oscillations of period two. This result was explained by a model in which a 2-electron carrier (R) is interposed between Q and the plastoquinone pool (A). R can accept a total of 2 electrons one at a time from Q, but can only donate electrons in pairs to A. Velthuys and Amesz (209) and van Best and Duysens (210) have used an identical model to explain their fluorescence measurements under conditions which indicate the interaction of Q with A in flash experiments. The transfer of electrons from photosystem II to the plastoquinone pool can then be represented in the following way (208):

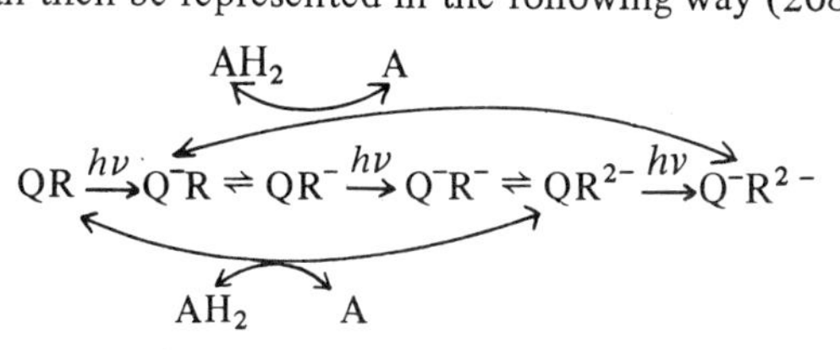

For formal purposes, the reactions have been represented as pure electron transfer. Two protons must be taken up in the sequence, but there is little information as to where this occurs. Data on the three equilibria involved, taken from the papers of Diner (211) and Bouges-Bocquet (208), are given in Table 1. According to van Best and Duysens (210), for the reaction $Q^-R^- \rightleftharpoons QR^{2-}$, $K \gg 1$ rather than $K = 1$, as suggested by Bouges-Bocquet (208).

There is little doubt that A represents plastoquinone, and according to evidence discussed above, Q is probably to be identified with a special bound molecule of plastoquinone. The suggestion made by several authors that the

Table 1. Equilibria within the Q complex[a]

Reaction (with approx. rate constants)	Equilibrium constant (K)	Relative oxidation-reduction potentials E (mV)
$Q^-R \underset{75\ s^{-1}}{\overset{1,500\ s^{-1}}{\rightleftharpoons}} QR^-$	20	+78
$Q^-R^- \underset{1,330\ s^{-1}}{\overset{660\ s^{-1}}{\rightleftharpoons}} Q^-R^{2-}$	0.5	−18
$R^{2-}A \underset{15\ s^{-1}}{\overset{15\ s^{-1}}{\rightleftharpoons}} RA^{2-}$	1.0	0

[a]Compiled from data given by Bouges-Bocquet (208). See text for explanation. Note that the equilibria as written should be regarded as including any equilibration with H^+ that occurs under normal conditions in the chloroplast.

intermediate R may also be a plastoquinone molecule bound to a specific protein is helpful, even though direct evidence is lacking.

The basic question to be answered is why the 1-equivalent reduction of A by photosystem II does not occur to a significant extent. The answer becomes apparent with the help of the potential diagram for the Q complex shown in Figure 2. $E_0'(Q/Q^-)$ has been taken as 0 V, and the relative potentials are from the data given in Table 1. This results in E_0' (A/AH_2) = +30 mV. Carrier (212) obtained E_0' = +110 mV for plastoquinone in solution, but the in vivo value is likely to be lower (cf. ubiquinone (39)). The key feature of the diagram is the pair of 2-equivalent couples, R and A, which in principle can be oxidized and reduced not only by pairs of electrons, but also in 1-equivalent steps via the semiquinone. The theory of semiquinone potentials (213) shows that the mid-point potential of the 2-equivalent couple lies midway between the potentials of the two 1-equivalent couples which involve the semiquinone. The diagram suggests that while $E_0'(R^-/R^{2-}) < E_0'(R/R^{2-}) < E_0'(R/R^-)$ (as shown in Table 1), for A the sequence is inverted, $E_0'(A^-/A^{2-}) > E_0'(A/A^{2-}) > E_0'(A/A^-)$, and the 1-equivalent reduction of A becomes thermodynamically very difficult.

The inverted sequence of potentials postulated for A is what is expected for quinones in solution at neutral pH when the semiquinone is very unstable (semiquinone formation constant $K_i \ll 1$). The degree of splitting shown (+260 mV) is in fact that which may be calculated from the data of Bishop and Tong (214) for trimethylquinone, a simple analogue of plastoquinone, at pH 7 (see Figure 3). The potentials would be expected to have a similar pattern in vivo if plastoquinone behaves as a diffusible carrier in solution in the hydrophobic phase of the membrane. The failure to detect an ESR signal attributable to $PQ^{\cdot-}$ in chloroplasts is consistent with $K_i \ll 1$. On the other hand, the sequence of potentials $R^-/R^{2-} < R/R^{2-} < R/R^-$ demands $K_i > 1$, and this could be achieved if the semiquinone is stabilized by binding to a protein. An alternative interpreta-

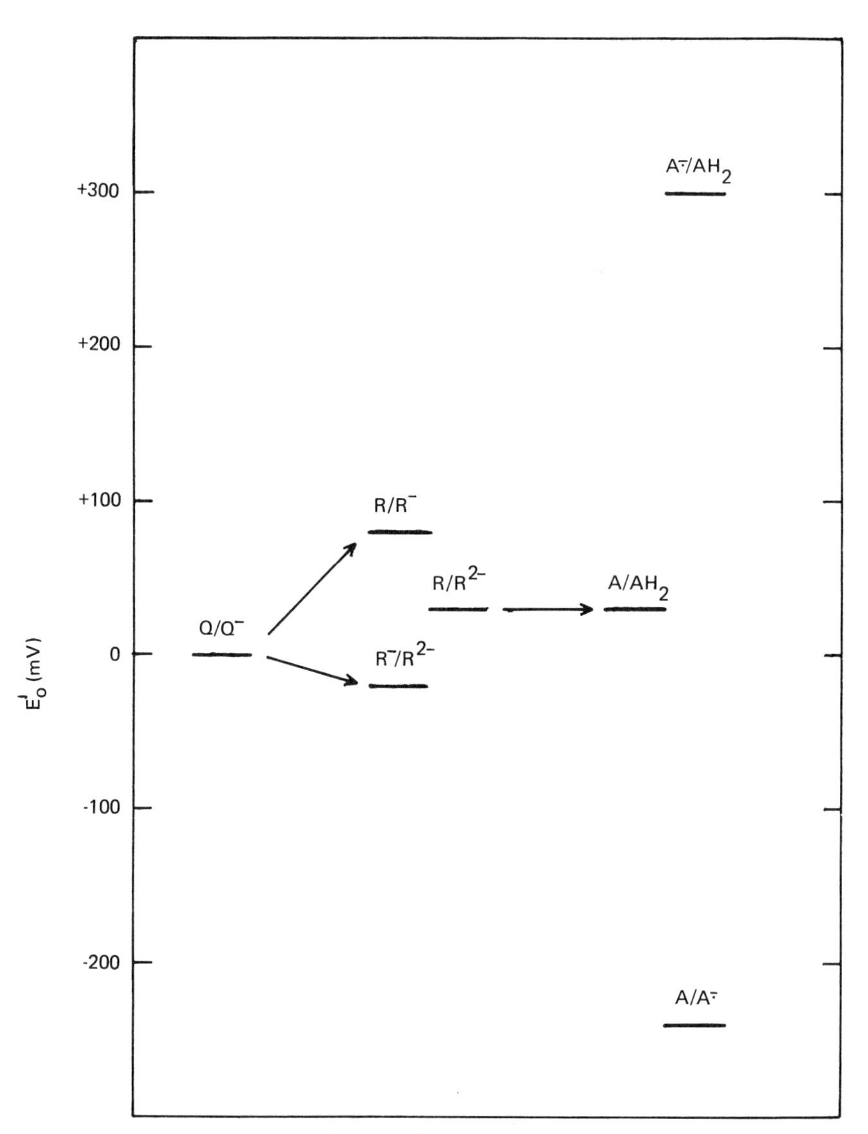

Figure 2. Postulated midpoint oxidation-reduction potentials for members of the Q complex. See text for explanation.

tion of R is that binding stabilizes the anionic forms of the quinol: R behaves much as expected of a quinone at high pH.

The primary acceptor Q could also be regarded as a quinone molecule in which the semiquinone is stabilized by binding. The couple $Q^{\cdot -}/Q^{2-}$ might have too low a potential to be reducible by photosystem II. However, there is a serious discrepancy between the spectroscopic data indicating that reduced Q is

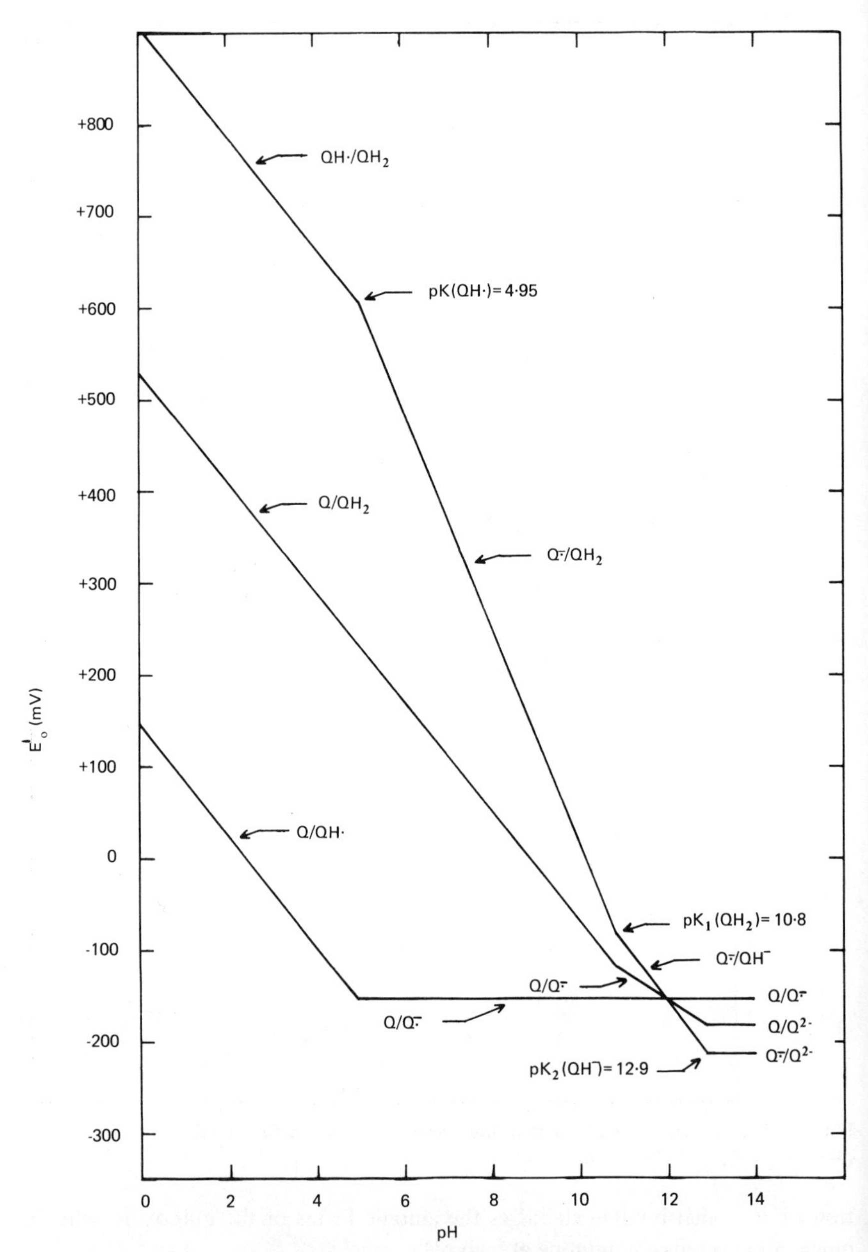

Figure 3. Midpoint oxidation-reduction potentials of the trimethylbenzoquinone system as a function of pH. The diagram has been constructed from the following data: E_0 (Q/QH_2) from Clark (213); K_i = 3.3 at pH 13, pK_1 (QH_2) = 10.8, pK_2 (QH^-) = 12.9 from Bishop and Tong (214); $pK(QH\cdot)$ = 4.95 from Patel and Willson (215). From the diagram it can be calculated that $K_i = 2.15 \times 10^{-9}$ at pH 7.

the plastosemiquinone anion and the oxidation-reduction titration results of Cramer and Butler (216) and Knaff (217), which indicated that Q is a hydrogen carrier. Knaff has determined a pK = 8.9 and $E_0(Q/Q^{\overline{\cdot}}) = -130$ mV; he suggested that protonation of $Q^{\overline{\cdot}}$ is slow and only apparent when slow equilibrations are deliberately taken into account, as in oxidation-reduction titrations. If Q does behave under normal functional conditions as a pure electron carrier, the observation of Bouges-Bocquet (218) that the reactions $Q^-R \rightarrow QR^-$ and $Q^-R^- \rightarrow QR^{2-}$ have the same rates at pH 5.2 and 8.0 suggests that the same is true of R also; proton uptake would then be associated with reduction of A, the plastoquinone pool.

The function of R as a gate precludes a genuine equilibrium between Q and A. The apparent equilibrium constant will be governed largely by the two equilibria $Q^-R \rightleftharpoons QR^-$ and $Q^-R^- \rightleftharpoons QR^{2-}$, and so will vary between about 0.5 and 20, according to the data of Bouges-Bocquet (208). Light would also influence the "equilibrium," but the distribution between the various states is not easy to predict without a computer analysis. If the oxidation-reduction potential of Q/Q^- is as low as −130 mV, as reported by Knaff, the apparent equilibrium constants would be much higher than estimated by Bouges-Bocquet and difficult to reconcile with fluorescence experiments (122, 200–204). The "equilibrium" would also be influenced by the rate of equilibration of Q^- with protons and might, therefore, be expected to show a shift from a large constant in the light to a small constant in the dark (the opposite of what is observed).

An attractive feature of the Q complex is its ability to explain the effects of DCMU and similar inhibitors. Velthuys and Amesz (209) showed that addition of DCMU to chloroplasts which had been subjected to a variable number of preilluminating flashes caused a fluorescence rise (reduction of Q) which oscillated according to the number of flashes with period two; odd numbers of flashes gave large increases and even numbers small ones. This could be explained by a lowering of the potential of the couple R/R^- by DCMU to bring about the reaction $QR^- \rightarrow Q^-R$. After an even number of flashes, R would be largely oxidized. Inhibition of electron transport would require the potential to be lowered to such an extent that reduction of R by Q^- became virtually impossible.

Knaff (219) has shown that *O*-phenanthroline, usually thought to inhibit in much the same way as DCMU, causes an increase in the potential of the primary acceptor (measured as C550). The relative potentials of Q and R would then change in the same manner as suggested by Velthuys and Amesz.

CONTROL OF ELECTRON TRANSPORT

The rate of noncyclic electron transport may be controlled by several physiologically important factors, but here our attention will be confined to the influence of phosphorylation substrates and uncouplers of photophosphorylation.

When envelope-free chloroplasts are suspended in a medium at pH 7–8 and exposed to a high light intensity, the rate of electron flow is usually limited by the activity of the phosphorylation machinery (220). A 2-fold stimulation of the basal rate can easily be demonstrated and factors of up to five or six have been claimed (221). This process of photosynthetic control, as West and Wiskich termed it, has its counterpart in the phenomenon of respiratory control exhibited by mitochondria. In chloroplasts, uncoupling agents will usually give a considerably greater rate of electron flow than can be obtained under phosphorylating conditions. The limiting factor then becomes the rate of electron transfer itself, i.e., the rate of reoxidation of plastoquinol (158).

Although Chance and Williams (222) viewed respiratory control as essentially a function of ADP concentration, it has usually been thought of as the consequence of an approach to equilibrium between the oxidation reactions and the phosphorylation system, i.e., the back pressure of the accumulating ATP causes back reactions to become significant. Indeed, Chance was the first to demonstrate energy-dependent reversed electron transport (223). The equilibrium model for mitochondria has recently received detailed experimental and theoretical support (224, 225).

Mitchell (6, 7) considered that electron transport would be controlled by the proton motive force, ΔpH acting on transmembrane hydrogen carriers and $\Delta\Psi$ on transmembrane electron carriers. In the case of chloroplasts, the electrogenic processes are the primary light reactions themselves and at saturating intensities these cannot be rate-limiting. We are left with the possibility of equilibration with the proton-translocating steps, oxidation of plastoquinol (site 1) and water (site 2). Although reversibility of oxygen evolution has not been demonstrated, the overpotential at which photosystem II operates is unknown and, as the midpoint potential for water splitting rises by 60 mV/pH unit, equilibration with the internal pH is possible. The effect of the membrane ΔpH on delayed light emission from photosystem II[121] supports this possibility. Site 1 can be discussed more concretely. For plastoquinone, $\Delta E/\Delta\text{pH} = -60$ mV, and for cytochrome *f* the midpoint potential is constant below pH 8, at least as far as pH 6 (183). If the midpoint potentials of cytochrome *f* and plastoquinone at pH 7 are taken as +360 mV and +110 mV, respectively (the in vivo value for plastoquinone is likely to be lower), it is found that equilibrium would be reached with plastoquinone and cytochrome *f* poised at their midpoints at pH 2.8. For various reasons, this is probably a conservative estimate of the internal acidity needed to control the rate of electron transport. These arguments lead to the general conclusion that although the chemiosmotic hypothesis encourages us to expect that the rate of electron flow would be controlled by the internal pH, in the case of site 1, at any rate, kinetic limitations would come into play long before equilibrium is reached if irreversible inactivation is to be avoided.

The main experimental evidence in favor of an equilibrium model for chloroplasts is the claim by Avron's group (226–228) to have demonstrated energy-dependent reversed electron transport from cytochrome *f* to Q, the fluorescence quencher of system II. However, it is difficult to accept the conclusion drawn from these experiments without further substantiation of the reduction of Q or plastoquinone by a means other than fluorescence, and with the presently available evidence that cytochrome *f* has actually been oxidized. The published difference spectrum shows that most of the measured absorbance change between 554 and 540 nm is certainly not due to cytochrome *f*, and the small peak at 554 nm does not correspond well with the spectrum of cytochrome *f* because of the absence of a shoulder on the short wave length side of the band.

Ideally, control in the site 1 and 2 regions should be investigated independently. Most of the attempts to do this rely upon the use of artificial electron donors or acceptors, and, as pointed out by Trebst and Reimer (229), the behavior of such partial reactions is likely to depend more on the properties (usually unknown) of the added reagents than on the properties of the chloroplast itself. A good example is provided by the effects of uncouplers on photosystem I reactions that depend on various artificial donor systems (in the presence of DCMU). Hauska (230) showed that at low donor concentrations (0.1 mM) NH_4Cl stimulates electron transport with phenolic donors such as dichlorophenolindophenol and has little or no effect with amines (diaminodurene, N,N,N′,N′-tetramethyl-*p*-phenylenediamine). The primary effect of NH_4Cl is to cause the internal pH to rise toward that of the external medium. Hauska suggested that in the case of dichlorophenolindophenol the phenolate anion reacts more rapidly with photosystem I than does the neutral form, whereas with the amines, on the contrary, the neutral form (free base) is a good donor but the protonated form is inactive. This suggestion is consistent with the known pK values of 7 for reduced 2,6-dichlorophenolindophenol (213) and 4.3 for diaminodurene (231). However, at higher concentrations of diaminodurene and higher rates of electron transport, the thylakoid interior might become sufficiently acid for an uncoupler to stimulate, as was found by Gould (232). To substantiate Hauska's hypothesis, the rates of reaction of dichlorophenolindophenol and diaminodurene with plastocyanin, cytochrome *f*, and P700 need to be determined as functions of pH.

The view that electron transport is controlled primarily by the internal pH was developed by Siggel (233–235) and by Rottenberg and Avron (236–239). The rate of electron transport at any given internal pH would depend not only on the intrinsic pH/activity relations of the system, but also on the leakiness of the membrane toward protons and whether or not phosphorylation is occurring. The permeability of the membrane toward protons increases 8-fold as the pH is raised from 6 to 8 (240). Siggel (235) measured the rate of overall electron transport by ferricyanide reduction, the activity of site 1 by the reduction of P700 after a flash, and site 2 activity by the reduction of plastoquinone, all as functions of pH; gramicidin was present to equalize internal and external pH. As the pH fell below 7, the activity of site 1 was limited as if the rate were determined by protonation of a group with pK = 5.7. The identity of this group is not known but, to quote Siggel, "it might be the proton pump molecule delivering the protons into the inner phase." Site 2 could be described as limited by an ionizing group with pK = 4.7. Thus, the rate of electron transport would normally be determined predominantly by site 1; the apparent pK for limitation of overall electron transport with ferricyanide as acceptor was 5.5.

Rottenberg et al. (236) showed, in agreement with the results of Siggel, that the rate of ferricyanide reduction depends on the internal pH with an optimum at 5–5.5. This dependence explains the shift in the optimum external pH to more acidic values that occurs in the presence of uncouplers (241). In a later paper, Bamberger et al. (238) concluded that ΔpH also has an important influence on the rate of electron transport. The rate of ferricyanide reduction was expressed as a function of internal pH at various values of ΔpH which were established with a range of concentrations of nigericin as uncoupler. The rate limitation on the acid side was fairly satisfactorily determined by the value of the internal pH alone, but on the alkaline side the pH at which electron transport declined depended on ΔpH. They concluded that the rate-controlling site is buried within the membrane and is influenced by an intramembrane pH which is unknown but dependent on both internal and external pH. A replot of the data of Bamberger et al. (Figure 4) suggests another interpretation. This is a composite diagram showing all the data from Figure 7 of the paper of Bamberger et al. as a function of either internal or external pH. Below pH 8.2, the rate is shown as determined by internal pH, above this value as a function of external pH. This diagram seems to show that, irrespective of ΔpH, the rate is limited *either* by an acid internal pH (as shown by Siggel) *or* by an alkaline external pH. The external site is probably associated with photosystem II (or plastoquinone reduction), as very high rates of electron transport can be achieved with photosystem I-dependent reactions at pH 9 (232).

In conclusion, work with envelope-free chloroplasts suggests that the state of the phosphorylation system controls the electron transport system mainly through kinetic effects of the internal pH. The coupling is fairly loose and, according to Heber and Kirk (242), this is also true of intact chloroplasts. Under normal conditions, when both photosystems are operating, the main restraint is placed on site 1 by a decrease in the rate constant for electron transfer from plastoquinol. Control of photosystem II may be by an approach to equilibrium rather than by a kinetic effect, but this is not normally rate-determining. Under most conditions, electron transport will be limited by the internal acidity, but at external pH $>$ 8.5 the rate is determined by an external region of the electron transport system on the acceptor side of photosystem II. The rate is also likely to decline if the internal pH rises above 7, but little is known about the mechanism of this effect. At high internal pH values, an irreversible inactivation of oxygen evolution may occur.

CONCLUSION

The strength of chemiosmotic ideas as applied to chloroplasts in the form of the vectorial Z scheme is their relative simplicity. Only one proton translocation

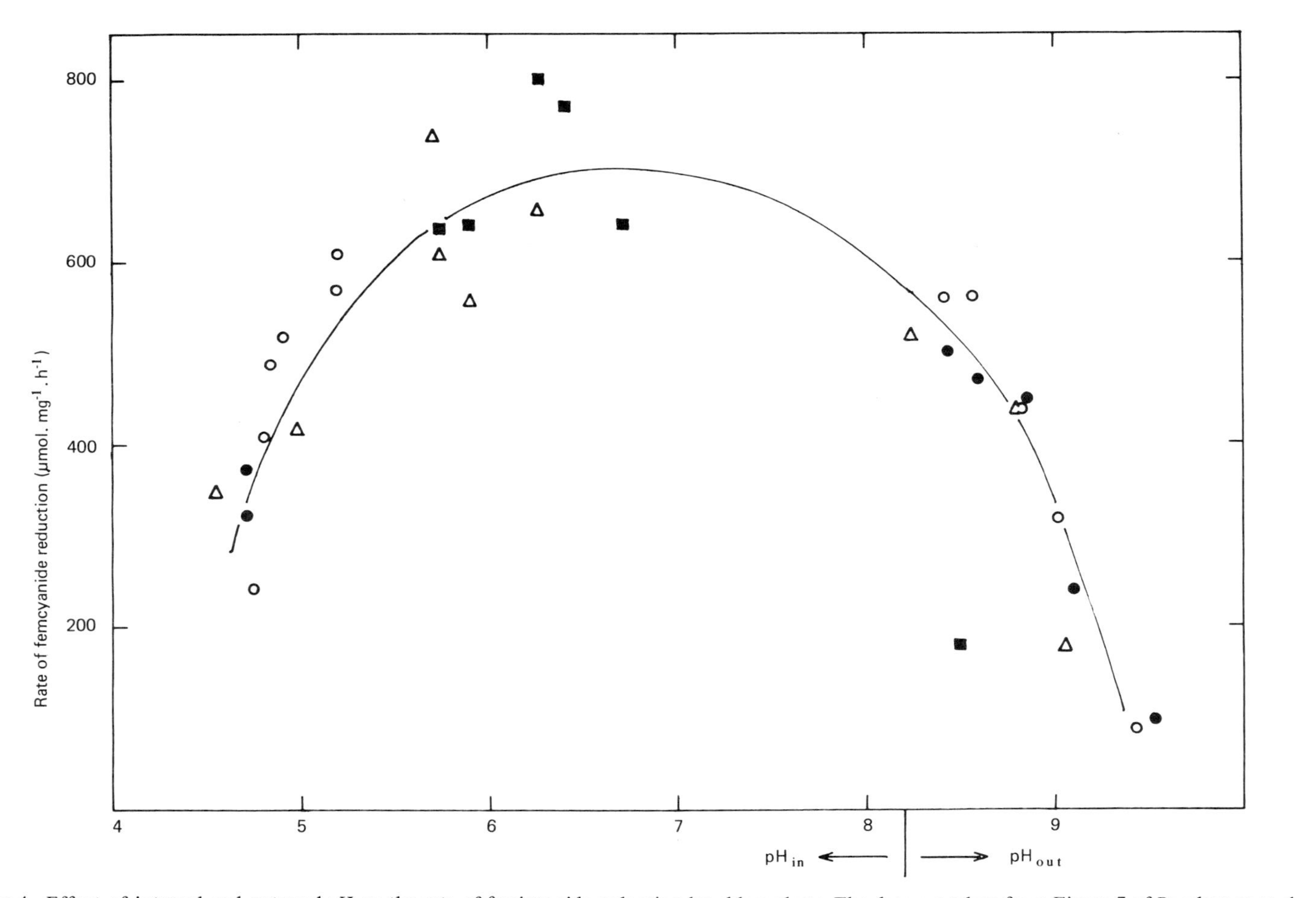

Figure 4. Effect of internal and external pH on the rate of ferricyanide reduction by chloroplasts. The data are taken from Figure 7 of Bamberger et al. (238). ΔpH values are ●, 3.4; ○, 2.8; △, 2.2; and □, 1.3.

site is required, and this is easily understood in terms of a diffusion of plastoquinol across the membrane. On the other hand, a fundamental difficulty for the chemiosmotic approach to oxidative phosphorylation in mitochondria has been the need to define three proton translocation sites; ubiquinol readily fills the role of H carrier at one site, but the other two have remained elusive. This difficulty might be partly overcome with the help of the ingenious quinone cycle recently proposed by Mitchell (243, 244). Basically, Mitchell has attempted to find a chemical explanation for two sites of proton translocation by invoking separate roles for the two 1-equivalent steps of ubiquinol oxidation. Gibson and Cox (245) had previously made similar suggestions for *Escherichia coli,* although without the catalytic electron cycle introduced by Mitchell. The purple photosynthetic bacteria also have a second site of proton translocation, and in this context Crofts et al. (246) have recently discussed various ways in which ubisemequinone could play a crucial role at two sites. An intensive study of the simpler but closely related chloroplast system may well contribute significantly to an eventual understanding of the more complex bacterial and mitochondrial systems.

At the molecular level, the thylakoid membrane is complex and progress is often hampered by the lability of some of the components. The major challenge now is to unravel the intricate interactions between individual membrane proteins and between proteins and lipids, in order to achieve an integrated account of electron and proton transfer in terms of structure at the molecular level. Only as a result of detailed work of this kind will it eventually be possible to evaluate the chemiosmotic view of chloroplast processes in relation to alternative ideas such as those propounded by Williams (247–249) and Dilley (164, 250).

REFERENCES

1. Hill, R., and Bendall, F. (1960). Nature 186:136.
2. McSwain, B. D., and Arnon, D. I. (1968). Proc. Natl. Acad. Sci. U.S.A. 61:989.
3. Anderson, J. M. (1975) Biochim. Biophys. Acta 416:191.
4. Singer, S. J. (1971). *In* L. I. Rothfield (ed.), Structure and Function of Biological Membranes, pp. 145–222. Academic Press, New York.
5. Mitchell, P. (1961). Nature 191:144.
6. Mitchell, P. (1966). Chemiosmotic Coupling in Oxidative and Photosynthetic Phosphorylation. Glynn Research, Bodmin, Cornwall, England.
7. Mitchell, P. (1966). Biol. Rev. 41:445.
8. Witt, H. T. (1971). Q. Rev. Biophys. 4:365.
9. Trebst, A. (1974). Annu. Rev. Plant Physiol. 25:423.
10. Hind, G., and Jagendorf, A. T. (1963). Proc. Natl. Acad. Sci. U.S.A. 49:715.
11. Neumann, J., and Jagendorf, A. T. (1964). Arch. Biochem. Biophys. 107:109.
12. Jagendorf, A. T., and Uribe, E. (1966). Proc. Natl. Acad. Sci. U.S.A. 55:170.
13. Crofts, A. R. (1974). *In* S. Estrada-O and C. Gitler (eds.), Perspectives in Membrane Biology, pp. 373–412. Academic Press, New York.
14. Clayton, R. K. (1973). Annu. Rev. Biophys. Bioeng. 2:131.
15. Parson, W. W., and Cogdell, R. J. (1975). Biochim. Biophys. Acta 416:105.
16. Clayton, R. K., and Wang, R. T. (1971). Methods Enzymol. 23:696.
17. Feher, G. (1971). Photochem. Photobiol. 14:373.
18. Sauer, K. (1975). *In* Govindjee (ed.), Bioenergetics of Photosynthesis, pp. 116–181. Academic Press, New York.

19. Prince, R. C., and Crofts, A. R. (1973). FEBS Lett. 35:213.
20. Noël, H., Van der Rest, M., and Gingras, G. (1972). Biochim. Biophys. Acta 275:219.
21. Okamura, M. Y., Steiner, L. A., and Feher, G. (1974). Biochemistry 13:1394.
22. Steiner, L. A., Okamura, M. Y., Lopes, A. D., Moskowitz, E., and Feher, G. (1974). Biochemistry 13:1403.
23. Straley, S. C., Parson, W. W., Mauzerall, D. C., and Clayton, R. K. (1973). Biochim. Biophys. Acta 305:597.
24. Sauer, K., Dratz, E. A., and Coyne, L. (1968). Proc. Natl. Acad. Sci. U.S.A. 61:17.
25. Reed, D. W., and Ke, B. (1973). J. Biol. Chem. 248:3041.
26. Sauer, K. (1972). Methods Enzymol. 24:206.
27. Norris, J. R., Uphaus, R. A., Crespi, H. L. and Katz, J. J. (1971). Proc. Natl. Acad. Sci. U.S.A. 68:625.
28. Katz, J. J., and Norris, J. R. (1973). Curr. Top. Bioenerg. 5:41.
29. Feher, G., Hoff, A. J., Isaacson, R. A., and Ackerson, L. C. (1975). Ann. N. Y. Acad. Sci. 244:239.
30. Feher, G., Okamura, M. Y., and McElroy, J. D. (1972). Biochim. Biophys. Acta 267:222.
31. Dutton, P. L., Leigh, J. S., and Reed, D. W. (1973). Biochim. Biophys. Acta 292:654.
32. Cogdell, R. J., Brune, D. C., and Clayton, R. K. (1974). FEBS Lett. 45:344.
33. Feher, G., Isaacson, R. A., McElroy, J. D., Ackerson, L. C., and Okamura, M. Y. (1974). Biochim. Biophys. Acta 368:135.
34. Loach, P. A., and Hall, R. L. (1972). Proc. Natl. Acad. Sci. U.S.A. 69:786.
35. Bensasson, R., and Land, E. J. (1973). Biochim. Biophys. Acta 325:175.
36. Slooten, L. (1972). Biochim. Biophys. Acta 275:208.
37. Clayton, R. K., and Straley, S. C. (1972). Biophys. J. 12:1221.
38. Wraight, C. A., Cogdell, R. J., and Clayton, R. K. (1975). Biochim. Biophys. Acta 396:242.
39. Urban, P. F., and Klingenberg, M. (1969). Eur. J. Biochem. 9:519.
40. Silberstein, B. R., and Gromet-Elhanan, Z. (1974). FEBS Lett. 42:141.
41. Govindjee, R., Smith, W. R., and Govindjee (1974). Photochem. Photobiol. 20:191.
42. Loach, P. A., Kung, C., and Hales, B. J. (1975). Ann. N. Y. Acad. Sci. 244:297.
43. McElroy, J. D., Feher, G., and Mauzerall, D. C. (1969). Biochim. Biophys. Acta 172:180.
44. McElroy, J. D., Mauzerall, D. C., and Feher, G. (1974). Biochim. Biophys. Acta 333:261.
45. Hsi, E. S. P., and Bolton, J. R. (1974). Biochim. Biophys. Acta 347:126.
46. Junge, W. (1975). Third International Congress on Photosynthesis, Vol. I, p. 273. Elsevier, Amsterdam.
47. Bearden, A. J., and Malkin, R. (1975). Q. Rev. Biophys. 7:131.
48. Thornber, J. P. (1975). Annu. Rev. Plant Physiol. 26:127.
49. Anderson, J. M., and Levine, R. P. (1974). Biochim. Biophys. Acta 357:118.
50. Machold, O. (1975). Biochim. Biophys. Acta 382:494.
51. Nelson, N., and Bengis, C. (1975). Third International Congress on Photosynthesis, Vol. I, p. 609. Elsevier, Amsterdam.
52. Bengis, C., and Nelson, N. (1975). J. Biol. Chem. 250:2783.
53. Malkin, R. (1975). Arch. Biochem. Biophys. 169:77.
54. Norris, J. R., Uphaus, R. A., and Katz, J. J. (1972). Biochim. Biophys. Acta 275:161.
55. Norris, J. R., Scheer, H., Druyan, M. E., and Katz, J. J. (1974). Proc. Natl. Acad. Sci. U.S.A. 71:4897.
56. Philipson, K. D., Sato, V. L., and Sauer, K. (1972). Biochemistry 11:4591.
57. Kok, B. (1961). Biochim. Biophys. Acta 48:527.
58. Ruuge, E. K., and Izawa, S. (1972). Fed. Proc. 31:901 (Abstr.).
59. Knaff, D. B., and Malkin, R. (1973). Arch. Biochem. Biophys. 159:555.
60. Strichartz, G. R., and Chance, B. (1972). Biochim. Biophys. Acta 256:71.
61. Jackson, J. B., and Crofts, A. R. (1969). FEBS Lett. 4:185.
62. Crofts, A. R., Holmes, N. G., and Crowther, D. (1975). Tenth FEBS Meeting, Vol. 40, p. 287. North-Holland/American Elsevier, Amsterdam.
63. Fowler, C. F., and Kok, B. (1974). Biochim. Biophys. Acta 357:308.
64. Hiyama, T., and Ke, B. (1971). Arch. Biochem. Biophys. 147:99.

65. Warden, J. T., and Bolton, J. R. (1974). Photochem. Photobiol. 20:251.
66. Ke, B., and Beinert, H. (1973). Biochim. Biophys. Acta 305:689.
67. Malkin, R., and Bearden, A. J. (1971). Proc. Natl. Acad. Sci. U.S.A. 68:16.
68. Evans, M. C. W., Reeves, S. G., and Cammack, R. (1974). FEBS Lett. 49:111.
69. Malkin, R., Aparicio, P. J., and Arnon, D. I. (1974). Proc. Natl. Acad. Sci. U.S.A. 71:2362.
70. Mayne, B. C., and Rubinstein, D. (1966). Nature 210:734.
71. Lozier, R. H., and Butler, W. L. (1974). Biochim. Biophys. Acta 333:465.
72. Visser, J. W. M., Rijgersberg, K. P., and Amesz, J. (1974). Biochim. Biophys. Acta 368:235.
73. Ke, B., Sugahara, K., Shaw, E. R., Hansen, R. E., Hamilton, W. D., and Beinert, H. (1974). Biochim. Biophys. Acta 368:401.
74. McIntosh, A. R., Chu, M., and Bolton, J. R. (1975). Biochim. Biophys. Acta 376:308.
75. Evans, M. C. W., and Cammack, R. (1975). Biochem. Biophys. Res. Commun. 63:187.
76. Warden, J. T., Moharty, P., and Bolton, J. R. (1974). Biochem. Biophys. Res. Commun. 59:872.
77. Visser, J. W. M., Amesz, J., and van Gelder, B. F. (1974). Biochim. Biophys. Acta 333:279.
78. Hauska, G. A., McCarty, R. E., Berzborn, R. J., and Racker, E. (1971). J. Biol. Chem. 246:3524.
79. Plesničar, M., and Bendall, D. S. (1973). Eur. J. Biochem. 34:483.
80. Brand, J., Baszynski, T., Crane, F. L., and Krogmann, D. W. (1972). J. Biol. Chem. 247:2814.
81. Brand, J., San Pietro, A., and Mayne, B. C. (1972). Arch. Biochem. Biophys. 152:426.
82. Selman, B. R., Johnson, G. L., Dilley, R. A., and Voegeli, K. K. (1975). Third International Congress on Photosynthesis, Vol. II, p. 897. Elsevier, Amsterdam.
83. Schmid, G. H., Radunz, A., and Menke, W. (1975). Z. Naturforsch. 30c:201.
84. Plesničar, M., and Bendall, D. S. (1970). Biochim. Biophys. Acta 216:192.
85. Malkin, R., and Bearden, A. J. (1973). Biochim. Biophys. Acta 292:169.
86. Joliot, P., Barbieri, G., and Chabaud, R. (1969). Photochem. Photobiol. 10:309.
87. Kok, B., Forbush, B., and McGloin, M. (1970). Photochem. Photobiol. 11:457.
88. Joliot, P., Joliot, A., Bouges, B., and Barbieri, G. (1971). Photochem. Photobiol. 14:287.
89. Forbush, B., Kok, B., and McGloin, M. (1971). Photochem. Photobiol. 14:307.
90. Joliot, P., and Joliot, A. (1968). Biochim. Biophys. Acta 153:625.
91. Radmer, R., and Kok, B. (1975). Annu. Rev. Biochem. 44:409.
92. Joliot, P., and Kok, B. (1975). *In* Govindjee (ed.), Bioenergetics of Photosynthesis, pp. 387–412. Academic Press, New York.
93. Vernon, L. P., Shaw, E. R., Ogawa, T., and Raveed, D. (1971). Photochem. Photobiol. 14:343.
94. Ke, B., Sahu, S., Shaw, E., and Beinert, H. (1974). Biochim. Biophys. Acta 347:36.
95. Wessels, J. S. C., Van Alphen-Van Waveren, O., and Voorn, G. (1973). Biochim. Biophys. Acta 292:741.
96. Butler, W. L., and Kitajima, M. (1975). Third International Congress on Photosynthesis, Vol. I, p. 13. Elsevier, Amsterdam.
97. Wessels, J. S. C., and Borchert, M. T. (1975). Third International Congress on Photosynthesis, Vol. I, p. 473. Elsevier, Amsterdam.
98. Levine, R. P., Burton, W. G., and Duram, H. A. (1972). Nature (New Biol.) 237:176.
99. Malkin, R., and Bearden, A. J. (1975). Biochim. Biophys. Acta 396:250.
100. Van Gorkom, H., Pulles, M. P. J., and Wessels, J. S. C. (1975). Biochim. Biophys. Acta 408:331.
101. Bendall, D. S., and Sofrová, D. (1971). Biochim. Biophys. Acta 234:371.
102. Kok, B., Radmer, R., and Fowler, C. F. (1975). Third International Congress on Photosynthesis, Vol. I, p. 485. Elsevier, Amsterdam.
103. Crofts, A. R., Wraight, C. A., and Fleischmann, D. E. (1971). FEBS Lett. 15:89.

104. Duysens, L. N. M., and Sweers, H. E. (1963). Studies on Microalgae and Photosynthetic Bacteria, p. 353. University of Tokyo Press, Tokyo.
105. Knaff, D. B., and Arnon, D. I. (1969). Proc. Natl. Acad. Sci. U.S.A. 64:715.
106. Erixon, K., and Butler, W. L. (1971). Photochem. Photobiol. 14:427.
107. Erixon, K., and Butler, W. L. (1971). Biochim. Biophys. Acta 234:381.
108. Amesz, J., Pulles, M. P. J., and Velthuys, B. R. (1973). Biochim. Biophys. Acta 325:472.
109. Malkin, R., and Knaff, D. B. (1973). Biochim. Biophys. Acta 325:336.
110. Cox, R. P., and Bendall, D. S. (1974). Biochim. Biophys. Acta 347:49.
111. Butler, W. L., and Okayama, S. (1971). Biochim. Biophys. Acta 245:237.
112. Okayama, S., and Butler, W. L. (1972). Plant Physiol. 49:769.
113. Stiehl, H. H., and Witt, H. T. (1968). Z. Naturforsch. 23b:220.
114. Witt, K. (1973). FEBS Lett. 38:116.
115. Van Gorkom, H. J. (1974). Biochim. Biophys. Acta 347:439.
116. Pulles, M. P. J., Kerkhof, P. L. M., and Amesz, J. (1974). FEBS Lett. 47:143.
117. Joliot, P., and Joliot, A. (1973). Biochim. Biophys. Acta 305:302.
118. Amesz, J., Pulles, M. P. J., De Grooth, B. G., and Kerkhof, P. L. M. (1975). Third International Congress on Photosynthesis, Vol. I, p. 307. Elsevier, Amsterdam.
119. Schliephake, W., Junge, W., and Witt, H. T. (1968). Z. Naturforsch. 23b:1571.
120. Joliot, P., and Delosme, R. (1974). Biochim. Biophys. Acta 357:267.
121. Wraight, C. A., and Crofts, A. R. (1971). Eur. J. Biochem. 19:386.
122. Joliot, P. (1965). Biochim. Biophys. Acta 102:135.
123. Malkin, S., and Michaeli, G. (1971). Second International Congress on Photosynthesis, Vol. I, p. 149. Dr. W. Junk N. V., The Hague.
124. Vermeglio, A., and Mathis, P. (1973). Biochim. Biophys. Acta 292:763.
125. Butler, W. L., Visser, J. W. M., and Simons, H. L. (1973). Biochim. Biophys. Acta 325:539.
126. Murata, N., Itoh, S., and Okada, M. (1973). Biochim. Biophys. Acta 325:463.
127. Mathis, P., and Vermeglio, A. (1974). Biochim. Biophys. Acta 368:130.
128. Mathis, P., and Vermeglio, A. (1975). Biochim. Biophys. Acta 396:371.
129. Fowler, C. F., and Kok, B. (1974). Biochim. Biophys. Acta 357:299.
130. Blankenship, R. E., and Sauer, K. (1974). Biochim. Biophys. Acta 357:252.
131. Girault, G., and Galmiche, J. M. (1974). Biochim. Biophys. Acta 333:314.
132. Giaquinta, R. T., and Dilley, R. A. (1975). Biochim. Biophys. Acta 387:288.
133. Lien, S., and Racker, E. (1971). J. Biol. Chem. 246:4298.
134. Ausländer, W., and Junge, W. (1974). Biochim. Biophys. Acta 357:285.
135. Babcock, G. T., and Sauer, K. (1975). Biochim. Biophys. Acta 396:48.
136. Wolff, C., Buchwald, H.-E., Rüppel, H., Witt, K., and Witt, H. T. (1969). Z. Naturforsch. 24b:1038.
137. Van Gorkom, H. J., and Donze, M. (1973). Photochem. Photobiol. 17:333.
138. Zankel, K. L. (1971). Biochim. Biophys. Acta 245:373.
139. Gläser, M., Wolff, C., Buchwald, H.-E., and Witt, H. T. (1974). FEBS Lett. 42:81.
140. Cox, R. P., and Bendall, D. S. (1972). Biochim. Biophys. Acta 283:124.
141. Horton, P., and Cramer, W. A. (1974). Biochim. Biophys. Acta 368:348.
142. Bearden, A. J., and Malkin, R. (1973). Biochim. Biophys. Acta 325:266.
143. Cramer, W. A., and Horton, P. (1975). Photochem. Photobiol. 22:304.
144. Floyd, R. A., Chance, B., and DeVault, D. (1971). Biochim. Biophys. Acta 226:103.
145. Izawa, S., and Ort, D. R. (1974). Biochim. Biophys. Acta 357:127.
146. Selman, B. R., and Bannister, T. T. (1971). Biochim. Biophys. Acta 253:428.
147. Selman, B. R., Bannister, T. T., and Dilley R. A. (1973). Biochim. Biophys. Acta 292:566.
148. Giaquinta, R. T., Dilley, R. A., Selman, B. R., and Anderson, B. J. (1974). Arch. Biochem. Biophys. 162:200.
149. Giaquinta, R., Dilley, R. A., Anderson, B. J., and Horton, P. (1974). J. Bioenerg. 6:167.
150. Giaquinta, R. T., Ort, D. R., and Dilley, R. A. (1975). Biochemistry 14:4392.
151. Lockau, W., and Selman, B. R. (1976). Z. Naturforsch. 31c:48.
152. Haehnel, W. (1976). Biochim. Biophys. Acta 423:499.

153. Ojakian, G. K., and Satir, P. (1974). Proc. Natl. Acad. Sci. U.S.A. 71:2052.
154. Joliot, P. (1968). Photochem. Photobiol. 8:451.
155. Radmer, R., and Kok, B. (1973). Biochim. Biophys. Acta 314:28.
156. Amesz, J. (1973). Biochim. Biophys. Acta 301:35.
157. Amesz, J., Visser, J. W. M., Van Den Engh, G. J., and Dirks, M. P. (1972). Biochim. Biophys. Acta 256:370.
158. Stiehl, H. H., and Witt, H. T. (1969). Z. Naturforsch. 24b:1588.
159. Siggel, U., Renger, G., Stiehl, H. H., and Rumberg, B. (1972). Biochim. Biophys. Acta 256:328.
160. Bouges-Bocquet, B. (1973). Biochim. Biophys. Acta 314:250.
161. Bouges-Bocquet, B. (1975). Biochim. Biophys. Acta 396:382.
162. Boardman, N. K., Anderson, J. M., Thorne, S. W., and Björkman, O. (1972). Carnegie Inst. Wash. Yearbook 71:107.
163. Boardman, N. K., Björkman, O., Anderson, J. M., Goodchild, D. J., and Thorne, S. W. (1975). Third International Congress on Photosynthesis, Vol. III, p. 1809. Elsevier, Amsterdam.
164. Dilley, R. A., and Giaquinta, R. T. (1975). Curr. Top. Membr. Transp. 7:49.
165. Junge, W., and Ausländer, W. (1974). Biochim. Biophys. Acta 333:59.
166. Vater, J., Renger, G., Stiehl, H. H., and Witt, H. T. (1968). Naturwissenschaften 55:220.
167. Avron, M., and Chance, B. (1967). Brookhaven Symp. Biol. 19:149.
168. Böhme, H., and Cramer, W. A. (1972). Biochemistry 11:1155.
169. Racker, E., Hauska, G. A., Lien, S., Berzborn, R. J., and Nelson, N. (1972). Second International Congress on Photosynthesis, Vol. II, p. 1097. Dr. W. Junk, The Hague.
170. Horton, P., and Cramer, W. A. (1974). Biochim. Biophys. Acta 368:348.
171. Hauska, G., Reimer, S., and Trebst, A. (1974). Biochim. Biophys. Acta 357:1.
172. Hauska, G. (1972). FEBS Lett. 28:217.
173. Hauska, G., Trebst, A., and Draber, W. (1973). Biochim. Biophys. Acta 305:632.
174. Hauska, G., and Prince, R. C. (1974). FEBS Lett. 41:35.
175. Witt, H. T. (1967). Fifth Nobel Symposium, p. 261. Almqvist and Wiksell, Stockholm.
176. Yamazaki, I., and Ohnishi, T. (1966). Biochim. Biophys. Acta 112:469.
177. Trebst, A., and Eck, H. (1963). Z. Naturforsch. 18b:694.
178. Wood, P. M., and Bendall, D. S. (1976). Eur. J. Biochem. 61:337.
179. Wessels, J. S. C. (1966). Biochim. Biophys. Acta 126:581.
180. Wessels, J. S. C., and Voorn, G. (1972). Second International Congress on Photosynthesis, Vol. I, p. 833. Dr. W. Junk, The Hague.
181. Wood, P. M. (1974). Biochim. Biophys. Acta 357:370.
182. Bennett, L. E. (1973). Progr. Inorg. Chem. 18:1.
183. Davenport, H. E., and Hill, R. (1952). Proc. Roy. Soc. Lond. B 139:327.
184. Harrison, J. E. (1974). Proc. Natl. Acad. Sci. U.S.A. 71:2332.
185. Duysens, L. N. M., and Amesz, J. (1962). Biochim. Biophys. Acta 64:243.
186. Knaff, D. B. (1973). Biochim. Biophys. Acta 292:186.
187. Kimimura, M., and Katoh, S. (1972). Biochim. Biophys. Acta 283:279.
188. Oitrakul, R., and Izawa, S. (1973). Biochim. Biophys. Acta 305:105.
189. Izawa, S., Kraayenhof, R., Ruuge, E. K., and DeVault, D. (1973). Biochim. Biophys. Acta 314:328.
190. Gorman, D. S., and Levine, R. P. (1966). Plant Physiol. 41:1643.
191. Biggins, J. (1967). Plant Physiol. 42:1447.
192. Wood, P. M., and Bendall, D. S. (1975). Biochim. Biophys. Acta 387:115.
193. Marsho, T. V., and Kok, B. (1970). Biochim. Biophys. Acta 223:240.
194. Haehnel, W. (1975). Third International Congress on Photosynthesis, Vol. I, p. 557. Elsevier, Amsterdam.
195. Haehnel, W. (1973). Biochim. Biophys. Acta 305:618.
196. Ramshaw, J. A. M., Brown, R. H., Scawen, M. D., and Boulter, D. (1973). Biochim. Biophys. Acta 303:269.
197. Cox, R. P. (1975). Eur. J. Biochem. 55:625.
198. Malkin, R., Knaff, D. B., and Bearden, A. J. (1973). Biochim. Biophys. Acta 305:675.

199. Knaff, D. B., and Malkin, R. (1973). Arch. Biochem. Biophys. 159:555.
200. Joliot, P. (1965). Biochim. Biophys. Acta 102:116.
201. Delosme, R. (1967). Biochim. Biophys. Acta 143:108.
202. Forbush, B., and Kok, B. (1968). Biochim. Biophys. Acta 162:243.
203. Joliot, A. (1968). Physiol. Vég. 6:235.
204. Malkin, S. (1971). Biochim. Biophys. Acta 234:415.
205. Levine, R. P., and Gorman, D. S. (1966). Plant Physiol. 41:1293.
206. Cramer, W. A., and Böhme, H. (1972). Biochim. Biophys. Acta 256:358.
207. Bouges-Bocquet, B. (1973). Biochim. Biophys. Acta 314:250.
208. Bouges-Bocquet, B. (1975). Third International Congress on Photosynthesis, Vol. I, p. 579. Elsevier, Amsterdam.
209. Velthuys, B. R., and Amesz, J. (1974). Biochim. Biophys. Acta 333:85.
210. van Best, J. A., and Duysens, L. N. M. (1975). Biochim. Biophys. Acta 408:154.
211. Diner, B. (1975). Third International Congress on Photosynthesis, Vol. I, p. 589. Elsevier, Amsterdam.
212. Carrier, J. -M. (1966). *In* T. W. Goodwin (ed.), Biochemistry of Chloroplasts, Vol. II, p. 551. Academic Press, London.
213. Clark, W. M. (1960). Oxidation-Reduction Potentials of Organic Systems. Baillière, Tindall and Cox, London.
214. Bishop, C. A., and Tong, L. K. J. (1965). J. Am. Chem. Soc. 87:501.
215. Patel, K. B., and Willson, R. L. (1973). J. Chem. Soc. Faraday Trans. (I) 69:814.
216. Cramer, W. A., and Butler, W. L. (1969). Biochim. Biophys. Acta 172:503.
217. Knaff, D. B. (1975). FEBS Lett. 60:331.
218. Bouges-Bocquet, B. (1974). Thèse de Doctorat d'État, Université de Paris.
219. Knaff, D. B. (1975). Biochim. Biophys. Acta 376:583.
220. West, K. R., and Wiskich, J. T. (1968). Biochem. J. 109:527.
221. Hall, D. O., Reeves, S. G., and Baltscheffsky, H. (1971). Biochem. Biophys. Res. Commun. 43:359.
222. Chance, B., and Williams, G. R. (1956). Adv. Enzymol. 17:65.
223. Chance, B., and Hollunger, G. (1961). J. Biol. Chem. 236:1534.
224. Owen, C. S., and Wilson, D. F. (1974). Arch. Biochem. Biophys. 161:581.
225. Wilson, D. F., Erecińska, M., and Dutton, P. L. (1974). Annu. Rev. Biophys. Bioenerg. 3:203.
226. Rienits, K. G., Hardt, H., and Avron, M. (1973). FEBS Lett. 33:28.
227. Rienits, K. G., Hardt, H., and Avron, M. (1974). Eur. J. Biochem. 43:291.
228. Shahak, Y., Hardt, H., and Avron, M. (1975). FEBS Lett. 54:151.
229. Trebst, A., and Reimer, S. (1973). Biochim. Biophys. Acta 325:546.
230. Hauska, G. (1975). Third International Congress on Photosynthesis, Vol. I, p. 689. Elsevier, Amsterdam.
231. Wepster, B. M. (1957). Rec. Trav. Chim. 76:335.
232. Gould, J. M. (1975). Biochim. Biophys. Acta 387:135.
233. Rumberg, B., and Siggel, U. (1969). Naturwissenschaften 56:130.
234. Rumberg, B., Reinwald, E., Schröder, H., and Siggel, U. (1969). *In* H. Metzner (ed.), Progress in Photosynthesis Research, Vol. III, p. 1374. I. U. B., Tübingen.
235. Siggel, U. (1975). Third International Congress on Photosynthesis, Vol. I, p. 645. Elsevier, Amsterdam.
236. Rottenberg, H., Grunwald, T., and Avron, M. (1972). Eur. J. Biochem. 25:54.
237. Schuldiner, S., Rottenberg, H., and Avron, M. (1972). Eur. J. Biochem. 25:64.
238. Bamberger, E., Rottenberg, H., and Avron, M. (1973). Eur. J. Biochem. 34:557.
239. Avron, M. (1972). Second International Congress on Photosynthesis, Vol. II, p. 861. Dr. W. Junk, The Hague.
240. Reinwald, E., Siggel, U., and Rumberg, B. (1968). Naturwissenschaften 55:221.
241. Good, N., Izawa, S., and Hind, G. (1966). Curr. Top. Bioenerg. 1:76.
242. Heber, U., and Kirk, M. R. (1975). Biochim. Biophys. Acta 376:136.
243. Mitchell, P. (1975). FEBS Lett. 59:137.
244. Mitchell, P. (1975). *In* E. Quagliariello, S. Papa, F. Palmieri, E. C. Slater, and N. Siliprandi (ed.) Electron Transfer Chains and Oxidative Phosphorylation, pp. 305–316. North Holland/American Elsevier, Amsterdam.
245. Gibson, F., and Cox, G. B. (1973). Essays Biochem. 9:1.

246. Crofts, A. R., Crowther, D., and Tierney, G. V. (1975). *In* E. Quagliariello, S. Papa, F. Palmieri, E. C. Slater, and N. Siliprandi (eds.), Electron Transfer Chains and Oxidative Phosphorylation, pp. 233–241. North Holland/American Elsevier, Amsterdam.
247. Williams, R. J. P. (1969). Curr. Top. Bioenerg. 3:79.
248. Williams, R. J. P. (1972). J. Bioenerg. 3:81.
249. Williams, R. J. P. (1974). Ann. N. Y. Acad. Sci. 227:98.
250. Dilley, R. A. (1971). Curr. Top. Bioenerg. 4:237.

International Review of Biochemistry
Plant Biochemistry II, Volume 13
Edited by D. H. Northcote

3
Riddle of Sucrose

H. G. PONTIS
Fundación Bariloche, Departamento de Biología, Argentina

GENERAL PROPERTIES OF SUCROSE 80

ENZYMES OF SUCROSE METABOLISM 83
Sucrose Synthetase 83
Historical Background 83
Occurrence 84
Methods of Assay 84
Isolation 85
Molecular Properties 85
General Characteristics of Catalytic Reaction 88
Sucrose Phosphate Synthetase 99
Historical Background 99
Occurrence 99
Methods of Assay 100
Isolation 100
Molecular Properties 100
General Characteristics of Catalytic Reaction 101
Sucrose Phosphate Phosphatase 105
Historical Background 105
Occurrence 105
Methods of Assay 105
Isolation 105
Molecular Properties 106
General Characteristics of Catalytic Reaction 106

PHYSIOLOGICAL ROLES OF SUCROSE SYNTHETASE AND SUCROSE PHOSPHATE SYNTHETASE 107

SUCROSE AND PLANT HORMONES 111
Hormonal Effects at Enzyme Level 111

Sucrose as Regulator of Plant Metabolism 112

CONCLUSION 113

Sucrose occupies a unique position in the plant kingdom, comparable only to glucose in the animal world. It is known, of course, that sucrose is not only one of the major products of photosynthesis and the main form of translocation from leaves to other organs in nearly all higher plants, but that it is also the principal form of carbohydrate storage and provides a ready source of fructose and glucose for synthesis and energy. The importance of sucrose is emphasized when it is realized that nearly all of the carbon compounds formed in the nonphotosynthetic parts of the plants are derived from it. Even more, most animal carbon compounds are somewhat derived from the sucrose carbon skeleton because they are not auxotropic and depend on the plant world for their subsistence and development.

This ubiquity of sucrose constitutes its riddle, because the reason for its occurrence in the plant is not very obvious. What are the advantages that the selection of sucrose rather than glucose brings to the plant kingdom?

There is no clear answer to this question. However, if one applies the principle of Optimal Design (1), the selection of sucrose indicates that this molecule possesses properties that particularly fit it to perform its function in plants. The properties of the sucrose molecule could perhaps explain why sucrose is especially important in the plant metabolism. In this review, the chemical and physical properties of sucrose will be analyzed from a biochemical and physiological point of view. Second, the properties of the enzymes which affect the sucrose level in the cell will be studied. Invertases will not be dealt with, because this subject has been well reviewed recently in this series (2). Third, the role that the sucrose level in the cell could play in the regulation of metabolic activities will be investigated, especially the possible interaction of sucrose with plant hormones.

Finally, no attempt has been made to cover the whole body of information available on sucrose. Several valuable reviews (2–8) on different aspects of sucrose metabolism have been published recently, and the reader is referred to them for the aspects not covered here, as well as for the extensive existing literature.

GENERAL PROPERTIES OF SUCROSE

The general properties of sucrose are familiar to most students of chemistry. It is nonreducing, but is exceedingly easily hydrolyzed by diluted acids, although it is reasonably stable in the presence of strong bases or alkalis.

Enzymatically, sucrose is hydrolyzed to glucose and fructose by β-fructofuranosidase (invertase), widely distributed in nature, and α-glucosidase, present in animal intestine and honey.

The ease of its hydrolysis comes from the nature of its glycosidic linkage. Sucrose is an α-glucopyranosyl-β-D-fructofuranoside, in which the α-glucopyranosyl moiety is in the chair conformation ($C_1{}^4$) and the β-D-fructofuranosyl possesses the envelope (E^4) conformation (9).

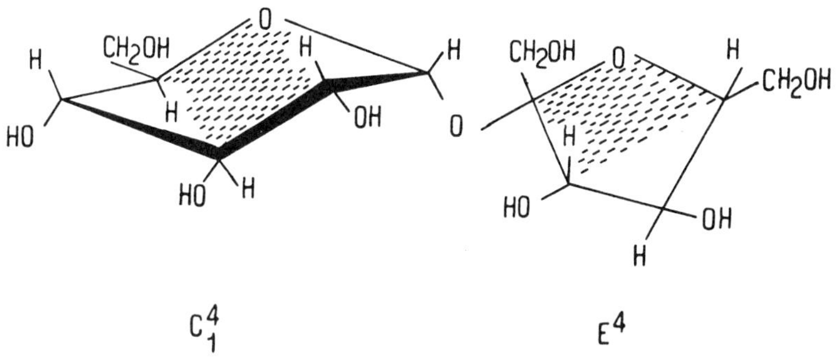

C_1^4 E^4

Chemically, this conformation contributes to a greater reactivity of the primary hydroxyls which are in a less hindered or more exposed position (10).

The other characteristic of sucrose is its β-fructofuranoside nature, which, as Edelman (11) has pointed out, is something of a rarity. It does occur in other compounds, but all these appear to be based on the sucrose molecule itself and can be considered derivatives of it.

The furanose configuration of the fructoside moiety in sucrose confers upon the linkage between the two sugars a very high free energy of hydrolysis. The ΔF° of hydrolysis of the glycosidic linkage of sucrose has been calculated as −6,600 cal (12) from the equilibrium constant (K = 0.053 at pH 6.6) of the reaction catalyzed by sucrose phosphorylase:

$$\alpha\text{-D-glucopyranosyl phosphate} + \text{D-fructose} \rightleftharpoons \text{sucrose} + \text{phosphate} \qquad (1)$$

According to Neufeld and Hassid (13) in the original calculation (14), a value of −4,800 cal was used for the ΔF° of hydrolysis of α-D-glucopyranosyl phosphate. This value was, however, obtained at pH 8.5. If a correction to pH 6.6 is applied, the ΔF° of hydrolysis of α-D-glucopyranosyl phosphate becomes −5,200 cal and that of sucrose is changed to −7,000 cal. This value may be contrasted with those of trehalose (15) (−5,700 cal/mol), maltose and lactose (16) (−3,000 cal/mol) and compared with those of UDP-glucose (16) (−7,600–8,000 cal/mol) and ATP (16) (−6,900 cal/mol).

This relatively high value for the glycosidic bond of sucrose has led Hassid to suggest that this might account for the distinctive role of sucrose in plant metabolism. Arnold (17) disregards this argument because sucrose seems to provide very little advantage over a glucose equivalent when taking into account the number of ATP molecules that each can yield potentially.

Arnold (17) considers only two enzymes as acting on sucrose: sucrose phosphorylase and invertase. He does not consider sucrose synthetase, which actually catalyzes a nucleoside pyrophosphorolysis of sucrose, thus allowing the cell to use the energy of the glycosidic linkage of sucrose for synthetic activities (see under "Substrate Specificity").

Considering the physical characteristics of sucrose, it is more appropriate to examine solutions of sucrose, as it is in this state that it occurs in plants. Arnold has compared water and solutions of sucrose and glucose with respect to several of their physical characters (see Table 1). He selected the level of sucrose (20%) that he used by taking into account the concentration of sucrose found in the translocation stream of plants, which ranges from 10–25% (w/v), depending upon the species and the time of the day.

Although there are some differences between the physical parameters illustrated in Table 1, Arnold concluded that these were insignificant in explaining why sucrose is used for translocation in plants and glucose is not.

Sucrose shares with glucose and other polyhydroxyl compounds the property of stabilization of isolated organelles and soluble proteins (18). Warner (19) has stated that such compounds could fit into the hydrated structure of proteins, and because of the inflexible nature of their hydroxyl bonds, relative to those of water, they supposedly could add to the stability of the system. However, sucrose structure does not seem to provide any advantage over that of monosaccharides or polyhydroxyl alcohols.

Thus, none of the properties of sucrose reviewed seems to account for its selection by plants as the main substance for translocation and storage. The nonreducing characteristic of sucrose might be important in understanding its function. Trehalose, the only other nonreducing disaccharide found in nature, occupies a place in fungi and insects similar to that of sucrose in plants. Therefore, Arnold has suggested that sucrose is a convenient and comparatively unreactive derivative of glucose, and has advanced the thesis that sucrose acts as a protective derivative of glucose.

Table 1. Physical parameters of water and 20% (w/w) solutions of sucrose and glucose

Parameter	Units	Temperature	Water	Sucrose	Glucose
Density	g/ml	d_4^{20}	0.998	1.081	1.080
Viscosity	millipoises	25	8.94	17.10	16.6
Surface tension	dyn/cm	25	72.0	73.0	73.4
Dielectric constant	$E_{vacuum} = 1$	20	81.9	75.3	55.7
Osmotic pressure	atm	0	0	17.7	32.0
Diffusion coefficient	$(cm^2/s) \cdot 10^7$	25	223	39.5	49.0

From W. N. Arnold (1968).

Arnold's point is that glucose, if it were a translocate, would be attacked by enzymes which catalyze its metabolism, whereas sucrose is less reactive. The disaccharide protects glucose residues from metabolic attack until they arrive at specific sites of growth or storage.

Nevertheless, one wonders whether the question "Why sucrose and not glucose?" has any meaning. Are we not too much influenced by the animal world? Could not sucrose be just the final product of photosynthesis–a process belonging to the plant kingdom–and as such more economical to plants as a translocate?

ENZYMES OF SUCROSE METABOLISM

The level of sucrose in the plant cell depends on four enzymes which are involved in its synthesis and degradation. They are sucrose phosphate synthetase (UDP-glucose: D-fructose-6-phosphate 2-glucosyltransferase, EC 2.4.1.14), sucrose phosphate phosphatase (sucrose-6-phosphate phosphohydrolase, EC 3.1.3.00), sucrose synthetase (UDP-glucose:D-fructose 2-glucosyltransferase, EC 2.4.2.13), and invertase (β-D-fructofuranoside fructohydrolase, EC 3.2.1.26). The latter may be divided into acid invertases and alkaline or neutral invertases. The claims that sucrose phosphorylase (EC 2.4.1.7) (200–21) is present in higher plants are not convincing. Furthermore, careful attempts (22) to confirm its existence in plants have not been successful.

Sucrose Synthetase

Historical Background The synthesis of sucrose was long of great interest because of its importance in plants. The discovery of sucrose synthesis from α-glucose-1-P and D-fructose catalyzed by extracts from *Pseudomonas saccharophila* (23) suggested a similar enzymic reaction in plants. Numerous attempts to demonstrate sucrose phosphorylase in plants were unsuccessful (22).

Evidence obtained in photosynthetic experiments with $^{14}CO_2$ led Buchanan et al. (24, 25) to formulate that UDP-glucose was the precursor of sucrose in green plants. It was left to Cardini et al. (26) to show first that extracts from wheat germ and other plants catalyze the synthesis of sucrose from UDP-glucose and fructose, according to the following reaction:

$$\text{UDP-glucose} + \text{D-fructose} \rightleftharpoons \text{UDP} + \text{sucrose} \tag{2}$$

In accordance with modern nomenclature, the enzyme is named UDP-glucose:D-fructose 2-glucosyltransferase. In Leloir's laboratory the enzyme was named saccharese, following a suggestion of von Euler that enzymes should receive the name of the substrates which they synthesize, with the ending changed to "-ese." Nevertheless, this name never caught the biochemists' attention, and the enzyme is commonly referred to as sucrose synthetase.

Occurrence The enzyme is widespread in plants, although a comprehensive systematic survey has not yet been reported. Evidence for the enzyme has been reported for the following plants: wheat, sugar cane, sweet sorghum, pine, fenugreek, maize, barley (26), mung bean (27), sweet potato (28), rice (29), tapioca (30), tomato (31), tobacco (32), grasses (33), Jerusalem artichoke (34), spinach (35), broad bean (36–38), castor bean (37), berries (39), pea (26, 40), sugar beet (26, 41), and potato (26, 42, 43).

The enzyme is not confined to a particular organ; it has been found in seeds, stems, tubers, roots, and leaves. Similarly, sucrose synthetase is found in endosperm, cotyledons, scutella, stele, callus tissue, mesophyll cells, and bundle sheath cells. On the other hand, no agreement has been reached regarding the presence of sucrose synthetase within the chloroplast (7, 35, 44, 45, 47). The present evidence seems to indicate that sucrose cannot penetrate the inner chloroplast envelope (46). Thus, the presence of sucrose-synthetizing enzymes in chloroplast preparations would only indicate their contamination with cytoplasmic proteins during normal methods of chloroplast isolation (44).

Methods of Assay Several assays for sucrose synthetase have been described for measuring the reaction in both directions. Among these, the procedure using the forward reaction is based on the estimation of sucrose or UDP. Sucrose has been determined colorimetrically after hydrolysis by invertase, followed by estimation of the glucose produced with a glucostat reagent (48), or after destruction of the remaining fructose by treatment with borohydride (49) or alkali (26). The latter method has been widely used, and it is quite suitable for kinetic determination, especially when sucrose is measured with the thiobarbituric acid procedure of Percheron (50). This method is not very suitable for testing the crude enzyme of some plants, however, owing to the presence of interfering substances. In these cases, it is advisable to use the methods that measure the formation of [^{14}C]sucrose from UDP-[^{14}C]glucose. Radioactive sucrose is separated from the labeled substrate by using anionic resins (51), paper electrophoresis (52), or by paper chromatography (53). In all cases, the separated [^{14}C]sucrose is determined by liquid scintillation counting. Radioactive techniques are particularly useful when studying the action of inhibitors that interfere with colorimetry.

The other product of the forward reaction, UDP, can be determined spectrophotometrically by coupling its formation to the pyruvate kinase-lactate dehydrogenase reaction and measuring the decrease in fluorescence due to oxidation of NADH (34, 48). The measurement of the pyruvate formed in kinase reaction with the method of dinitrophenylhydrazine of Cabib and Leloir (15) cannot be used because sucrose interferes with pyruvate kinase (54).

Procedures for measuring the back reaction are based on the estimation of fructose or UDP-glucose. Fructose formed is determined colorimetrically by the Nelson-Somogyi reducing sugar assay (55) or spectrophotometrically by coupling hexokinase, phosphoglucose isomerase, and glucose 6-phosphate dehydrogenase following the appearance of NADPH (34). The estimation of

UDP-glucose formed can be performed spectrophotometrically with the use of UDP-glucose dehydrogenase and following the reduction of NAD (34).

Radioactive techniques measure the incorporation of [^{14}C]glucose into UDP-glucose from [^{14}C]sucrose. The sugar nucleotide formed is separated from the excess [^{14}C]sucrose by paper chromatography (56) or by ion exchange paper (57). The degree of contamination of enzyme solution obtained from different plant material by endogenous substances, invertases, and phosphatases should determine the choice of method to be employed.

Isolation Sucrose synthetase has been prepared from many different materials. The enzyme seems to be most abundant in ripening seeds (27) and young etiolated seedlings (27). Procedures for its purification vary, but generally, tissue homogenization is followed by an ammonium sulfate fractionation, chromatography on DEAE-cellulose, and, usually, a step to gel filtration. Sucrose synthetase is a sulfhydryl enzyme (see under "Chemical Properties" in next section), and it is very sensitive to inhibition by substances formed during the oxidation of phenolic compounds present in plant extracts. Slack (58) found that this inhibition could be prevented by addition of cysteine or sodium diethyldithiocarbamate to the extracting buffers.

Molecular Properties

Purity Many studies on sucrose synthetase have been performed with relatively crude preparations. The enzyme has been prepared to various degrees of purification from potato (42), wheat germ (26), sugar beet (48), sweet potato roots (59), pea seedlings (60), maize endosperm (61), and tapioca tuber (30).

Wolosiuk and Pontis (62) purified sucrose synthetase from Jerusalem artichoke tubers to only one band by gel electroisofocusing. The specific activity of this preparation was not modified by rechromatography of the enzyme in Sepharose 6B or DEAE-cellulose, which seemed to indicate that it was homogeneous. However, when the enzyme was submitted to disc electrophoresis, it showed two protein bands with very close mobility (63). Nomura and Akazawa (29) obtained a sucrose synthetase from ripening rice seeds, which seems to be homogeneous as examined by polyacrylamide gel electrophoresis, with a specific activity similar to that of Jerusalem artichoke. On the other hand, Delmer (27) has purified sucrose synthetase from etiolated *Phaseolus aureus* seedling to homogeneity by electrophoretic immunological and ultracentrifugation criteria. Nevertheless, the isolation of a crystalline enzyme has not been reported so far.

Physical Properties In regard to optical properties, the absorption spectrum of the *P. aureus* enzyme is typical of a protein lacking any other chromophore groups with ultraviolet or visible spectrum. It shows absorption maxima at 215 and 276 nm. The $A_{280}:A_{260}$ absorption ratio is 1.43. The extinction coefficient of a 1% solution in 0.01 M ammonium bicarbonate is 9.5 measured at 280 nm.

In studying mobility, Nomura and Akazawa (29) determined the sedimentation constant of sucrose synthetase by glycerol gradient centrifugation, following the method of Martin and Ames (64) except for the replacement of sucrose by glycerol. With catalase (s_{20w} 11.4) and spinach leaf ribulose-1,5-diphosphate carboxylase ($s_{20,w}$ 18.6) as markers, sucrose synthetase from ripening rice seeds was found to have an $s_{20,w}$ of 13.3×10^{-13} s^{-1}.

Delmer (27), using sucrose as the gradient medium, with catalase as marker, found the sucrose synthetase from *P. aureus* seedlings to have an $s_{20,w}$ of 12.4 $\times 10^{-13}$ s^{-1}.

Delmer (27) calculated the partial specific volume of the homogeneous protein from the amino acid composition as 0.733 cm^3/g. Previous determinations by Grimes et al. (52) for a partially purified enzyme measured in sucrose gradients ($\nu = 0.83$) or in $CsCl_2$ gradients ($V = 0.79$) seem to be inaccurate due to solvent interactions, particularly in the case of sucrose gradients. Delmer also calculated the diffusion constant ($D_{20,w} = 3.03 \times 10^{-7}$ cm^{-2} s^{-1}) and the corresponding frictional ratio (f:fo = 1.49) for the *P. aureus* enzyme.

The molecular weight of sucrose has been determined for enzymes prepared for various sources, as can be seen in Table 2. The determinations were carried out by different methods and indicate a molecular weight of approximately 400,000. This is considerably smaller than the value estimated by gel filtration by Grimes et al. (52) for the *P. aureus* enzyme, but in this case only a void volume marker was run. Also the data accumulated for the more purified preparations of the enzyme show that there is no basis for the claim that sucrose

Table 2. Purification and properties of sucrose synthetase

Source	Specific activity[a] (μmol/min/ mg protein)	Apparent purification	Molecular weight	Reference
Potato	2.5	35	290,000[b]	41
Wheat germ	0.08	48		26
Sugar beet	0.9[c]	91		48
Sweet potato roots	3.2	40	540,000[b]	59
Maize endosperm	180	29		61
Tapioca tuber	2.8	96		30
Jerusalem artichoke tuber	4	153	420,000[d]	63
Mung bean seedling	0.1	252	375,000[e] 405,000[f]	27
	17.3	79	900,000[g]	52
Rice seeds	5.7	11	410,000[h] 440,000[b]	29

[a]Activity measured on the direction of sucrose synthesis.
[b]Gel filtration on Sephadex G-200.
[c]Cleavage, fructose formed.
[d]Gel filtration on Sepharose 6B.
[e]Equilibrium sedimentation.
[f]Gel filtration on Bio-Gel A-1.5 M.
[g]Sucrose gradient.
[h]Polyacrylamide gel electrophoresis.

synthetase is a lipoprotein. Moreover, the claim for a high lipid content was not based on analysis but on the density determined for the partially purified enzyme. As pointed out earlier, it has been found that these density determinations were inaccurate duc to solvent interaction. Finally, analyses support the conclusion that *P. aureus* sucrose synthetase contains no significant amounts of nonprotein components (27).

The subunit size of sucrose synthetase, as determined by sodium dodecylsulfate (SDS)-gel electrophoresis (29), is 90,000–100,000, indicating that the enzyme is a tetramer of identical protein subunits. Furthermore, low speed equilibrium sedimentation of the enzyme in 6 M guanidine-HCl indicated an average molecular weight of 94,000, which is in good agreement with the results of SES gels (27).

Chemical Properties The reported stability of sucrose synthetase varies with the enzyme source and the purity attained. Nevertheless, freezing and thawing cause a marked or local loss of activity regardless of the enzyme source. The *P. aureus* enzyme is quite stable at 4°C in the presence of 0.1 mM dithiothreitol and EDTA, losing less than 50% of its activity over a 1-month period (27). A similar stabilizing effect of mercaptoethanol and EDTA was found for the sugar beet enzyme (48). Wolosiuk (63) found that addition of 50 mM sucrose to buffer already containing mercaptoethanol and EDTA allows the conservation of 70% of the enzyme activity for at least 8 months at 4 °C.

Amino acid analyses have only been reported for the enzyme from *P. aureus* seedlings (Table 3). This enzyme shows an average distribution of the common amino acids. NH_2-terminal analysis by dansylation yielded only one conclusive NH_2-terminal amino acid, threonine (27).

In studies of thiol groups, all sucrose synthetase preparations examined gave evidence of essential sulfhydryl groups. In most cases, the enzymes are inhibited by mercurial reagents such as *p*-chloromercuribenzoate (58, 65), but they may be reactivated by the addition of a thiol such as cysteine, dithiothreitol, or mercaptoethanol (58). There is no evidence demonstrating direct participation of thiol groups in the reaction mechanism. The thiol group may be involved in maintaining a conformation state or facilitating subtle conformation changes necessary to the catalytic process. The action of mercaptoethanol on the forward and reverse reactions catalyzed by sucrose synthetase was investigated by Pressey (42), who found that sucrose cleavage activity was increased 4-fold, whereas sucrose synthesis was only activated by 30%. No titration studies of the thiol groups present in sucrose synthetase have been carried out, so no information exists at present that would allow speculation on the existence of different reactive thiol groups. Wolosiuk (63) found no difference in the inhibition of the forward and reverse reactions by different concentrations of *p*-chloromercuribenzoate.

In regard to the reactivity toward proteolytic enzymes, limited proteolysis of some enzymes has produced a change in their properties (66–68). Sucrose synthetase from Jerusalem artichoke has been submitted to the action of three

Table 3. Amino acid analysis of *P. aureus* sucrose synthetase

Amino acid	Residues/94,000 MWt[a]
Lysine	44.9
Histidine	26.2
Arginine	49.9
Tryptophan	16.4
Aspartic acid	70.6
Threonine	50.7
Serine	40.6
Glutamic acid	109.8
Proline	34.2
Glycine	59.1
Alanine	56.2
Half-cystine	3.1
Valine	53.0
Methionine	10.2
Isoleucine	45.0
Leucine	93.6
Tyrosine	29.8
Phenylalanine	40.5

[a]From D. P. Delmer (1972). J. Biol. Chem. 247:3822. Values are given as number of residues for subunit of 94,000 molecular weight.

proteolytic enzymes of varied specificity: papain, chymotrypsin, and trypsin (69, 70). No difference in the rate of inactivation for the forward or reverse reactions was observed when the enzyme was incubated with papain or chymotrypsin. On the other hand, there was a clear difference in the rate of inactivation of the synthetase on cleavage activities of the enzyme when it was submitted to the action of trypsin. After 15 min, 70% of the synthetase activity remains, whereas only 30% of the cleavage activity can be measured.

Results similar to those found with the Jerusalem artichoke enzyme were obtained when ammonium sulfate-purified sucrose synthetase from mung beans and peas was submitted to the proteolytic action of trypsin.

General Characteristics of Catalytic Reaction

Substrate Specificity The reaction catalyzed by sucrose synthetase is distinguished by two very special features: 1) it is the only sugar nucleotide transglucosylation reaction which is readily reversible, and 2) it exhibits a wide specificity for the nucleoside base, unlike the enzymes of sugar nucleotide metabolism which are specific for a particular base (71–73).

Thus, sucrose synthetase from various sources can accept UDP-glucose, ADP-glucose, TDP-glucose, CDP-glucose, and GDP-glucose (52, 74, 75) as glucosyl donors for the forward reaction. However, Budowsky et al. (76) have determined that for UDP-glucose the presence of an acylamido grouping $-C^2(X)-N^3H-C^4(X)-$ (where X = *O* or *S*) in the uracyl nucleus seems necessary.

The corresponding nucleoside diphosphates are substrates for the reverse reaction, but the affinity of the enzyme for the different sugar nucleotides and nucleosides phosphate is not the same. The point is discussed under "Kinetics" below.

The enzyme appears to be very specific for fructose or sucrose. According to Slabnik et al. (65), the potato enzyme also uses L-sorbose and D-tagatose, although at a very low rate. Similarly, Avigad and Milner (48) found that sugar beet sucrose synthetase uses 5-keto-D-fructose, fructose 6-phosphate, and levanbiose, as well as L-sorbose and D-tagatose, as glucosyl acceptors to a small, detectable extent. Raffinose and the sucrose analogs, D-xylsucrose and D-galsucrose, could not replace sucrose as aldosyl donors.

Equilibrium This aspect of sucrose synthetase has attracted attention because of the ready reversibility of the reaction. The equilibrium constant of reaction (2) defined as:

$$K = \frac{(\text{sucrose})\,(\text{UDP})}{(\text{UDP-glucose})\,(\text{fructose})}$$

was originally estimated by Cardini et al. (26) to be between 2 and 8 at 37° C and pH 7.4, although the determination was not considered very exact due to interference by hydrolytic enzymes. Years later, Milner and Avigad (77), working with more purified preparations of the sugar beet enzyme found an apparent equilibrium constant of 1.3 at 37°C and pH 7.4. Similarly, for the Jerusalem artichoke enzyme, the apparent *K* was found (34) to be 1.4–1.8 at 30°C and pH 7.6, whereas Akazawa et al. (78) found *K* to equal 2 at 37°C and pH 7.4 for sucrose synthetase from rice.

In contrast, Delmer (27), measuring the reaction in the direction of UDP-glucose synthesis, found a value of 0.149 for the equilibrium constant at 25°C and pH 7.5. This value is equal to a *K* of 6.7 in the sucrose synthesis direction, which is much higher than the values determined by Avigad (34). By using a *K* value of 1.6, ΔF° can then be estimated at −283 cal/mol at 30°C and pH 7.6. Thus, the equilibrium of the reaction starting with UDP-glucose and D-fructose is in favor of sucrose synthesis. Also, the small ΔF° of reaction (2) is an indication of the similarity in ΔF° of hydrolysis of sucrose and UDP-glucose.

The above determination of the apparent constant of equilibrium does not take into consideration that during the reaction a secondary phosphoric acid group becomes unmasked.

Thus, the reaction may be written as follows:

$$\text{UDP-glucose}^{2-} + \text{fructose} \rightleftharpoons \text{UDP}^{-3} + \text{sucrose} + H^+ \qquad (3)$$

It is apparent, then, that hydrogen ions participate in the reaction and that the pH should affect the final equilibrium.

Kinetics The substrate affinities for sucrose synthetase from various sources are indicated by the Michaelis constants shown in Table 4. It can be seen that in contrast to the high degree of specificity for fructose and sucrose, sucrose synthetase can use a wide variety of nucleoside diphosphates glucose and nucleoside diphosphates. However, with the homogeneous enzyme from mung bean, the K_m for UDP-glucose is 10 times smaller than that for any other nucleoside diphosphate glucose. Furthermore, the K_m for fructose is also 10 times smaller in the presence of UDP-glucose than in the presence of any other nucleoside diphosphate glucose.

Enzymes from other sources show nearly the same affinity for ADP-glucose as for UDP-glucose. The maximum rate of sucrose synthesis is measured in nearly every case when ADP-glucose is the glycosyl donor.

The fact that fructose affinity for the enzyme varies very little when UDP-glucose is the other substrate but changes with other nucleoside diphosphates glucose, might suggest that the conformation of the enzyme could be affected differently, leading to the variations observed in fructose K_m(52).

Analysis of the data shown in Table 4 for the reverse reaction presents a different picture. No large differences in K_m are observed between UDP and other nucleoside diphosphates, from any enzyme source. Maximum rates of sucrose cleavage are always observed when UDP is the substrate. On the other hand, K_m for sucrose varied from 10–400 mM. Moreover, the affinity of the enzyme for sucrose was changed on treatment with trypsin (see under "Sucrose Synthetase," section "Chemical Properties"). Saturation curves for sucrose were no longer hyperbolic but sigmoidal in shape. The K_m increased from 56 mM for the native enzyme to 200 mM for the trypsin-treated one (70). No change in UDP K_m could be observed.

The difference that the enzyme presents in the affinity toward nucleotides for the forward and reverse reactions is surprising. It would have been expected that nucleoside diphosphate and nucleoside diphosphate sugar would attach to the same site and, therefore, the affinity relationship between different nucleosides would have been the same for the forward and the reverse reactions.

Kinetic data (56) for the Jerusalem artichoke enzyme actually substantiate the above assumption, making the interpretation more difficult.

In regard to pH optimum, sucrose synthetase is characterized by presenting different pH optima for the forward and reverse reactions. Sucrose synthesis generally occurs at slightly alkaline pH, with a broad maximum from pH 7.5–8.5. Sucrose cleavage takes place on the other side, with a tendency to present a sharp maximum centered around pH 6–6.5. The pHs indicated in Table 4 correspond to the optimum for each enzyme source.

Sucrose synthetase from any source does not show an absolute requirement for metal ions. On the other hand, it has been shown that enzyme isolated from various sources (Table 5) is activated in the direction of sucrose synthesis by divalent cations (Mg^{2+}, Mn^{2+}, Ca^{2+}, and Ba^{2+}) while inhibiting sucrose cleavage (34, 42, 57, 61, 63, 70). It has been shown by Fekete and Cardini (81) that UDP is a strong inhibitor of sucrose synthesis. This inhibition can be reversed by Mg^{2+}. As can be seen in Figure 1, UDP at 10 mM produces at 70–80% inhibition. Activity was regained almost to control values when 20 mM $MgCl_2$ was added. Wolosiuk and Pontis (70) have shown also that if the ratio $MgCl_2$:UDP is kept at 2, the activity of sucrose synthetase remains constant even in the presence of UDP concentrations that would normally produce 90% inhibition. A similar effect could be demonstrated for $CaCl_2$. The effect of divalent cations on the enzyme activity may be considered a consequence of a cation-nucleotide complex formation or of the interaction between the divalent cation and the protein or both. The fact that $MgCl_2$, as well as $MnCl_2$, produces a change in the K_m for fructose, increasing its affinity, gives some support to the latter possibility.

In studies of activators and inhibitors, sucrose synthesis is inhibited by UDP, one of the reaction products (81). Kinetic studies carried out by Wolosiuk and Pontis (56) and by Gabrielian et al. (60) indicate that UDP is a competitive inhibitor with respect to UDP-glucose. The same inhibition pattern appears in the reverse direction, that is, UDP-glucose acts as competitive inhibitor with respect to UDP. The results obtained by Wolosiuk and Pontis seem to indicate that UDP-glucose, like UDP, binds to only one enzyme form. These may be the same or different enzyme forms which are separated by freely reversible paths.

UDP not only acts as an inhibitor toward UDP-glucose, but also toward ADP-glucose, when this nucleotide is used as substrate for sucrose synthesis (81). The inhibition produced by ADP when UDP-glucose is the glucosyl donor is very weak.

Pridham et al. (36) were the first to detect an inhibitory action of fructose on sucrose synthetase activity. This inhibition has been shown by Wolosiuk and Pontis (56) to be competitive toward sucrose but uncompetitive to the other substrate UDP.

Several compounds have been tested for possible activation or inhibition effects. According to Delmer (57), glucose-1-P, glucose-6-P, fructose-6-P, fructose-1,6-diP, ribose-5-P, ribulose-1,5-diP, P-enolpyruvate, 3-P-glycerate, Pi, citrate, chondroitin sulfate, and kinetin, at a final concentration of 2 mM, have no effect upon the rates of either the forward or reverse reactions. Similar results were reported by Shukla and Sanwal (30) with respect to the direction of sucrose synthesis.

Delmer (57) found that pyrophosphate in the presence of $MgCl_2$ activated sucrose cleavage. These results conflict somewhat with those of Pontis et al. (82), who found an inhibitory action of pyrophosphate alone or in the presence of Mg^{2+}, for the Jerusalem artichoke enzyme. Furthermore, Pontis et al. (82), in contrast to Delmer's results, found that glucose-1-P, glucose-6-P, and fructose-1,6-diP, at a final concentration of 2–5 mM, inhibited sucrose cleavage without affecting sucrose synthesis. Fekete (83), working with *Vicia faba* enzyme, also observed the inhibitory action of glucose 1-phosphate and glucose 6-phosphate in the cleavage of sucrose.

Nucleoside mono- and triphosphates, in particular adenosine and uridine

Table 4. Kinetic constants for sucrose synthetase

Plant	Synthesis						Cleavage					Reference
			NDP-glucose		Fructose				NDP		Sucrose	
	pH	Base	K_m (mM)	V_m (μmol/min/mg protein)	K_m (mM)	V_m (μmol/min/mg protein)	pH	Base	K_m (mM)	V_m (μmol/min/mg protein)	K_m (mM)	
Jerusalem artichoke	8.3	U[a]	0.64		1.6		8.3	U	0.32		33	34
	8.0	U	1.0[b]		4.7[b]		6.2	U	0.9[b]		56[b]	82
		U	0.4[c]		2.5[c]			U	0.3[c]		100[c]	
Pea	7.3	U	1.9									79
		A	3.2									
Rice	8.0	U	5.3	6.3	6.9	7.1	6.0	U	0.8		290	29
		A	3.8	3.3	40	15.4		A	3.3		400[d]	
Mung bean	7.3	U	0.21	140	2							52
		A	1.8	350	23							
		T	1.7	170	35							
		C	2.5	48	35							
		G	2.5	41	35							
Mung bean							7.5	U	0.19	100[e]	17	27
								A	0.19	28[e]	29	
								T	0.30	5.8[e]		
								C	0.44	3[e]		
								G	0.17	3[e]		

Tomato	8.0	U			3.8							31
Tapioca	7.5	U	8.3	6.2	2.6	1.5	7.5	U	6.6	3.7	10	
		A	6.7	6.2	2.5	2.6		A	33	6.1	12.5	30
Sugar beet	7.2	U	0.3		3.1		6.0	U	0.06	100[e]	110	
		T	0.09					T	0.09	52[e]		48
								A		16[e]		
Potato	8.0	U	1.65	0.01	5.9							
		A	86	0.006								
		T	1.0	0.003								65
		G	2.8	0.002								
	7.0	U	1.6		6.2		7.0	U	1.7		130	
								A	1.3			42
	8.7	U	0.6		6.6							80
Wheat	7.2	U			2.3							26
Corn	8.0	U	1.1		2.1		6.5	U	0.14		40	
								A	1.25			61
Sweet corn	7.4	U	1.1				7.4	U	0.06			
		A	2.0					A	2.1			81
Sweet potato	7.5	U	1.7		1.5		6.0	U	0.13[d]		31[d]	
		U	0.7[f]		1.0[f]			A	0.44		125[d]	59
		A	5.0									

[a]U, uracil; A, adenosine; T, thymine; C, cytosine; G, guanine.
[b]Corresponding to sucrose synthetase form A, pI 5.85.
[c]Corresponding to sucrose synthetase form B, pI 5.4.
[d]Sigmoidal curves, $A_{0.5}$.
[e]Relative numbers.
[f]In the presence of 5 mM $MnCl_2$.

Table 5. Effect of divalent cations on sucrose synthetase

Source	Synthesis: Ion concentration (mM)		Synthesis: Activation (%)	Cleavage: Ion concentration (mM)		Cleavage: Inhibition (%)	Reference
Potato	Mn^{2+}	5	200	Mn^{2+}	10	55	42
	Mg^{2+}	20	100		20	84	42
	Ba^{2+}	8	50				
Jerusalem artichoke				Mg^{2+}	8	58	34
				Mn^{2+}	8	46	34
	Mg^{2+}	10	40	Mg^{2+}	20	10	63, 70
Mung bean	Mg^{2+}	2	57	Mg^{2+}	2	54	57
Maize	Mg^{2+}	4	27	Mg^{2+}	4	21 39[a]	61
	Ca^{2+}	4	26	Ca^{2+}	4	10 27[a]	61

[a]Measured at pH 8.0.

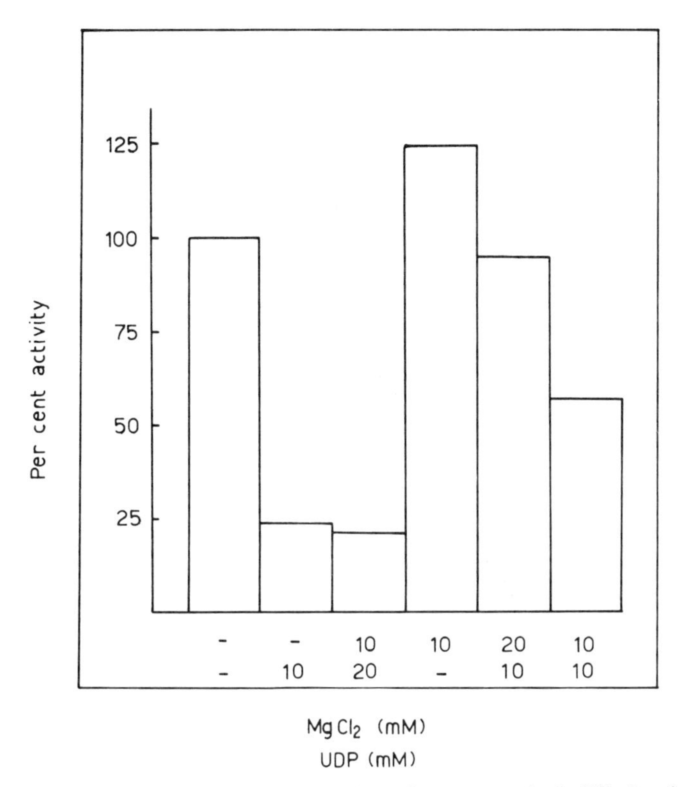

Figure 1. Effect of $MgCl_2$ on the UDP inhibition of sucrose synthesis (70). Reprinted with permission of Mol. Cell Biochem. 4:115.

derivatives, also belong to the group of metabolites that affect only one of the directions of the reaction.

At optimum pH, Tsai (61) found that the synthetic activity of the *Zea mays* endosperm enzyme is unaffected by 4 mM ATP, ADP, and AMP, and only slightly affected by UTP. Shukla and Sanwal (30) also found no inhibitory action toward sucrose synthesis at ATP up to a concentration of 20 mM. These nucleotides inhibit sucrose cleavage at pH 6.5 to the extent of 80% for UTP, 30% for ATP, and 50% for ADP and AMP. Similar results have been found by Wolosiuk (63) for the Jerusalem artichoke enzyme and by Avigad and Milner (48) for the sugar beet enzyme. Dermer reported no action by ATP at 2 mM in any direction for the homogeneous mung bean enzyme. Thus, except for these last observations, all the data indicate that nucleotides only affect the cleavage reaction, regardless of the enzyme source.

How is it that these metabolites seem to differentially alter the reaction rates of the forward and reverse reactions?

Frieden (84), discussing the effects of modifiers on the kinetic parameters of single substrate enzyme, pointed out that "the effect of a particular modifier need not be the same for the forward and reverse directions of a particular enzyme. Not only is it possible for the type of effect to differ, but it is apparent that in certain cases a modifier could enhance the activity in one direction and inhibit the enzyme in the reverse direction." The theoretical treatment of Frieden might also be applicable to enzymes which use more than one substrate. If this were the case, it would readily explain the results obtained with sucrose synthetase. Moreover, this type of control on a reversible reaction may produce a means of regulation.

In order to study this possibility, Wolosiuk and Pontis (70) tested phenolic glycosides. These are among normal metabolites which are very widespread in plants and which Slabnik et al. (65) have reported as affecting sucrose synthetase activity.

Wolosiuk and Pontis showed that β-phenylglycoside inhibited the forward and reverse reactions of sucrose synthetase differentially (Figure 2). In the presence of 5 mM β-phenylglycoside, cleavage activity is practically suppressed whereas 50% of synthesis activity remains. A detailed study of the inhibition showed that β-phenylglycoside is a competitive inhibitor for UDP-glucose with K_i 5.2 mM, and it has the same type of inhibition for sucrose with K_i 0.8 mM. It would be assumed then that β-phenylglycoside competes for the same form of the enzyme with UDP-glucose or sucrose. When the interaction of β-phenylglycoside with fructose and UDP was studied, the results pointed to a mixed type of inhibition. Replots of slopes and intercepts versus inhibitors were linear, indicating that β-phenylglycoside combines with only one form of the enzyme, as fructose or UDP does.

These kinetic data indicate, therefore, that in the presence of β-phenylglycoside the K_m of the four substrates may be modified, although the alteration of enzyme activity need not be the same for each substrate. The result is the differential modification of the relative rates of cleavage and synthesis, supporting Frieden's hypothesis.

The activity of sucrose synthetase is also affected by glucose. Glucose not only inhibits cleavage (70), but also sucrose synthesis (65). The action of sugars on the activity of the enzyme has been mainly studied in connection with the cleavage reaction. Besides glucose, the cleavage reaction is also inhibited by fructose, β-D-thioglucose, glucosamine, gluconic acid, and δ-gluconolactone (70). It is interesting that the products of the action of another enzyme which acts on sucrose, invertase, are some of the few sugars that affect sucrose cleavage catalyzed by sucrose synthetase (82).

The most detailed kinetic study of the mechanism of sucrose synthesis by sucrose synthetase used the Jerusalem artichoke enzyme (56). A plot of the reciprocal initial velocity versus reciprocal substrate concentration gave a series of intersecting lines indicating a sequential mechanism. Product inhibition studies showed that UDP-glucose was competitive with UDP, wherever fructose was competitive with sucrose and uncompetitive with UDP. A dead-end inhibitor, salicine, was competitive with sucrose and uncompetitive with UDP.

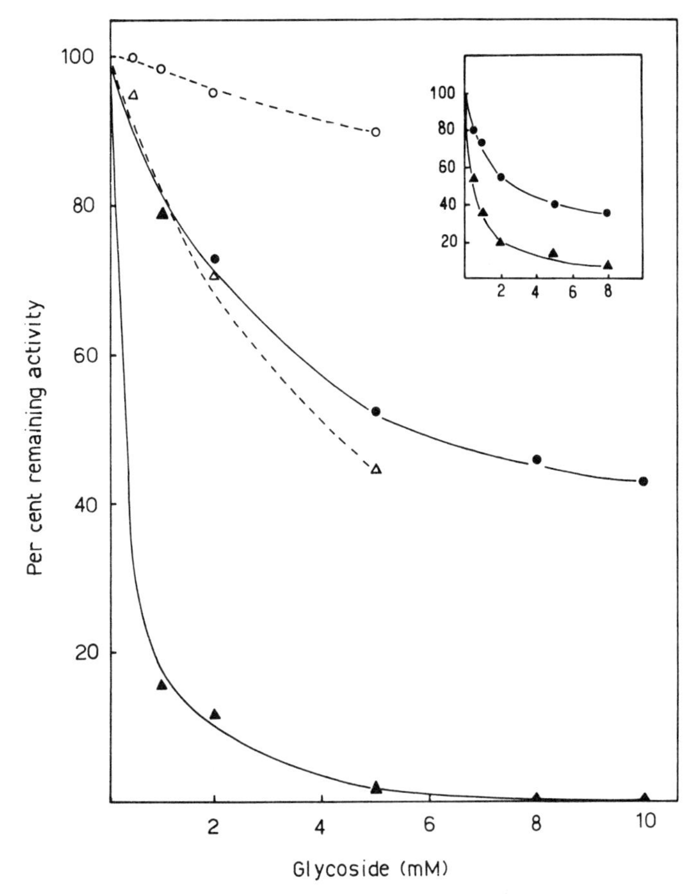

Figure 2. Inhibition of sucrose synthetase activity by glycosides. ○, salicine synthesis; △, salicine cleavage; ●, β-phenylglycoside synthesis; ▲, β-phenylglycoside cleavage. The inset shows the inhibition of β-phenylglycoside of the activity assayed at pH 7.4. Reprinted with permission of Mol. Cell Biochem. 4:115.

The results of initial velocity product and dead-end inhibitor studies suggested that the addition of substrates to the enzyme follows an ordered mechanism. In contrast, a "ping-pong" mechanism was proposed by Sung et al. (85) for pea seedling enzyme. The initial velocity studies illustrated in Figure 3 show an intersecting pattern characteristic for ordered mechanism. Failure to show isotopic exchange between sugars in the absence of nucleoside diphosphate and sugar nucleotide adds further evidence that the reaction is sequential rather than ping-pong.

Gabrielian et al. (60) arrive at a similar conclusion in studying the effect of UDP inhibition on the rate of the forward reaction at different concentrations of UDP-glucose and fructose, with the use of a purified sucrose synthetase from pea seedling.

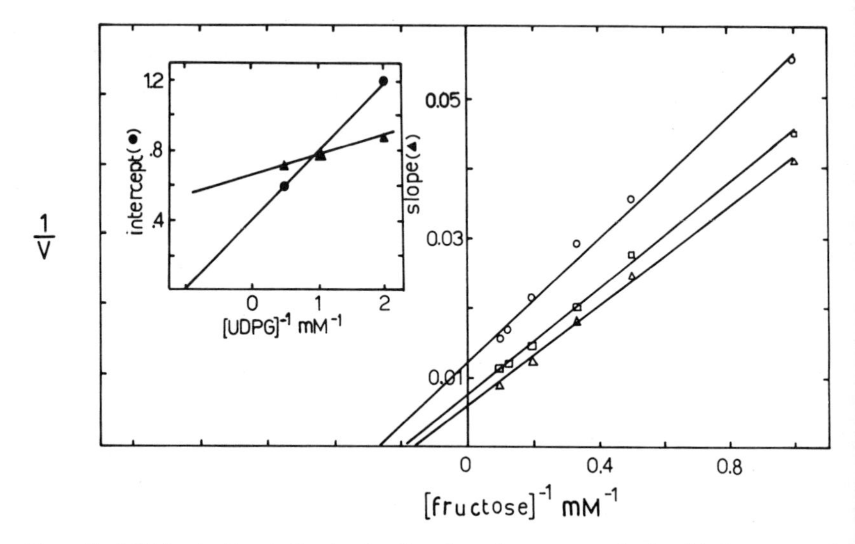

Figure 3. Initial velocity studies in the direction of sucrose synthesis with fructose as the varied substrate. Sucrose formed was determined by the thiobarbituric acid method. The concentrations of UDP-glucose were ○, 0.5 mM; □, 1 mM; △, 2 mM. The *slopes* and *intercepts* of the primary plot are replotted as a function of the reciprocal of the concentration of UDP-glucose in the secondary plot. Reprinted by permission of Arch. Biochem. Biophys. 165:140.

Delmer (27) has suggested a random mechanism for *P. aureus* sucrose synthetase. Her suggestion derives basically from two results. First, the K_m for UDP is independent of the concentration of sucrose. Second, the experimentally determined equilibrium constant fits with kinetic constants in the Haldane relationship. Cleland (86) pointed out that the first result arises from kinetic constants involved in the mechanism itself. Particularly in the Bi-Bi reaction mechanism, the crossover point depends on the K_{ia}:K_a ratio. Several authors draw attention to the Haldane relationship, indicating that it is not very sensitive to errors involved in individual determination of constants. The kinetic data reported by Wolosiuk and Pontis for the Jerusalem artichoke enzyme did not fit into a random mechanism. The product and dead-end inhibition seems to agree more with an ordered mechanism than with a random one.

An ordered reaction mechanism indicates that contrary to the case of the sucrose phosphorylase reaction, a glycosyl enzyme is probably not formed.

Inhibition studies with sugars shed some light on the mechanism of the transglycolation reaction between sucrose and UDP-glucose. In both compounds, the configuration of the anomeric carbon is α, which means that the reaction proceeds with retention of configuration. This rules out a single displacement mechanism, strongly favoring a double displacement one in which two Walden inversions lead to a product with the same configuration as the substrate, such as has been shown with sucrose phosphorylase (87).

A double displacement reaction is unlikely, because attempts to demonstrate the presence of a glucosyl enzyme have failed. No exchange was detected either between [^{3}H]UDP and UDP-glucose when these substrates were incubated in the absence of fructose, and no exchange between free glucose and sucrose was found. Similar studies provided unequivocal evidence for a double displacement reaction in sucrose phosphorylase, and the glucosyl enzyme intermediate was ultimately isolated (87).

Glycogen phosphorylase (88) and β-glycosidase (89), in which the transglycosidation reaction proceeds with retention of configuration, are inhibited by δ-gluconolactone (90). The lactone has a half-chair conformation similar to that of a cyclic carbonium-oxonium ion. Thus, the lactone apparently competes for the glucosyl transfer site, and the normal transition state is likely to be a glucose-enzyme complex in which the glucosyl unit has the half-chair conformation (88). This mechanism can also be envisaged for sucrose synthetase. In this case, the enzyme would facilitate a displacement reaction followed by a front side attack by fructose on a stabilized carbonium ion.

This conclusion is supported by the data reported by Avigad (34) on the exchange of [^{14}C]fructose between sucrose and free D-fructose, which occurs only in the presence of UDP.

Sucrose Phosphate Synthetase

Historical Background During the isolation of sucrose synthetase, Leloir and Cardini (91) found that while some preparations of the enzyme were almost devoid of the action of fructose 6-phosphate, other extracts catalyzed the formation of free sucrose either from fructose or from its phosphate. Because the extracts contained phosphatase, it was thought initially that the formation of sucrose was due to the action of sucrose synthetase on the fructose liberated from its ester. Moreover, further work showed that it was possible to obtain enzyme preparations which catalyze the reaction:

$$\text{UDP-glucose} + \text{fructose 6-phosphate} \rightarrow \text{UDP} + \text{sucrose phosphate} \qquad (4)$$

At that time, the enzyme isolated from wheat germ by Leloir and Cardini was purified approximately 5- to 10-fold. The preparation contained significant quantities of sucrose synthetase and phosphatases which hydrolyzed fructose 6-phosphate, as well as sucrose phosphate. Nevertheless, the evidence presented conclusively proved the above reaction, and the phosphoric ester of sucrose was isolated and its structure determined, showing the phosphate group at position 6 of the fructose moiety.

Occurrence The enzyme is widespread in plants, but a systematic survey has not yet been reported. Evidence for the enzyme has been reported for the following plants: wheat (35, 91, 92), sugar beet (49), broad bean (37, 93), rice (37), sugar cane (94), pea (94, 95), maize (37, 96), sweet potato (97), barley (97), rape (97), potato (65, 97), ladino clover (97), spinach (92), field bean (35), grasses (33), castor bean (37), grape (39), tobacco (94), and mung bean (98).

The enzyme is not confined to green tissues, as has been suggested (98); it has also been found in seeds (37), stems (22), tubers (65, 97), roots (99), and leaves (94).

As with sucrose synthetase, no agreement has been reached regarding the presence of sucrose phosphate synthetase within the chloroplast (44, 45). The present view is that its presence in chloroplast preparations is due to cytoplasmic contamination (44) (see under "Sucrose Synthetase," section "Occurrence").

Methods of Assay There are several assays for sucrose phosphate synthetase that are based on the estimation of sucrose phosphate or UDP formed. Sucrose phosphate has been determined colorimetrically after destruction of the remaining fructose 6-phosphate by treatment with borohydride (46) or alkali (91). The latter method has been widely used, and it is quite suitable for kinetic determinations. When radioactive UDP-glucose is used, the labeled sucrose phosphate formed is separated from the substrate by paper chromatography (94), or ion exchange resin (100) after submitting the ester to the action of alkaline phosphatase. If radioactive fructose 6-phosphate is used instead of UDP-glucose, the separation must be carried out after phosphatase action by paper chromatography (94).

The other product of the reaction, UDP, can be determined spectrophotometrically by coupling its formation to the pyruvate kinase lactate dehydrogenase reaction and measuring the decrease in fluorescence due to oxidation of NADH (101).

Isolation Sucrose phosphate synthetase has been prepared from different materials, but there is no report indicating which material would be most favorable for extracting the enzyme. Procedures for its purification vary, but tissue homogenization is followed by ammonium sulfate fractionation and chromatography on DEAE-cellulose or protamine precipitation. Sucrose phosphate synthetase seems to be a sulfhydryl enzyme (65), so care should be taken to avoid enzyme inhibition during the preparation of plant extracts (see under "Sucrose Synthetase," section "Chemical Properties").

Molecular Properties

Purity The enzyme has been purified from wheat germ (92), broad bean cotyledons (93), rice scutellum (102), and potato tubers (65). However, most studies have been limited to the detection of enzyme activity. No homogeneous preparations have been described. Moreover, an analysis of the purification procedures described seems to indicate that in many cases sucrose phosphate synthetase preparations are not completely free of sucrose synthetase.

Physical Properties In mobility studies, Nomura and Akazawa (102) determined the sedimentation constant of sucrose phosphate synthetase by glycerol gradient centrifugation following the method of Martin and Ames (64) except for the replacement of sucrose by glycerol. With catalase ($s_{20,w}$ 11.4) and urease ($s_{20,w}$ 18.6) as markers, sucrose phosphate synthetase from rice scutellum was found to have an $s_{20,w}$ of 10.4×10^{-13} s^{-1}.

This value is rather low, taking into account the estimated molecular weight found by gel filtration on Sepharose 6B102. On the other hand, Salerno and Pontis (103), using the same technique, found that the wheat germ enzyme presents more than one species, judging from the pattern obtained after the high speed centrifugation.

The molecular weight of sucrose phosphate synthetase has been determined only for the rice scutellum and wheat germ enzymes. In both cases, the same method, gel filtration through Sepharose 6B, was used. Nomura and Akazawa (102) found a molecular weight of 450,000 for the rice scutellum enzyme, whereas for the wheat germ enzyme it was calculated to be about 360,000 (103). Such a large molecular weight indicates that the protein is probably composed of a number of subunits. However, no determination of subunit size has been carried out.

Chemical Properties Regarding stability, sucrose phosphate synthetase seems to be quite stable when kept frozen at temperatures ranging from −10–20°C. Thus, Mendicino (92) has reported only a 10% loss of activity after 2 months at −10°C. Salerno (104), for the same wheat germ enzyme, found that it retained at least 40–50% of its activity after 1 year at −20°C. Nomura and Akazawa (102) could keep their preparation from rice scutellum 1 week at 10°C without apparent loss of activity.

Fekete (93) found for the broad bean enzyme that freezing and thawing caused loss of enzyme activity. Simultaneously, a very faint precipitate appeared which was separated by centrifugation. The full activity could be found again if both supernatant fluid and precipitate were mixed. No such behavior has been described for enzymes prepared from other sources. Murata (97) has reported that enzyme prepared from plant leaves is quite unstable and loses half of its activity after 4–5 days when kept at −20°C. Similarly, Slabnik et al. (65) found that the enzyme from potato is also very unstable.

In studies of thiol groups, sucrose phosphate synthetase has been found to be inhibited by mercurial reagents. Thus, Slabnik et al. (65) reported that their enzyme from potato was inhibited 25% by 0.7 mM *p*-chloromercuribenzoate. Salerno (104) found an inhibition of 60% with 0.1 mM *p*-chloromercuribenzoate for wheat germ enzyme and could recover full activity upon addition of 1 mM mercaptoethanol.

No titration studies of the thiol groups present in sucrose phosphate synthetase have been carried out, so no information exists on different reactive thiol groups.

General Characteristics of Catalytic Reaction

Substrate Specificity Sucrose phosphate synthetase is very specific toward its substrates. UDP-glucose cannot be replaced by other sugar nucleotides like ADP-glucose, GDP-glucose, or TDP-glucose (65). Similarly, fructose 1-phosphate or fructose-1,6-diP cannot serve as substrates (92). This is a very striking difference from sucrose synthetase. In this respect, sucrose phosphate synthetase behaves like all other sugar nucleotide transglycosylases, which are highly specific toward their substrates.

Equilibrium The apparent equilibrium constant for the sucrose phosphate synthetase reaction was determined at pH 7.5 by Mendicino (92) for the wheat germ enzyme. As the reaction proceeded to the right to near completion and the amount of UDP-glucose and fructose 6-phosphate remaining was very small, Mendicino measured the equilibrium with the use of UDP and sucrose 6-phosphate as substrates. Radioactive sucrose 6-phosphate was used so that the formation of very small amounts of fructose 6-phosphate could be detected. The fructose 6-phosphate concentration at equilibrium was 4.67×10^{-4} M at pH 7.5 and 6.7×10^{-5} M at pH 5.5. The apparent equilibrium constant was determined by the following equation:

$$K = \frac{(\text{UDP})\,(\text{sucrose-6-P})}{(\text{UDP-glucose})\,(\text{fructose-6-P})} \tag{5}$$

where the concentrations are expressed as total concentrations of all molecular species. The average value of K is 53 at pH 5.5 and 3,250 at pH 7.5. These values indicate a surprisingly low ΔF° of −2,700 cal for the hydrolysis of the glycosidic bond of sucrose 6-phosphate.

The reaction catalyzed by sucrose phosphate synthetase may be written as follows:

$$\text{UDP-glucose}^{2-} + \text{fructose-6-P}^{2-} \rightleftharpoons \text{UDP}^{3-} + \text{sucrose-P}^{2-} + \text{H}^{+} \tag{6}$$

It is apparent that hydrogen ions participate in the reaction and, therefore, that the pH should affect the final equilibrium. Thus, a shift of 2 pH units (which is equivalent to a 100-fold change in the H^+ concentration) altered the value of the sucrose phosphate synthetase reaction by about 60-fold.

Kinetics The substrate affinities for sucrose phosphate synthetase from various sources are indicated by the Michaelis constants shown in Table 6. There seem to be no large differences between the values found for the various enzymes.

Some of the enzymes seem to exhibit sigmoidal saturation curves. In these cases, the reported values for the Hill coefficient are included in the table, as well as an indication of the change from K_m to $A_{0.5}$ or the substrate concentration needed for half-maximum velocity. It should be pointed out that in every case only a very slight cooperativity could be observed, as indicated by the values obtained for the Hill coefficient.

Sucrose phosphate synthetase presents a pH optimum around 6.5. Depending on the enzyme source, pH curves tend to show a slight plateau around the maximum, which could explain some reports of a higher pH optimum. The activity of the enzyme is also influenced by the buffers used. It should be mentioned that phosphate and Tris-maleate have an inhibitory action (104) on the activity, so they are not very suitable for determining either the activity or

Table 6. Kinetic constants for sucrose phosphate synthetase

Plant	pH	UDP-glucose K_m (mM)	UDP-glucose $\bar{n}$	Fructose-6-phosphate K_m (mM)	Fructose-6-phosphate $\bar{n}$	Reference
Wheat	7.4	7.4		3.0		92
	6.5			2.2		91
	6.4	0.67[a,b]	1.5	2.6[a,b]	1.8	101
		1.9[b]	1.7	3.4[b]	1.4	
	6.5	5		2.2		104
Barley	6.5	2.7		1.8[b]	1.4	97
Rape	6.5	4.2		1.3[b]	1.4	97
Ladino clover	6.5	3.0		0.6[b]	1.5	97
Potato	6.0	2.0[c]		0.9[b]	1.4	97
		2.5				
	7.5	7.4		5.5		65
Sweet potato	6.0	5.4		0.6[b]	1.4	97
Sugar cane	6.7			2.4		94
Broad bean	6.4	3.9	1.0	1.4	1.0	93
		19.8[d]	1.0	6.3[d]	1.1	
		9.8[e]	1.5	4.3[e]	1.6	
		4.6[f]	1.0	1.4[f]	1.0	
Rice	7.5	25.0		2.0–5.9		102

[a]In the presence of 22 mM $MgCl_2$.
[b]Sigmoidal curves, $A_{0.5}$.
[c]In the presence of 5 mM $MnCl_2$.
[d]In the presence of 40 mM citrate.
[e]Devoid activator.
[f]In the presence of 100 mM citrate.

the pH curve. The pH indicated in Table 6 corresponds to the optimum for each enzyme source.

In studies of metal requirements, sucrose phosphate synthetase from any source does not show an absolute requirement for metal ions. It has been shown that enzymes isolated from various sources are activated by divalent cations. According to Murata (97), Mg^{2+} (2.5 mM) has a slight stimulative effect (about 10–20% with the enzyme from sweet potato roots and potato tubers), but is somewhat inhibitory with barley leaf enzyme. With the wheat germ enzyme, Mg^{2+} has a more pronounced effect. At 20 mM, Mg^{2+} produced a 60% activation. Preiss and Greenberg (101) obtained similar values depending on the concentration of fructose 6-phosphate used. Mg^{2+} also has a stimulating effect for sweet potato, potato, and wheat enzymes. Murata has pointed out that the effect seems to depend on the pH. Salerno and Pontis (105) have shown that the pH has no special significance for the activation produced by Mn^{2+} on the wheat germ enzyme. The latter authors have also demonstrated that product inhibition produced by UDP is reverted by Mg^{2+} or Mn^{2+}. Monovalent ions such as Na^+ and K^+ have been tested, but no effect on the enzyme activity was found (104).

In activator and inhibitor studies, sucrose phosphate synthetase is inhibited by UDP, one of the reaction products, but not by sucrose 6-phosphate, the other product. The inhibition for the wheat germ enzyme has been shown by Mendicino (92) to be competitive with an apparent K_i of 3.6 mM. Similar values have been reported for the enzymes from rice (102), potato (65), and sweet potato (97).

Several metabolites have been tested for possible activities or inhibition effects. Nomura and Akazawa (102) found no significant effect on the enzyme reaction by sucrose, fructose, glucose, glucose-1-P, fructose-1-P, fructose-1,6-diP, 3-phosphoglycerate, glyceraldehyde-3-P, phosphoenolpyruvate, citrate, succinate, malate, oxaloacetate, and glycolate (5 and 10 mM), UMP (4.4 mM), UTP (2.7 mM), ADP (6.3 mM), ATP (5.1 mM), and ADP-glucose (5 mM), confirming Mendicino's data regarding UMP and UTP.

Fekete (93), as a result of her studies with the enzyme from *V. faba,* claims that sucrose phosphate synthetase has a bound activator. The presence or absence of this activator in the enzyme preparation would explain the contradictory information concerning the inhibition of sucrose phosphate synthetase by nucleosides mono- and triphosphates. Thus, although she found inhibition of the enzyme with UTP, AMP, ATP, UMP, 3PG (3-phosphoglyceric acid), and PEP (phosphoenolpyruvate) in the presence of activator, in its absence she could not detect any inhibition, but in contrast found a stimulatory effect by the same compounds.

There has not been any confirmatory or contradictory evidence regarding the presence of an activator bound to sucrose phosphate synthetase from any other source. However, Salerno (104), working with a very purified preparation of the wheat germ enzyme, found an inhibitory action by sucrose, which neither Nomura and Akazawa (102) nor Hawker (94) could detect in their preparations. The inhibitory action of sucrose is quite interesting, as sucrose could be considered the end product of the action of sucrose phosphate synthetase. The fact that sucrose also inhibits sucrose phosphate phosphatase points to a possible feedback regulatory mechanism. In this respect, the sucrose inhibition found by Salerno is of the noncompetitive type, suggesting the presence of a second site in the enzyme. The stimulation of the enzyme by analogues of fructose 6-phosphate found by Preiss and Greenberg (101) points in that direction. They showed that whereas both 2-deoxyglucose-6-P and 1,5-anhydroglucitol-6-P could not bind to the fructose-6-P substrate site, they appear to replace fructose-6-P at an activator site or second site.

In regard to reaction mechanisms with the use of experiments performed with the wheat germ enzyme highly labeled [^{14}C] UDP-glucose and [^{3}H] UDP failed to detect a stable glucosyl intermediate (104). On the other hand, δ-gluconolactone strongly inhibits (104) sucrose phosphate synthetase. Thus, it seems that the reaction of transglucosylation between UDP-glucose and fructose

6-phosphate takes place through a mechanism similar to that of sucrose synthetase. See under "Sucrose Synthetase," section "Kinetics," for a complete discussion.

Sucrose Phosphate Phosphatase

Historical Background The pioneering work of Leloir and Cardini (26, 91), which led to the discovery of sucrose synthetase and sucrose phosphate synthetase, showed that sucrose 6-phosphate was hydrolized in crude extracts to sucrose and phosphate. This effect was attributed to the current phosphatases existing in plant tissues. Mendicino (92) was the first to indicate that the hydrolysis of sucrose 6-phosphate was apparently due to a specific phosphatase, but it was left to Hawker and Hatch (106) to prove conclusively the existence of sucrose phosphate phosphatase. The enzyme should be referred to more correctly as sucrose-6-phosphate phosphohydrolase (EC 3.1.3.00), but the first name is more currently used.

Occurrence The enzyme has been found and measured in so many plants that it seems certain that it is as widespread as sucrose synthetase and sucrose phosphate synthetase. Evidence for its presence has been presented for the following plants: wheat (91), sugar cane (106), carrot (106, 107), tobacco (108), spinach (107), barley (106), oat (106), potato (106), parsnip (106), pea (106), castor bean (37), rice (37), maize (37), grape (39), and broad bean (37). The enzyme is not confined to a special organ, having been found in all tissues of the plants examined. On the other hand, there is no conclusive evidence regarding the location of sucrose phosphate phosphatase in the cell. Initial studies by Hawker and Hatch (106) indicated that the enzyme was associated with mitochondria. However, later work by Hawker (107) seems to indicate that the enzyme is a soluble one, being present in the supernatant fraction after centrifugation at 100,000 *g*.

Methods of Assay The assay used for sucrose phosphate phosphatase is based on the measurement of the [^{14}C]sucrose liberated from labeled sucrose phosphate. The radioactive sugar is separated by chromatography (109). Nevertheless, it should be possible to follow the reaction by estimating the appearance of sucrose or inorganic phosphate, especially with purified preparations. Sucrose can be estimated specifically by invertase, as this enzyme does not act on sucrose phosphate (91). The limited amount of sucrose phosphate available has probably been the reason for the selection of the radioactive assay.

The assay of sucrose phosphate phosphatase in homogenates is helped by some of its properties. The enzyme is completely inactive in systems with no added magnesium and in the presence of EDTA (107), whereas nonspecific phosphatases like those acting on fructose 6-phosphate are not affected by the addition of EDTA. Thus, any phosphatase activity on sucrose phosphate, in the presence of EDTA, corresponds to nonspecific phosphatases.

Isolation The enzyme has been partially purified from carrot roots and immature tissue of sugar cane. The preparation procedure consisted of an ammonium sulfate fractionation after homogenization, followed by chromatography on Sephadex G-100 and calcium phosphate gel (106).

Molecular Properties All existing information on the properties of sucrose phosphate phosphatase corresponds to enzymes partially purified; no homogeneous preparations have been reported.

There are no data on its optical properties, mobility, or molecular weight. The enzyme elution pattern from columns of Sephadex G-100 seems to indicate that its molecular weight is below 100,000 (106).

The preparations of sucrose phosphate phosphatase isolated by Hawker and Hatch (106) from sugar cane mitochondria remained stable when stored for 10 days at −15°C. Upon further purification, losses of up to 50% were observed during storage for 7 days at −15°C. Activity was completely destroyed by heating the partially purified enzyme at 40°C for 10 min. Exposure to pH 5.2 for 15 min at 20°C also destroyed the activity.

General Characteristics of Catalytic Reaction Although the partially purified preparations of sucrose phosphate phosphatase were not totally free of other nonspecific phosphatases, it has been possible to determine the substrate specificity for the carrot and sugar cane enzymes. The rates of hydrolysis of several other phosphorylated compounds relative to the rate of sucrose phosphate hydrolysis are shown in Table 7. It is apparent that the enzyme is very specific for sucrose phosphate.

Several determinations of the Michaelis constant for the sugar cane (106) and carrot (107) enzyme gave values between 0.13 and 0.17 mM. In contrast, enzymes extracted from broad bean, maize, and castor bean gave values between 0.045 and 0.065 mM (37). Both enzymes exhibit a very similar pH curve with a pH optimum between 6.4 and 6.7. The activity falls rather sharply on the acid

Table 7. Substrate specificity of sucrose phosphate phosphatase

Substrate	Relative activity	
	Carrot enzyme (106)	Sugar cane enzyme (107)
Sucrose phosphate	100[a]	100[a]
Fructose-6-phosphate	1.9	2.1
Fructose-1-phosphate	0	1.4
Fructose-1,6-diphosphate	1.5	0
Glucose-6-phosphate	1.0	1.1
Glucose-1-phosphate	0.4	0.5
UDP	1.7	3.8
AMP	0.9	
IMP	0	
Phosphoenol pyruvate	5.0	
α-Glycerophosphate	2.5	2.8
β-Nitrophenylphosphate	3.9	10.2

[a]The activity toward sucrose-6-phosphate taken as 100%.

side of the pH optimum. This effect seems to be due, according to Hawker and Hatch (106), to enzyme inactivation rather than to low initial rates.

The activity of the partially purified enzyme for sugar cane was increased about 4-fold by addition of 10 mM $MgCl_2$. EDTA (10 mM) in the presence of 5 mM $MgCl_2$ inhibited activity by 81% (106) and in its absence by 100%. It seems, then, that the enzyme shows an absolute requirement for Mg^{2+}. In contrast, $CaCl_2$ (10 mM), $MnCl_2$ (10 mM), and KF (50 mM) inhibited activity by 89, 37, and 66%, respectively. Inorganic phosphate (50 mM) and pyrophosphate (30 mM) inhibited sucrose phosphate phosphatase activity by 75 and 87%, respectively (107).

The enzyme is inhibited by sucrose with K_i about 10 mM. The inhibition was shown by Hawker (108) to be partially competitive. Maltose, melizitose, and 6-ketose also inhibited enzyme activity by 45, 86, and 34%, respectively. It should be noticed that these sugars were assayed at 100 mM concentration, whereas the reaction mixtures contained 71 mM sucrose phosphate. No inhibitory effect was found for 1-ketose, furanose, glucose, fructose, lactose, cellooliose, trehalose, gentiobiose, or raffinose, also tested at 100 mM concentration.

The inhibitory effect of maltose and melizitose was studied further with the use of the sugar cane and tobacco leaf enzymes. The inhibition constant K_i for melizitose was about 7 mM and for maltose 23 mM. In the latter case, the inhibition is of the mixed type.

The inhibition of sucrose phosphate phosphatase by sucrose could act as a control mechanism, limiting the concentration of sucrose. Because sucrose exhibits a partially competitive type of inhibition, the enzyme is never completely inhibited. Thus, sucrose phosphate phosphatase could still be partially active even in the presence of a very high concentration of sucrose.

PHYSIOLOGICAL ROLES OF SUCROSE SYNTHETASE AND SUCROSE PHOSPHATE SYNTHETASE

The discovery of sucrose synthetase and sucrose phosphate synthetase presented plant biochemists with two pathways leading to sucrose. The existence of two separate mechanisms for the synthesis of sucrose raises the question of their respective roles in vivo. Experiments of fixation of $^{14}CO_2$ (25, 111–113) and [^{14}C]glucose infusion studies (114, 115) have shown that the fructose moiety of sucrose becomes highly labeled before any label appears in the free fructose pool, suggesting that this monosaccharide is not an intermediate in sucrose synthesis, or, at least, that a direct synthesis of sucrose from UDP-glucose and free fructose does not represent a significant portion of sucrose synthesis. It was also found that fructose 6-phosphate becomes labeled before sucrose, and small amounts of sucrose 6-phosphate have been detected among the labeled products. These results strongly indicate that sucrose 6-phosphate is synthesized first and then hydrolyzed to free sucrose.

Furthermore, the reaction catalyzed by sucrose phosphate synthetase has an equilibrium constant which has been measured as 3,250, with a corresponding

ΔF° of −2,700 cal at pH 7.5, clearly showing that the reaction is fully displaced toward the synthesis of sucrose 6-phosphate (see under "Sucrose Phosphate Synthetase," section "Equilibrium").

The equilibrium constant of the reaction catalyzed by sucrose synthetase was found to be approximately 1.6, a value corresponding to a ΔF° of −280 cal at pH 7.5 (see under "Sucrose Synthetase," section "Equilibrium"), which indicates that although the equilibrium of the reaction is somewhat in favor of synthesis, complete cleavage of sucrose could occur if the resulting UDP-glucose or fructose were removed by being used up in various other metabolic reactions.

Moreover, supporting this conclusion, Pressey (42) found that sucrose synthetase from potato tuber catalyzes sucrose cleavage more rapidly than sucrose synthesis at physiological pH. These results have been confirmed by several investigators working with enzymes from other sources (28, 69, 116). At the same time, the ability of the enzyme to use several nucleoside diphosphates as glucosyl acceptors provides an alternative way to the pyrophosphorylase reaction for the formation of nucleoside diphosphate sugars. Such a "nucleotide pyrophosphorolysis" of sucrose would allow the cell to use the energy of the glycosidic linkage of sucrose for the synthesis of other saccharides, such as aryl-D-glucosides, starch, callose, or cell wall polysaccharides (13).

Nevertheless, this interpretation, attractive as it might be, does not quite fit with the kinetic data regarding the affinities of sucrose synthetase for UDP and the nucleoside diphosphates of adenine, guanine, thymine, and cytosine. Grimes et al. (52) have shown that when UDP is available, sucrose synthetase preferentially synthesizes UDP-glucose rather than any of the other nucleoside diphosphate glucoses, even if ADP, TDP, CDP, or GDP is present in amounts equivalent to or greater than the concentration of UDP.

Similar results have been presented by Murata et al. (117) in their study of the enzymic mechanism of starch synthesis in ripening rice grains. They found that ADP-glucose was scarcely detectable in a system containing sucrose synthetase and both ADP and UDP (in the ratios 1:1/5 and 1:1), whereas a small amount of the sugar nucleotide was formed in a system containing the above two nucleotides at the ratio 1:1/25. In the former two cases, a marked formation of UDP-glucose was demonstrated, whereas the magnitude of its synthesis was much less in the latter case. The inhibitory effect of UDP in the ADP-glucose sucrose transglucosylase reaction was originally reported by Cardini and Recondo (74), who used the wheat germ enzyme, and later in more detail by Fekete and Cardini (81), using an enzyme from sweet corn.

The energy of the glycosidic linkage of sucrose could still be used for the synthesis of the needed nucleoside diphosphate sugars if UDP-glucose were used for the formation of glucose-1-phosphate by the following steps:

$$\text{Sucrose} + \text{UDP} \rightarrow \text{UDP-glucose} + \text{fructose} \qquad (7)$$

$$\text{UDP-glucose} + \text{PP} \rightarrow \text{UTP} + \text{glucose-1-P} \qquad (8)$$

The glucose 1-phosphate produced in the second reaction could be converted to the different nucleoside diphosphates glucose by the respective pyrophosphorylases. This is actually the scheme previously suggested by Turner and Turner (118), Fekete and Cardini (81), Murata et al. (117), Pressey (42), and Fekete (83) to explain how the sucrose translocated into the endosperm would be made available for starch synthesis.

A difficulty of this scheme is the presence of inorganic pyrophosphatase in plant cells, which has been shown to appear in endosperm after anthesis during seed development (119).

It would be easier to imagine the above pathway if the whole process occurs within a compartment that would separate pyrophosphate from pyrophosphatases. Delmer and Albersheim (98), as a way to circumvent the difficulty, have suggested that the entire reaction sequence could be considered cyclic by using the UTP generated in the reaction as a source of pyrophosphate and by using the UMP thus generated in conjunction with another UTP to produce two UDPs as substrate for sucrose synthetase. However, an enzyme system which catalyzes this reaction sequence has not been described.

Nevertheless, considerable evidence linking the level of activity of sucrose synthetase with the rate of starch synthesis cannot be ignored. Thus, Pressey (42) and Sowokinos (43) have shown that the activity of sucrose synthetase in potato tubers declines as the sucrose content falls and starch accumulates. Tsai et al. (120) have also demonstrated that in the developing endosperm of maize the onset of starch accumulation coincides with an appreciable rise in sucrose synthetase activity. These two examples are only an illustration of many results pointing in the same direction, reported by several investigators (81, 83, 117, 119, 121–124).

All these studies strongly suggest a relationship between sucrose level, sucrose synthetase, and starch synthesis. However, a scheme that explains all data available without postulating compartmentation remains to be proposed.

This discussion so far stresses the view that the physiological role of sucrose synthetase would be sucrose cleavage, whereas the role of sucrose phosphate synthetase coupled to sucrose phosphate phosphatase would correspondingly be that of sucrose synthesis. If this were the case, it would be expected that according to the physiological situation one enzyme or the other will be acting or even present. This is actually what seems to be happening during the development of seeds. Here, sucrose translocated to the endosperm or storage cotyledon is converted to glucose and fructose, or possibly to nucleotide sugars, for use in respiration or synthetic processes. According to Fekete (83), at this state the synthesis of sucrose is quantitatively unimportant. Sucrose synthetase activity is relatively high in tissues during development such as maize endosperm, young developing broad bean cotyledons, barley endosperm, rice endosperm, and young castor bean seeds (37). In contrast, the activity of sucrose phosphate synthetase was very low (see Table 8).

The opposite situation is encountered during germination, when sucrose synthesis becomes quantitatively important (125). Hawker (37) has found that in broad bean cotyledons sucrose phosphate synthetase reached a maximum during germination, whereas sucrose synthetase activity decreased. Similarly,

Table 8. Activities of sucrose enzymes in different tissues[a]

Tissue	Enzyme activity (μmol product formed h^{-1} g^{-1} fresh wt)					Reference
	Sucrose synthetase	Sucrose phosphate synthetase	Sucrose phosphate phosphatase	Acid invertase	Alkaline invertase	
Vicia faba developing cotyledons	9.0	15.6	132	0.18	4.2	37
Pennisetum purpureum leaves	0.7–13.8	77–161				129
Muhlenbergia montana leaves grown 30–35°C high light	2.6	150				129
Maize leaves, mesophyll cells	0.03[b]	0.2[b]	6.9[b]	0.9[b]		123
Maize leaves, bundle sheath cells	0.14[b]	0.19[b]	4.2[b]	0.5[b]		123
Sugar cane, mesophyll chloroplasts	5930[b]	1350[b]		2.9[b]	1.4[b]	7
Sugar cane, bundle sheath chloroplasts	4810[b]	1660[b]		2.4[b]	0.9[b]	7
Rice, mature seeds; endosperm	10.7	0.5	4.2	1.0		37
Castor bean, germinating endosperm	18.6	16.9	165	8.4		37
Castor bean, germinating cotyledons	57.6	43.2	389	201		37
Pea root, cortex from tip	0.086			0.46	0.1	2
Pea root, stele from tip	0.44			0.14	0.3	2
Maize, developing endosperm, midmilky stage	41	12	41	2.6		37
Maize, developing scutella, midmilky stage	7.1	6.8	104	0.8		37
Maize, 3 days germinating endosperm	4.2	6.8	15.9	2.6		37
Maize, 3 days germinating scutella	6.3	30.0	165.5	4.1		37

[a]The values were calculated from data in the papers quoted.
[b]Activities expressed as μmol of product formed h^{-1} mg^{-1} of protein.

Nomura and Akazawa (126) have demonstrated the de novo synthesis of sucrose phosphate synthetase in rice scutellum, accompanied by very low activities of sucrose synthetase. Nevertheless, the situation is not too clear, because Hawker (37) also found in maize scutella and castor bean endosperm that the activities of both enzymes and sucrose phosphate phosphatase increased during germination.

The presence of the three enzymes in the same tissue raises the question whether, as in some animal cells, a futile cycle (127, 128) is in existence. This will arise if in the same cells the following reactions are taking place:

$$
\begin{aligned}
\text{UDP-glucose} + \text{fructose-6-P} &\rightarrow \text{UDP} + \text{sucrose-6-P} \qquad (9)\\
\text{Sucrose-6-P} &\rightarrow \text{sucrose} + \text{Pi}\\
\text{Sucrose} + \text{UDP} &\rightarrow \text{UDP-glucose} + \text{fructose}\\
\text{Fructose} + \text{ATP} &\rightarrow \text{fructose-6-P} + \text{ADP}\\
\hline
\text{Sum} \quad \text{ATP} &\rightarrow \text{ADP} + \text{Pi}
\end{aligned}
$$

Thus, a recycling of sucrose will have as its only consequence a dissipation of energy through hydrolysis of ATP. On the other hand, if the enzymes are present in separate cells, the problem will be one of opposing fluxes, in which case a futile cycle might not exist.

Invertases are not considered to participate, as their levels have been shown by Hawker (37) to be low in general, in many of the tissues he studied. Nevertheless, their presence will tend to bleed some sucrose from the cycle, increasing the demand for ATP.

A cycle would be avoided if sucrose synthetase fulfills a role toward sucrose synthesis in those cells, like castor bean endosperm, in which the above situation is encountered. Thus, the cell may use the easy reversibility of the sucrose synthetase reaction, directing the enzyme to catalyze sucrose synthesis, thus reinforcing the action of sucrose phosphate synthetase. Alternatively, sucrose cleavage would take place when sucrose concentration is too high. In this respect, it should be remembered that sucrose has an inhibitory effect on the activities of both sucrose phosphate synthetase and sucrose phosphate phosphatase.

SUCROSE AND PLANT HORMONES

The interaction of sucrose and plant hormones can be considered from two points of view–hormonal effects that are seen as a modification of the enzyme levels and reinforcement or synergic effects of hormones and sucrose.

Hormonal Effects at Enzyme Level

There are few reports on the action of plant hormones on sucrose synthetase and sucrose phosphate synthetase. Lavintman and Cardini (130), studying the activities of both enzymes in aging potato tuber slices in water, found that the presence of 2,4-dichlorophenoxyacetic acid (2,4-D) increased sucrose synthetase activity, whereas sucrose phosphate synthetase was not significantly modified. Plant growth substances such as gibberellic acid, indoleacetic acid (IAA), and kinetin have no effect. A similar action of 2,4-D on the levels of sucrose synthetase was reported by Pontis (53) on Jerusalem artichoke explants. Addition of the hormone to the growing media of the explants resulted in a 10-fold increase in sucrose synthetase.

Huber and Sankhla (131) have studied the influence of abscisic and gibberellic acids on the activities of various enzymes related to carbohydrate metabolism in leaves of *Pennisetum typhoide* seedlings. They found that abscisic acid promoted the activity of sucrose synthetase and sucrose phosphate synthetase. Gibberellic acid alone did not produce any enhancement of activity, but its addition, together with abscisic acid, canceled the stimulation effect of the latter.

In contrast to these in vivo studies, Delmer (57) has investigated the regulation of sucrose synthetase in vitro and found that IAA and gibberellic acid activate sucrose cleavage, whereas kinetin has no effect. However, hormones were tested at a rather higher concentration than the enzyme encounters in vivo.

It should be noticed that in all the in vivo studies, no conclusive proof that hormones were directly responsible for the increased enzyme activity was presented, as usually a rather long time had elapsed between addition of the hormone and the measurement of the enzyme activity.

Sucrose as Regulator of Plant Metabolism

It is difficult to visualize a metabolite which is usually in a rather high concentration (132–135) fulfilling a role as a controlling factor. Evidence of this role comes from in vitro experiments in which sucrose concentration in culture media affects the degree of tissue differentiation (136–143). Wetmore and Rier (136), as well as Jeffs and Northcote (137) have induced differentiation in callus tissue by gradient techniques and have shown that the xylem differentiation is favored by IAA together with a relatively low concentration of sucrose (1–2%), whereas IAA and a higher sucrose concentration (3–4%) favored phloem induction. Similarly, Wright and Northcote (138) have reported that *Acer pseudoplatanos* was induced to differentiate xylem, phloem, and roots when the culture medium contained naphthalene acetic acid (1 mg/liter), kinetin (0.05–0.5 mg/liter), and sucrose (3 or 5%). If the concentration of sucrose were allowed to fall below 2%, no roots were differentiated even though xylem and phloem elements were formed. The amount of sucrose present in a tissue has been shown to be correlated to the levels of sucrose, sucrose phosphate synthetase, sucrose phosphate phosphatase, and acid and neutral invertases.

Thus, as Wright and Northcote have pointed out, there is a fine control over the concentration of sucrose, both intracellularly and extracellularly. They have also shown that in callus tissue sucrose is degraded to glucose and fructose but that sucrose is also synthesized. Therefore, local conditions in the callus may control variation in the sucrose level, with the possibility that this might exert some physiological effect. It is very difficult, as Wright and Northcote emphasized, to distinguish an essential nutritional requirement from a hormone.

The effect of sucrose has not only been observed in connection with induced differentiations. Pontis et al. (81) have shown that growing explants of Jerusalem artichoke in the dark in a medium supplemented with sucrose promoted not only a rise in the level of starch synthetase, but also the appearance of amylopectin. The level of enzyme attained depended on the sucrose concentration in the medium. The effect of sucrose could not be replaced by adding glucose, fructose, maltose, or raffinose to the medium.

The addition to the medium of glucose (250 mM) and 2,4-D (10 mg/liter) also resulted in an increase of starch synthetase. This result is of particular interest as Pontis (53) has shown earlier that the addition of auxin and glucose raised the level of sucrose synthetase in growing tuber tissue. Thus, here again the control of sucrose level through the action of enzymes seems to be important in triggering some phenomena.

The experiments of Edelman and Hanson (144, 145) also point in this direction. They found that sucrose can suppress chlorophyll synthesis in carrot callus tissue containing no invertase. No such effect was seen if sucrose was replaced by glucose, fructose, or a mixture of both. This sucrose effect was shown by the substrate rather than the hormonal level of the sugar in the medium, the effect being marked at 100 mM and measurable at 30 mM. The effect could not be observed in a different callus strain that had a high invertase activity. Nevertheless, Wright and Northcote (138) pointed out that since sucrose might be regarded as the end product of chloroplast metabolism, the apparent control of chlorophyll synthesis may be a feedback mechanism. That this may be the case has been shown by Pamplin and Chapman (146), who, working with the same carrot tissue as Edelman and Hanson, found that the major effect of sucrose upon chlorophyll synthesis occurs at the stage controlling 5-amino-levulinic acid synthesis.

All these cases indicate once more that sucrose exerts an influence on cell metabolism that surpasses its value as a carbon source.

CONCLUSION

No concise concluding remarks can usefully or adequately summarize this review, because it will be seen from the above discussions that understanding of the sucrose riddle is far from complete. Not only is the physiological role of sucrose in the differentiation process obscure and our knowledge of the properties of the enzymes related to sucrose incomplete, but also the transformation of sucrose into starch is not yet fully explained. This review will, however, have fulfilled its purpose if it has presented the many facets with which the ubiquitousness of sucrose in the plant kingdom confronts us.

REFERENCES

1. Rosen, R. (1967). Optimality Principles in Biology, p. 7. Plenum Press, New York.
2. apRees, T. (1974). *In* D. H. Northcote (ed.), Plant Biochemistry, 11:89. Butterworths, London.
3. Hassid, W. Z. (1967). Annu. Rev. Plant Physiol. 18:253.
4. Glasziou, K. T. (1969). Annu. Rev. Plant Physiol. 20:63.
5. Preiss, J., and Kosuge, T. (1970). Annu. Rev. Plant Physiol. 21:433.
6. Yudkin, J., Edelman, J., and Hough, L. (1971). Sugar. Butterworths, London.
7. Davies, D. R. (1974). *In* J. B. Pridham (ed.), Plant Carbohydrate Chemistry, pp. 61–81. Academic Press, New York.
8. Turner, J. F., and Turner, D. H. (1975). Annu. Rev. Plant Physiol. 26:159.
9. Rodd, E. H. (1968). Chemistry of the Carbon Compounds, pp. 114–125. Elsevier, Amsterdam.
10. Hough, L. (1971). *In* J. Yudkin, J. Edelman, and L. Hough (eds), Sugar, pp. 49–59. Butterworths, London.
11. Edelman, J. (1971). *In* J. Yudkin, J. Edelman, and L. Hough (eds.), Sugar, pp. 95–102. Butterworths, London.
12. Hassid, W. Z. (1951). *In* W. D. McElroy and B. Glass (eds.), Phosphorus Metabolism, Vol. 1, pp. 11–42. Johns Hopkins Press, Baltimore.
13. Neufeld, E. F., and Hassid, W. Z. (1963). Adv. Carbohyd. Chem. 18:309.
14. Hassid, W. Z., and Doudoroff, M. (1950). Adv. Enzymol. 10:123.
15. Cabib, E., and Leloir, L. F. (1958). J. Biol. Chem. 231:259.
16. Leloir, L. F., Cardini, C. E., and Cabib, E. (1960). *In* M. Florkin and H. S. Mason (eds.), Comparative Biochemistry, Vol. 2, pp. 97–138. Academic Press, New York.
17. Arnold, W. N. (1968). J. Theor. Biol. 21:13.
18. Levitt, J. (1972). Responses of Plants to Environmental Stresses, p. 197. Academic Press, New York.
19. Warner, D. T. (1962). Nature 196:1055.
20. Panya, K. P., and Ramakrishnan, C. V. (1956). Naturwissenschaften 43:85.
21. Shukla, J. P., and Prabhu, K. A. (1959). Experientia 16:202.
22. Hatch, M. D., Sacher, J. A., and Glasziou, K. T. (1963). Plant Physiol. 38:338.
23. Hassid, W. Z., and Doudoroff, M. (1950). Adv. Carbohyd. Chem. 5:29.
24. Buchanan, J. G., Barsham, J. A., Benson, A. A., Bradley, D. F., Calvin, M., Daus, L. L., Goodman, M., Hayes, P. M., Lynch, V. H., Norris, L. T., and Wilson, A. T. (1952). *In* W. D. McElroy and B. Glass (eds.), Phosphorus Metabolism, Vol. 2, p. 440. Johns Hopkins Press, Baltimore.
25. Buchanan, J. G., Lynch, V. H., Benson, A. A., Bradley, D. F., and Calvin, M. (1953). J. Biol. Chem. 203:935.
26. Cardini, C. E., Leloir, L. F., and Chiriboga, J. (1955). J. Biol. Chem. 214:149.
27. Delmer, D. P. (1972). J. Biol. Chem. 247:3822.
28. Murata, T. (1971). Nippon Nogei Kagaku Kaishi 45:441.
29. Nomura, T., and Akazawa, T. (1973). Arch. Biochem. Biophys. 156:644.
30. Shukla, R. N., and Sanwal, G. G. (1971). Arch. Biochem. Biophys. 142:303.
31. Chin, C. K., and Weston, G. D. (1975). Phytochemistry 14:69.
32. Thorpe, T. A., and Meier, D. D. (1973). Phytochemistry 12:493.
33. Bucke, C., and Oliver, I. R. (1975). Planta 122:45.
34. Avigad, G. (1964). J. Biol. Chem. 239:3613.
35. Bird, I. F., Cornelius, M. J., Keys, A. L., and Whittingham, C. P. (1974). Phytochemistry 13:59.
36. Pridham, J. B., Walter, M. W., and Worth, H. G. (1969). J. Exp. Botany 20:317.
37. Hawker, J. S. (1971). Phytochemistry 10:2313.
38. Nakamura, M. (1959). Bull. Agr. Chem. Soc. Jap. 23:398.
39. Hawker, J. S. (1969). Phytochemistry 8:9.
40. Bean, R. C., and Hassid, W. Z. (1955). J. Am. Chem. Soc. 77:5737.
41. Burma, D. P., and Mortimer, D. C. (1956). Arch. Biochem. Biophys. 62:16.
42. Pressey, R. (1969). Plant Physiol. 44:759.
43. Sowokinos, J. R. (1971). Am. Potato J. 48:37.

44. Walker, D. A. (1974). *In* D. H. Northcote (ed.), Plant Biochem. 11:1–49. Butterworths, London.
45. Bird, I. F., Porter, H. K., and Stocking, C. R. (1965). Biochim. Biophys. Acta 100:366.
46. Heldt, H. W., and Sauer, F. (1971). Biochim. Biophys. Acta 234:83.
47. Haq, S., and Hassid, W. Z. (1965). Plant Physiol. 40:591.
48. Avigad, G., and Milner, Y. (1966). *In* E. F. Neufeld and V. Ginsburg (eds.), Methods in Enzymology, Vol. 8, pp. 341–345. Academic Press, New York.
49. Rorem, E. S., Walker, H. G., and McCready, R. M. (1960). Plant Physiol. 35:269.
50. Percheron, F. (1962). C.R. Acad. Sci. 255:2521.
51. Pontis, H. G. (1976). Manuscript in preparation.
52. Grimes, W. J., Jones, B. L., and Albersheim, P. (1970). J. Biol. Chem. 245:188.
53. Pontis, H. G. (1970). Plant Physiol. 23:1089.
54. Salerno, G. L., and Pontis, H. G. Manuscript in preparation.
55. Ashwell, G. (1966). *In* S. P. Colowick and N. O. Kaplan (eds.), Methods in Enzymology, Vol. 3, pp. 73–105. Academic Press, New York.
56. Wolosiuk, R. A., and Pontis, H. G. (1974). Arch. Biochem. Biophys. 165:140.
57. Delmer, D. P. (1972). Plant Physiol. 50:469.
58. Slack, C. R. (1966). Phytochemistry 5:397.
59. Murata, T. (1971). Agr. Biol. Chem. 35:1441.
60. Gabrielian, N. D., Komaleva, R. L., and Shibaev, V. N. (1973). Moldavskaya Biol. 7:337.
61. Tsai, C. Y. (1974). Phytochemistry 13:885.
62. Wolosiuk, R. A., and Pontis, H. G. (1971). FEBS Lett. 16:237.
63. Wolosiuk, R. A. (1974). Ph.D. thesis, Buenos Aires University.
64. Martin, R. G., and Ames, B. N. (1961). J. Biol. Chem. 236:1372.
65. Slabnik, E., Frydman, R. B., and Cardini, C. E. (1968). Plant Physiol. 43:1063.
66. Drechsler, E. R., Boyer, P. D., and Kowalsky, A. G. (1959). J. Biol. Chem. 234:2627.
67. Belocopitow, E., García Fernandez, M. C., Birnbaumer, L., and Torres, N. H. (1967). J. Biol. Chem. 242:1227.
68. Meyer, W. L., Fischer, E. H., and Krebs, E. G. (1964). Biochemistry 3:1033.
69. Pontis, H. G., and Wolosiuk, R. A. (1972). FEBS Lett. 28:86.
70. Wolosiuk, R. A., and Pontis, H. G. (1974). Mol. Cell Biochem. 4:115.
71. Salo, W. L., Nordin, J. H., Peterson, D. R., Bevill, R. D., and Kirkwood, S. (1968). Biochim. Biophys. Acta 151:484.
72. Furlong, C. E., and Preiss, J. (1969). J. Biol. Chem. 244:2539.
73. Hassid, W. Z. (1972). *In* R. Piras and H. G. Pontis (eds.), Biochemistry of the Glycosidic Linkage. An Integrated View, pp. 315–335. Academic Press, New York.
74. Cardini, C. E., and Recondo, E. (1962). Plant Cell Physiol. 3:313.
75. Milner, Y., and Avigad, G. (1965). Nature 206:825.
76. Budowsky, E. I., Drushinina, T. N., Eliseeva, G. I., Gabrielyan, N. D., Kochetkov, N. K., Novikova, M. A., Shibaev, V. N., and Zhdanov, G. L. (1966). Biochim. Biophys. Acta 122:213.
77. Milner, Y., and Avigad, G. (1964). Israel J. Chem. 2:316.
78. Akazawa, T., Minamikawa, T., and Murata, T. (1964). Plant Physiol. 39:371.
79. Gabrieljan, N. D., Lapina, E. B., Spiridonova, S. M., Shibaev, V. N., Budowsky, E. I., and Kochetkov, N. K. (1969). Biochim. Biophys. Acta 185:478.
80. Jaarma, M., and Rydström, J. (1969). Acta Chem. Scand. 23:3443.
81. Fekete, M. A. R., and Cardini, C. E. (1964). Arch. Biochem. Biophys. 104:173.
82. Pontis, H. G., Wolosiuk, R. A., Fernandez, L. M., and Bettinelli, B. (1972). *In* R. Piras and H. G. Pontis (eds.), Biochemistry of the Glycosidic Linkage. An Integrated View, pp. 239–266. Academic Press, New York.
83. Fekete, M. A. R. (1969). Planta 87:311.
84. Frieden, C. (1964). J. Biol. Chem. 239:3522.
85. Sung, H. Y., and Su, J. C. (1973). Abstract of the Ninth International Congress of Biochemistry, July 1–7, Stockholm.
86. Cleland, W. W. (1963). Biochim. Biophys. Acta 67:188.
87. Voet, J. G., and Abeles, R. H. (1970). J. Biol. Chem. 245:1020.
88. Tu, J., Jacobson, G. R., and Graves, D. J. (1971). Biochemistry 10:1229.

89. Levvy, G. A., and Snaith, S. M. (1972). Adv. Enzymol. 36:151.
90. Bentley, R. (1972). Annu. Rev. Biochem. 41:953.
91. Leloir, L. F., and Cardini, C. E. (1955). J. Biol. Chem. 214:157.
92. Mendicino, J. (1960). J. Biol. Chem. 235:3347.
93. Fekete, M. A. R. (1971). Eur. J. Biochem. 19:73.
94. Hawker, J. S. (1967). Biochem. J. 105:943.
95. Lyne, R. L., and apRees, T. (1972). Phytochemistry 11:2171.
96. Huber, W., Fekete, M. A. R., and Ziegler, H. (1973). Planta 112:343.
97. Murata, T. (1972). Agr. Biol. Chem. 36:1877.
98. Delmer, D. P., and Albersheim, P. (1970). Plant Physiol. 45:782.
99. Vieweg, G. H. (1974). Planta 116:347.
100. Salerno, G. L., and Pontis, H. G. Manuscript in preparation.
101. Preiss, J., and Greenberg, E. (1969). Biochem. Biophys. Res. Commun. 36:289.
102. Nomura, T., and Akazawa, T. (1974). Plant Cell Physiol. 15:477.
103. Salerno, G. L., and Pontis, H. G. Manuscript in preparation.
104. Salerno, G. L. Ph.D. thesis, Buenos Aires University.
105. Salerno, G. L., and Pontis, H. G. FEBS Lett. 64:415.
106. Hawker, J. S., and Hatch, M. D. (1966). Biochem. J. 99:102.
107. Hawker, J. S. (1966). Phytochemistry 5:1191.
108. Hawker, J. S. (1967). Biochem. J. 102:401.
109. Hatch, M. D. (1964). Biochem. J. 93:521.
110. Calvin, M., and Benson, A. A. (1949). Science 109:140.
111. Edelman, J., Ginsburg, V., and Hassid, W. Z. (1955). J. Biol. Chem. 213:843.
112. Bean, R. C., Barr, B. K., Welch, H. V., and Porter, G. C. (1962). Arch. Biochem. Biophys. 96:524.
113. Bean, R. C., Porter, G. G., and Barr, B. K. (1963). Plant Physiol. 38:280.
114. Putnam, E. W., and Hassid, W. Z. (1954). J. Biol. Chem. 207:885.
115. Vittorio, P. V., Krotkov, G., and Reed, G. B. (1954). Can. J. Botany 32:369.
116. Hardy, P. J. (1968). Plant Physiol. 43:224.
117. Murata, T., Sugiyama, T., Minamikawa, T., and Akazawa, T. Arch. Biochem. Biophys. 113:34.
118. Turner, D. H., and Turner, J. F. (1957). Aust. J. Biol. Sci. 10:302.
119. Baxter, E. D., and Duffus, C. M. (1973). Phytochemistry 12:1923.
120. Tsai, C. Y., Salamini, F., and Nelson, O. E. (1970). Plant Physiol. 46:229.
121. Nomura, T., Nakayama, N., Murata, T., and Akazawa, T. (1967). Plant Physiol. 42:327.
122. Rollit, J., and Maclachlan, G. A. (1974). Phytochemistry 13:367.
123. Downton, W. J. S., and Hawker, J. S. (1973). Phytochemistry 12:1551.
124. Perez, C. M., Perdon, A. A., Resurrecciόn, A. P., Villareal, R. M., and Juliano, B. O. (1975). Plant Physiol. 56:579.
125. Edelman, J., Shibko, S. I., and Keys, A. (1959). J. Exp. Botany 10:178.
126. Nomura, T., and Akazawa, T. (1973). Plant Physiol. 51:979.
127. Clark, D. G., Rognstad, R., and Katz, J. (1973). Biochem. Biophys. Res. Commun. 54:1141.
128. Rognstad, R., and Katz, J. (1972). J. Biol. Chem. 247:6047.
129. Bucke, C., and Oliver, I. R. (1975). Planta 122:45.
130. Lavintman, N., and Cardini, C. E. (1968). Plant Physiol. 43:434.
131. Huber, W., and Sankhla, N. (1974). Planta 116:55.
132. Edelman, J., and Jefford, T. G. (1964). Biochem. J. 93:148.
133. Chrispeels, M. J., Tenner, A. J., and Johnson, K. D. (1973). Planta 113:35.
134. Humphreys, T. E. (1973). Phytochemistry 12:1211.
135. Humphreys, T. E. (1972). Phytochemistry 11:1311.
136. Wetmore, R. H., and Rier, J. P. (1963). Am. J. Botany 50:418.
137. Jeffs, R. A., and Northcote, D. H. (1967). J. Cell Sci. 2:77.
138. Wright, K., and Northcote, D. H. (1972). J. Cell Sci. 11:319.
139. Greenwood, M. S., and Berlyn, G. P. (1973). Am. J. Botany 60:42.
140. Butcher, D. N., and Street, H. E. (1960). J. Exp. Botany 11:206.
141. Michel, B. E. (1968). Am. J. Botany 55:1126.

142. Jacobs, W. P., and Morrow, I. B. (1958). Science 128:1084.
143. Torrey, J. G. (1963). Symp. Soc. Exp. Biol. 17:285.
144. Edelman, J., and Hanson, A. D. (1971). Planta 98:150.
145. Edelman, J., and Hanson, A. D. (1971). Planta 101:122.
146. Pamplin, E., and Chapman, J. (1975). J. Exp. Botany 26:212.

International Review of Biochemistry
Plant Biochemistry II, Volume 13
Edited by D. H. Northcote

4 Biochemistry of Osmotic Regulation

H. KAUSS
Department of Biology
University of Kaiserslautern, Kaiserslautern, Federal Republic of Germany

FEEDBACK LOOP FOR OSMOTIC REGULATION 121

NATURE OF OSMOTIC AGENTS 123

SYSTEMS EMPLOYING ION TRANSPORT IN ALGAE 123
Transport Mechanisms 123
Pressure-sensing Mechanism 125

OSMOTIC REGULATION WITH ORGANIC COMPOUNDS IN LOWER PLANTS 126
Ochromonas 127
Dunaliella 133
Other Organisms 134

OSMOTIC REGULATION IN HIGHER PLANTS 135
Expanding Cells 136
Stomata and Pulvini 137
Role of Proline 137

The work discussed under "Ochromonas" was supported by the Deutsche Forschungsgemeinschaft.

An excellent review by J. A. Hellebust on osmoregulation has recently appeared (Annu. Rev. Plant Physiol. 27:485–505, 1976). Readers should consult this for more references and physiological (e.g., *Nitella*) and ecological aspects not treated in this chapter due to their complexity.

Living cells represent an osmotic system in which the sum of all substances solubilized in the cellular water results in an osmotic value. The cellular membranes are permeable to most solutes only to a limited extent, but they are permeable to water. In accordance with the laws of entropy, water will enter so long as the internal osmotic value is higher than the external value (Figure 1), the cell volume tending to increase with the occurence of an inflow of water. In most plant cells, the rigidity of the walls allows for only relatively small volume changes; in such cases a relatively high turgor pressure (*P*) will be built up and will counteract the water inflow. In cells without walls (such as plant gametes, protists, or animal cells), water inflow results in dramatic volume changes although these are accompanied by only small resultant turgor values. In balance, water is held within the cells by the forces encountered. Osmotic relations provide an important method of regulating water content and water flow of the cell.

Physiological studies have provided many examples which clearly demonstrate that the osmotic balance of cells is closely regulated. Different investigators studying different tissues have variously used the terms osmoregulation, turgor regulation, and volume regulation. It is not yet known which of these parameters is the principal basis of water regulation so criticism of any of these terms (1) is not justified at the present time. The more general term "osmotic regulation" will be used in this chapter to mean regulation of an osmotic system regardless of whether it is held in a state of balance or is continuously adjusted

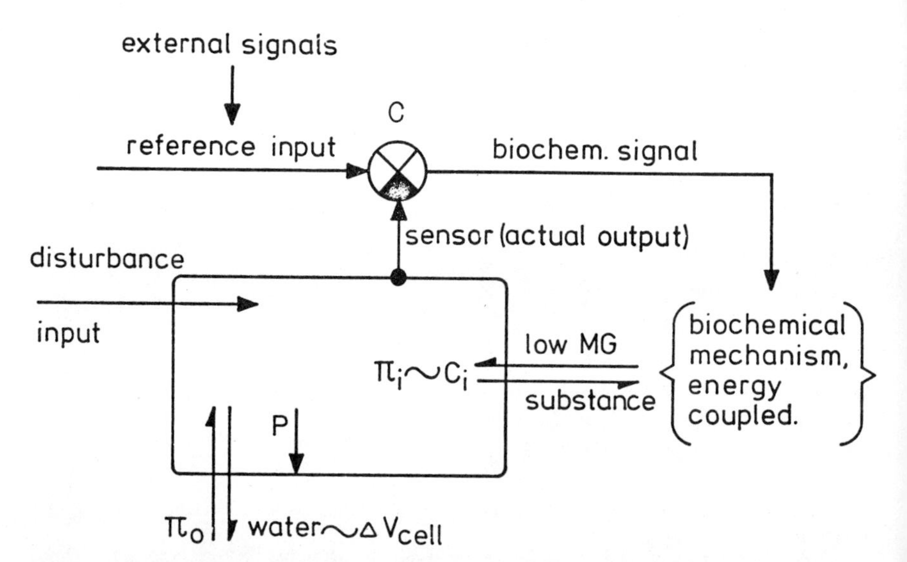

Figure 1. Block diagram of a system for osmotic regulation. π_i, cellular osmotic value; π_0, external osmotic value; c_i, solute concentration; *P*, turgor pressure; ΔV, volume change; *C*, controller.

to a new desired equilibrium. The discussion is restricted to osmotic regulation at the cellular level; subcellular organelles will not be considered.

Osmotic, as well as nonosmotic water relations in general, covering both osmotic and nonosmotic aspects in higher plants, have recently been expertly reviewed (2). In most cases, the processes involved are quite complex, and are difficult to correlate with biochemical events. Therefore, this chapter will deal only with systems showing at least some evidence that the concentration of cellular solutes is manipulated in order to adjust the osmotic balance or to perform work. It is quite difficult, however, to measure exactly and simultaneously the turgor, osmotic value, and proportion of the cell water content which is osmotically bound in intact tissues of higher plants. The biochemistry of osmotic regulation has, therefore, been studied to date with some success only in systems of single-celled organisms or in specialized cell lines at a certain stage of differentiation.

The aim of an investigation into the problems in question would be a better understanding of osmotic regulation in terms of distinct definable biochemical processes. One of the principal difficulties in studying the biochemistry of osmotic regulation is that the entire osmotic system operates only in living cells, requiring intact membranes or membrane-wall arrangements to maintain hydrostatic pressures while the biochemical steps are investigated. Such studies, therefore, must comprise many observations from such diverse fields as physical chemistry, membrane ultrastructure and function, enzymology, physiology, and ecology. Unfortunately, because our present knowledge comes from such diverse sources, generalization is valid only in a few cases. Nevertheless, an attempt is made in this chapter to show features common to the various systems. Comments on pertinent data from animal and bacterial cell systems will be included where relevant.

FEEDBACK LOOP FOR OSMOTIC REGULATION

The osmotic balance in a given cellular system is determined by the concentration of substances in a given volume, the turgor pressure, and the external osmotic value. Flow of water in or out of the cell then establishes an equilibrium. The events occurring in osmotic regulation can best be clarified by using the terms of system theory (3); experts in this field are asked to tolerate the pragmatic oversimplification of Figure 1 and the mixing of symbols used in cybernetics with those used in biochemistry, this being necessary to find an approach to the biochemical aspects of the problem. Furthermore, detailed treatments of the control systems and ion fluxes involved in osmotic regulation can be found in references 1 and 4.

The status of an osmotic system is measured continuously by a sensor, and the measured value must then be subsequently processed to provide the required information. The actual output is compared with a reference input in the controller (Figure 1). When alteration of any one of the cited parameters occurs (disturbance input), the actual output will deflect from the reference input. This deflection is converted into a signal, which represents information that can be related to the biochemical mechanism responsible for alteration in the concentration (c) of the internal solute. This biochemical mechanism, which uses the cellular energy supply, causes an increase or decrease in the amount of low molecular weight substances within the cell, until the optimum concentration for osmotic equilibrium is established. In the simple case of an integral controller, in which the actual output of the sensor equals the reference input, the absence of a biochemical signal will allow the energy-coupled biochemical mechanism to stop the change of production rate of low molecular weight substances, so that the system remains osmotically balanced.

Such a negative feedback loop would be adequate only if a balance was to be held. In physiological situations, however, regulation by indirectly related processes is often necessary to maintain life in a cell; this is easily seen in examples of osmotic regulation designed for the performance of work. For example, stomata opening and closing is controlled by carbon dioxide concentration, light, hormones (4), and nyctinastic movement of leaflets by the phytochrome system, endogenous oscillators, or seismic stimuli (5). Such external signals may affect the loop, as an alteration of the reference input as indicated in Figure 1. In addition, external signals may exert some influence at the level of the osmoregulatory biochemical mechanism itself, as a disturbance input or by an alteration of the sensitivity of the controller. In the following paragraphs, examples will be given of systems exhibiting osmotic regulation; and indications of the nature of the solutes, the sensor, the controller, the biochemical signal, and the biochemistry of the energy-coupled mechanisms will be suggested.

At this point, some remarks are relevant in regard to the reference input. Anyone who has experienced that his thirst can be appeased by the addition of a few percentages of water to the body will agree that organisms and cells obviously have an optimum water content. It is understandable that in plants under *severe* water stress ultrastructural organization is injured both mechanically and by loss of hydration water from macromolecules. However, there are in addition numerous reports that a *small* water stress accompanied by changes in the cell water of only 10% also has pronounced effects on various metabolic activities such as growth, respiration, photosynthesis, protein synthesis, and enzyme production (2). Possible mechanisms underlying the responses to small water deficits may be seen in a reduction of the water activity, a direct influence of hydrostatic pressure changes on macromolecules, dehydration of macromolecules, spatial effects on ultrastructure, or changes in the concentration of molecules. Hsiao (2) has recently carefully discussed the literature and points out that neither of these possibilities is likely to offer a satisfying explanation for the alteration of metabolism by mild water deficit and that these changes are most probably regulatory responses of physiological mechanisms. This author fully agrees with this idea and feels that, at the present time, it is best to regard the reference input—that is, the optimum value for volume, turgor pressure, or water content of the cell—as a prerequisite for the integrity of the physiological system. For example, erythrocytes of different animal species and the cells of different species of the same flagellate genus always tend to maintain a species-specific volume. In the context of this chapter, it would appear that the optimum values can be considered as features advanced for the given cells of a species during evolution. On this basis, the different biochemical steps underlying the physiological mechanisms available in each system for the regulation of the osmotic balance can be understood.

NATURE OF OSMOTIC AGENTS

Numerous solutes contribute to the osmotic value of the cell sap, but substances of the fundamental metabolic pathway are not likely to be subject to the great alteration in concentration that is necessary to effect osmotic regulation. To find which substance out of the many soluble cellular constituents does indeed play a role in osmotic regulation, properly designed experiments fulfilling several prerequisites must be performed. The cellular concentration of very many constituents must be measured before and after an alteration in external osmotic value or after a volume increase. There must also be a simultaneous determination of that part of the cell volume constituting the solubilization volume for the various substances. In addition, it must be established that the experiments are done within the proportional range of the system in question. The fulfillment of such exacting requirements is difficult, so these methods can only be properly applied at present to simple systems, preferably wall-less single cells rather than tissues of higher plants.

Table 1 gives some examples of cellular constituents reported to play a role in osmotic regulation. The substances employed are either inorganic ions, mainly potassium salts, or low molecular weight organic molecules. The observed changes in concentration reported definitely suggest that these agents contribute considerably toward the maintenance of osmotic balance. In most cases, however, the published data, due to the experimental difficulties cited above, do not allow the calculation of an exact balance.

A discussion is presented in detail for *Ochromonas* (see under "*Ochromonas*") to illustrate that in addition to the easily recognizable major osmotic agents, a number of other solutes do contribute to a minor extent to osmotic regulation. The latter are to be recognized only if the experiments fulfill all of the above-mentioned criteria. Therefore, it seems likely that additions of minor solutes to the main osmotic agents listed in Table 1 will have to be made in the future.

SYSTEMS EMPLOYING ION TRANSPORT IN ALGAE

Transport Mechanisms

Marine algae have attracted the interest of plant physiologists for a long time because they offer many advantages for the study of ion permeability and

Table 1. Some systems suitable for investigation of osmotic regulation

Organism[a]	Main cellular osmotic agents studied	Notes
Bacteria		
Bacillus (6)	Proline	Δc_i roughly $\Delta\pi_0$
Clostridium (6)	Proline, glutamic acid, γ-aminobutyric acid	Δc_i roughly $\Delta\pi_0$ (+ counterion)
Algae		
Chaetomorpha (7, 8)	K^+, Na^+, Mg^{2+}, Cl^-, SO_4^{2-}	Mainly K^+ flow, P const.
Valonia (9, 10)	K^+, Na^+, Cl^-	K^+ flow, P const.
Codium (11, 12)	K^+, Na^+, Cl^-	Prop. of K^+-Na^+-Cl^- const.
Dunaliella (13, 14)	Glycerol	Partial leakage, c_i high
Phaeodactylum (15)	Proline	Contribution to π_i uncertain
Platymonas (16, 17)	Mannitol	Δc_i roughly 40% of $\Delta\pi_0$
Monochrysis (18)	Cyclohexanetetrol	Cyclitol, contribution to π_i uncertain
Ochromonas (19, 20)	α-Galactosyl-(1,1)-glycerol	Δc_i about 80% $\Delta\pi_0$, 20% $\Delta\pi_0$: amino acids + K^+
Fungi		
Dendryphiella (21)	Arabitol, mannitol	c_i const.
Saccharomyces (22–24)	Alditols, trehalose, K^+, organic acids	Variable with strain and species
Higher Plants		
Helianthus (25)	K^+, Na^+, Cl^-, ? organic acid, sugars	Growing sunflower hypocotyl
Avena (30)	K^+, malate	IAA-induced extension of coleoptiles
Hordeum (26–28)	Glucose, maltose	Inhibition of α-amylase synthesis in aleuron cells
Vicia, Zea (4)	K^+, malate, Cl^-	Stomata, proportion malate:Cl^- different with species
Gossypium (29)	K^+, malate	Cotton fibers growing in vitro and in vivo
Albizzia (5)	K^+	Motor cells of leaflet pulvini

[a]Species name is not given as this may vary with different investigators. For discussion and more references see text.

bioelectrical phenomena (11). A few recent papers pertinent to the subject of the present chapter are discussed here. The multinucleate giant cells of some siphonal species can be punctured to allow the positioning of electrodes in the vacuole of single cells, the withdrawal of samples, exchange of vacuolar sap, and application of the hydrostatic pressure inside the cell. It has been found that the cells maintain a positive turgor pressure by maintaining the osmolarity of their intracellular fluid above the osmolarity of the external sea water. The observed turgor values are typical for each species, but in most cases are more or less constant over a wide range of external salinities. The main osmotically active substances in the vacuolar sap are K^+, Na^+, and Cl^-. In *Chaetomorpha* (7,8) and *Valonia* (9,10,31), the concentration of KCl is much higher than that of NaCl. During turgor regulation, the internal concentration of KCl changes markedly, whereas NaCl concentration shows only minor fluctuations. The regulatory mechanism appears to work by active inward transport of K^+, whereas chloride is passively transported as the counter ion. In *Codium* (12), K^+ and Na^+ are present in approximately equal concentrations, and turgor regulation occurs by the alteration of the concentrations of both ionic species.

The overall regulation mechanisms in these algae strongly resemble the situation in animal red blood cells. In the latter case, osmotic regulation (volume regulation) employs KCl as an osmotic agent in "high potassium cells" such as duck (32,33) and turkey (34), whereas "high sodium cells" (e.g., dog (35)) employ NaCl in addition to KCl. As in algae, the actual concentrations of the ions are regulated by a "pump and valve" system which combines active transport with passive fluxes in opposite directions to obtain the desired concentrations. Both directions seem to be controlled by the osmotic regulation mechanism in erythrocytes, and hormonal control is superimposed (32–34). Other common features of red blood cells and algae are a marked influence of Ca^{2+} ions (7,35) on osmotic regulation and a striking dependence of the influx, and outflux, of K^+ and Na^+ on the relative external concentrations of K^+ (8,36).

Our knowledge of the biochemistry of ion transport mechanisms in plants is far below that for animal cells. However, the similarities between plant and animal systems cited above do suggest that there are features in common. At present, plant physiologists seem to be mainly interested in K^+-H^+ exchange mechanisms (4,22,30,37). In several investigations using plant cells, a small but definite transport of Na^+ in an opposite direction to the observed K^+ transport has been reported (10,37,38). This observation may be an indication that the Na^+-K^+-ATPase system is also involved in the K^+ transport observed in plant membranes. In erythrocytes, the participation of this enzyme system in K^+-mediated osmotic regulation is generally accepted (39,40). Plant physiologists should bear in mind that in erythrocyte systems the stoichiometry between the observed apparent rates of Na^+ and K^+ transport can also be subject to great variation (40).

Pressure-sensing Mechanism

One feature which makes the giant algae very attractive for studies of osmotic regulation is the fact that they represent the only example in which experimental evidence exists in regard to the nature of the parameter measured by the sensor (see Figure 1). Cells of *Valonia* were perfused and connected to a mercury manometer, and a hydrostatic pressure was thereby applied to the inside of the cells (9). Potassium influx was high when no pressure was applied and dropped to about 30% when the pressure was raised to 1 atm., i.e., that pressure which approximately equals the turgor pressure normally observed in *Valonia.* Decreases in hydrostatic pressure resulted in the opposite effect. Later the experiments were repeated in a pressure chamber which allowed the additional application of an outside hydrostatic pressure (41). The results of the experiments clearly indicate that the sensor of the feedback system does not respond to an alteration in a chemical property such as concentration, but rather to the pressure gradient across the plasmalemma and the cell wall. It can be assumed on a hypothetical basis that a decrease in membrane thickness (42) or a stretching of membrane components leads to a change in the conformation of critical proteins, which may in turn lead either to a direct "permeability change," or to a chemical signal regulating membrane functions. In *Valonia* (10), hydraulic conductivity and electrical membrane resistance, two directly measurable membrane properties, have been shown to be affected by small changes in turgor pressure. As yet however, such observations cannot be understood in terms of the ultrastructure and the known biochemical properties of membranes. It is reasonable to assume that they are related in some way to osmotic regulation processes, such as alterations of passive pressure-sensitive fluxes and active fluxes. It is of interest to note that Kregenow (32), in a discussion of completely different experiments performed with duck red blood cells, assumes that changes in properties related to membrane elasticity are probable candidates for the role of a volume-measuring indicator.

In the pressure-sensitive regulation of K^+ transport in *Valonia* described above (9,41), a serious problem arises from the observation that the pressure change does not exert an observable influence until about 30 min after its onset. This would indicate that unknown chemical events occur between the recognition of the pressure change and the realization of the effect, and that the "pump mechanism" is not identical with the pressure sensor. Experimental manipulation in duck erythrocytes (36) also demonstrated a functional separation between pump mechanism and volume-controlling mechanism, while in the "high sodium" red blood cells of the dog (35) the volume-sensitive parameter appears to be mainly the Na^+ leak.

OSMOTIC REGULATION WITH ORGANIC COMPOUNDS IN LOWER PLANTS

In contrast to the difficulties encountered in the understanding of the biochemistry of systems employing ion transport for osmotic regulation (see under "Systems Employing Ion Transport in Algae"), systems involving organic molecules offer greater opportunities to explore biochemical mechanisms. The respective enzymatic steps and principles of their regulation are often known either directly from metabolic pathways or indirectly from an appreciation of similar reactions already studied in connection with metabolic events.

Ochromonas

The osmotic balance in the golden-brown wall-less alga *Ochromonas malhamensis* is mediated by a series of reactions involving carbohydrate metabolism. When suspensions of the flagellate were allowed to assimilate $^{14}CO_2$ in the light for some hours in the presence or absence of phosphate buffer, the composition of the end products differed markedly. At low buffer concentration, the water-soluble high molecular weight reserve β-(1→3)-glucan was the most prominent labeled cell constituent. In contrast, at higher buffer concentrations, a low molecular weight carbohydrate was found to contain high proportions of radioactivity (19). It was demonstrated that this was not due to the chemical composition of the suspension fluid but to the osmotic value, because solutions of various chemical composition resulted in similar alterations. The substance in question was identified by enzymatic and hydrolytic cleavage and by cochromatography as *O*-α-D-galactopyranosyl-(1→1)-glycerol (isofloridoside, IF). Enzymatic determination of the IF content of cells showed that high levels of the substance could be present and that the internal concentration was directly related to the external osmotic value. If ^{14}C-labeled IF were allowed to accumulate in high concentration, followed by a decrease of the osmotic pressure, IF rapidly disappeared and all the carbon of both the galactose and glycerol parts of the molecule could be recovered in the reserve β-(1→3)-glucan. This suggested that the reversible conversion of polysaccharide to IF may be an important factor in osmotic regulation of *Ochromonas.*

Further investigations (43) showed that this mechanism may be of ecological significance, because the level of IF changed in parallel with alterations in the osmotic value of the synthetic nutrient solution during growth of the cultures. In physiological experiments, suspensions of the cells in solutions containing mannitol, polyethylene glycol, sucrose, KCl, and NaCl as osmotic agents, resulted in the same IF levels if concentrations of equal osmolarity were used. Above concentrations of 75 mosM, the IF content of the cells is proportional to the outside osmotic value (π_0), whereas at values of π_0 lower than 75 mosM, the inward flow of water is balanced by means of the contractile vacuole (20).

After addition of an osmotic agent to suspensions of *Ochromonas,* the cells shrink within 1–3 min to a smaller volume which is indirectly proportional to the osmotic stress applied and which can be as low as 40% of the initial value at a π_0 of about 300 mosM (43). The initial volume (but a slightly different cell shape) is fully regained in parallel with the accumulation of IF (24, 43) which will take, for example, about 2 hr after a $\Delta\pi_0$ of 110 mosM (see Figure 2). In order to answer the question as to whether the measured amounts of IF are sufficient for osmotic balance in its entirety in *Ochromonas,* the solubilization volume in the cells was estimated. Careful determinations of total cell volume were made, using [^{14}C]inulin to correct for intercellular space in the pellets, and that proportion of the cell volume which does not respond osmotically was

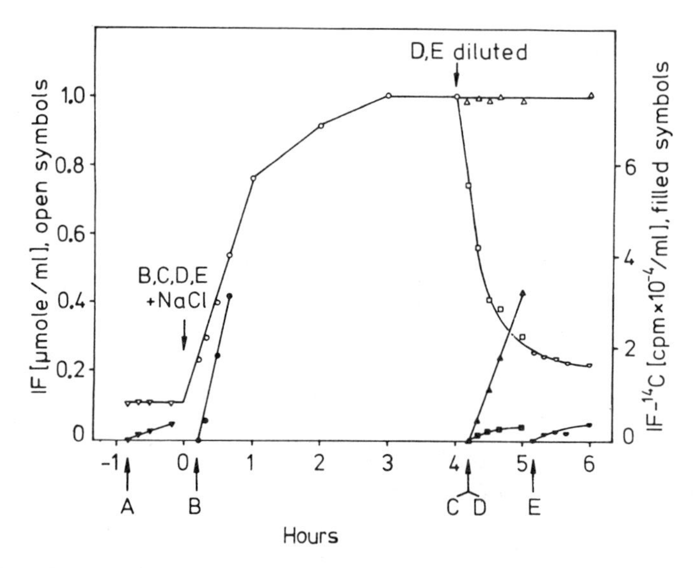

Figure 2. Pool size and labeling of IF in cells of *Ochromonas* under various osmotic conditions. Five identical suspensions were aerated in dilute nutrient solution containing 4 mg/ml of glucose; A, ▽– –▽ and ▼– –▼; B, ○——○ and ●——●; C, △——△ and ▲——▲; D, □——□ and ■——■; E, ○——○ and ●——●. At the time indicated by the *arrow,* NaCl was added to increase the π_0 in samples B, C, D, and E from 30 to 185 mosM. Samples D and E were diluted 4 hr later to bring π_0 from 185 to 30 mosM. [^{14}C]glucose was added at the times indicated by the *lower arrows* to the respective samples. Aliquots were drawn, inactivated by heating in water, and the radioactivity in IF was determined. Reprinted with permission of Plant Physiol. 52:613 (1973).

determined. For this purpose, aliquots of cell suspensions were subjected to solutions of increasing π, and the altered cell volumes were measured after completion of shrinkage. If $1/\pi$ is plotted against total cell volume, the intercept of the resulting straight line with the ordinate represents that proportion of the cell volume which remains after the cells have lost all osmotically-held water. As much as 40–50% of the apparent total cell volume was found to be nonosmotic, and this value could be used to calculate a balance. It was found that IF production accounted for about 70–80% of the observed volume regained by cells of *Ochromonas* after stress. Another 10% was due to an increase in free amino acids (mainly alanine) (43) and about 10–20% of the volume increase was accounted for by K^+ and its unknown counterion (44). The mechanisms responsible for the handling of the three types of osmotic agents appear to operate in parallel, but in proportions which vary with different batches of cells (44). The mechanism which operates by production and degradation of IF was always found to be by far the most prominent mechanism, and the known biochemical details are described below.

Some features of the underlying IF pathway were already evident from in vivo studies. About 1–3 min after addition of an osmotic agent to suspensions of *Ochromonas,* IF is produced at a constant rate. This, together with studies using inhibitors of protein synthesis (43), shows that the regulation does not involve de novo synthesis of enzymes, but occurs instead by activation of pre-existing enzymes. The rate of formation, as well as the time at which increased production is terminated (Figure 2), depends on the extent of the osmotic stress (43). This observation indicates that not only can gradual differences in the external osmotic pressure be sensed, but that the status of osmotic balance already reached can be measured as well. Signals of variable extent and of unknown nature must exist leading to a temporary increased production of IF (see Figure 1).

The operation of the control mechanism can best be demonstrated with chase experiments (45) using [^{14}C] glucose in cell suspensions which produce IF from [^{12}C] glucose (Figure 2). Incorporation of radioactivity into IF occurs under any osmotic condition and even when a fully constant pool size is observed; these results clearly indicate an intensive turnover of IF. The changes in the pool size of IF occurring shortly after $\Delta\pi_0$ and the amounts of [^{14}C] IF produced from [^{14}C] glucose permit calculations of the amounts of carbon which flow into and out of the IF pool (Table 2). There are both formation and degradation of IF at any osmotic pressure and at any pool size, the turnover rate being higher the greater the pool size. There is a significant increase in the formation of IF after an increase in π_0, while there is an increased degradation of IF after a decrease in π_0. Regulation obviously occurs by means of enzymatic steps which are involved in the formation as well as the degradation of IF. It has been shown previously (19) that the labeled IF which disappears from the cells following a decrease of π_0, is fully transformed to the reserve glucan. The degradation products of the IF from the turnover at constant pool size are, therefore, probably converted to reserve polysaccharides as well.

The combination of the physiological studies described above with enzymatic studies allows us to propose the participation of the following reactions in the IF pathway (Figure 3). The substance can be formed from any available carbon material by using classical glycolytic reactions. Ultimate precursors are UDP-galactose and glycero-3-phosphate. The key enzyme of the IF pathway is a galactosyltransferase (46) (I in Figure 3) which transfers galactose to glycerophosphate, forming a phosphorylated form of IF. This intermediate has been isolated from cells of *Ochromonas* and has been tentatively identified. The respective galactosyltransferase was studied in some detail in vitro (46). Dephosphorylation of IF-phosphate (IFP) and hydrolysis of IF by an α-galactosidase can be demonstrated in extracts of *Ochromonas* (H. Kauss, unpublished results); however, details on the specifity of the respective enzymes are still awaited. The turnover rates, in addition to the observation that [^{14}C] IF is fully transformed to the reserve polysaccharide following a decrease in π_0, strongly suggest that the enzymatic reactions necessary for the conversion of glycerol and galactose to the glucan are in operation in the cell.

The scheme in Figure 3 provides the basis for a crude estimation of the amount of energy necessary for the osmotic regulation effected by the IF pathway. Under conditions of constant high π_0 (185 mosM (Figure 2, Table 2)), formation and degradation of IF are equal. Therefore, the produced and used reduction equivalents are the same. The formation of 1 mol of IF from 1.5 mol of glucose will require about 3 mol of ATP; about the same amount of ATP is necessary for the conversion of 1 mol of IF back to the reserve glucan. Therefore, a complete passage through the IF pool requires about 6 mol of ATP, which are equivalent to 1 mol of O_2/mol of IF. The turnover rate for IF at 185 mosM (Table 2) was 0.34 μmol/hr/ml of suspension; 0.34 μmol of O_2/hr/ml is then used to produce the required amount of ATP. The respiration rate of a similar suspension of *Ochromonas* amounts to about 8–10 μmol of O_2/hr (H. Kauss, unpublished results); thus, less than 5% of the respiration of the cells is employed in maintaining the turnover of the IF pool. The stoichiometry of IF to ATP used in the above calculation is not fully established, and a relatively small amount of carbon material used for IF synthesis is not derived from exogenous glucose but from other unlabeled precursors (45), so the required energy may roughly be estimated to be 10% at most of the total energy derived from respiration. This appears economically feasible.

Regulation of the enzymes in metabolic pathways involves special mechanisms such as de novo synthesis, allosteric interaction with effectors, or chemical modification. It appears probable on the basis of that knowledge that the regulation of enzymes involved in osmotic regulation also employs these principles. The rather short time of 1–3 min necessary to establish a constant rate of IF production after a rise in π_0, as well as the failure of the antimetabolites of protein synthesis to interfere with the initiation of that process, may be an indication for regulation of the IF pathway, not by de novo synthesis, but by activation of preformed enzymes. Some experimental evidence exists to suggest that at least part of the regulation takes place at the steps indicated by roman numbers in Figure 3. A phosphorylase specific for β-(1→3)-glucans (enzyme IV in Figure 3) has been purified and crystallized from *Ochromonas* (47). The partial activation of this enzyme by addition of AMP may indicate an allosteric type of regulation, although details on the actual role of this enzyme and its regulation in IF metabolism are still lacking. Studies (48) on changes in the relative pool sizes of some intermediates of the IF pathway shortly after an increase or decrease of π_0 indicate that an additional regulation may occur at the level of the phosphatase (II in Figure 3) which is responsible for the dephosphorylation of IFP, and also at the level of IF degradation (III). However, the enzymes involved in the two latter processes have not been studied in detail as yet. The experiments (48) on changes in relative pool size induced by changes in π_0 indicate that the degradation of IF may even occur by transgalactosylation rather than by the action of an α-galactosidase.

Galactosyltransferase (IFP-synthase, I in Figure 3) is the most intensively explored enzyme with respect to osmotic regulation. An increased activity of

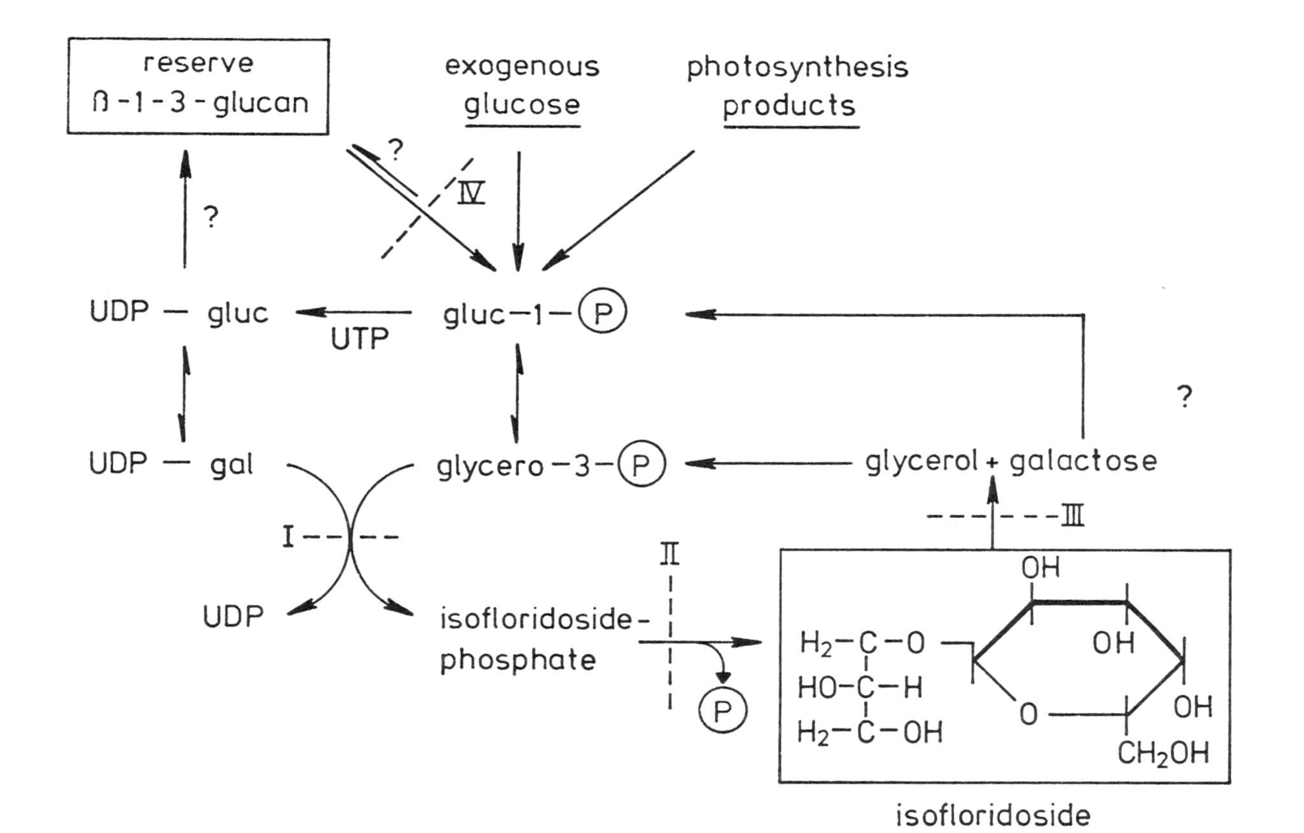

Figure 3. Pathway of isofloridoside in *Ochromonas.* For explanation see text. The enzymatic steps most probably subject to regulation are marked with Roman numbers.

Table 2. Flow of material into and out of isofloridoside pool in *Ochromonas* under various osmotic conditions

External osmotic condition	Isofloridoside		
	Pool size	Formed	Degraded
(+ NaCl or dilution)	(μmol per ml)	(μmol/hr/ml)	
30 mosM, constant	0.12	0.07	0.07
10 min after 30 → 185 mosM	0.23	0.64	?
185 mosM, constant	1.01	0.34	0.34
10 min after 185 → 30 mosM	0.70	0.04	0.85
70 min after 185 → 30 mosM	0.28	0.04	0.08

From H. Kauss (1973). Plant Physiol. 52:613.

that enzyme can be demonstrated in crude extracts of cells which were exposed to an increased π_0 *before* homogenization (46). Addition of the osmotic agent *after* homogenization, however, did not increase its activity. It was concluded, therefore, that the enzyme was not directly affected by the higher π_0, but metabolic activity of the living cell was required to effect some change in the activity of the IFP-synthase. The enzyme is rather unstable, and 20-fold purification at most could be achieved by ion exchange chromatography (49). The resulting preparation showed strongly sigmoid kinetics when its activity was assayed with varying concentrations of its substrate, UDP-galactose. However, an allosteric effector could not be found.

More recent experiments (50) suggest an alternative mechanism for the regulation of the IFP-synthase. The enzymatic activity in the crude extract can be increased about 5- to 20-fold by incubation at pH 6.2, whereas the optimum for the transferase activity is at pH 7.8. The activation is time-dependent and shows a distinct pH optimum, as well as Q_{10} value of 1.5–3.0. These features strongly indicate an activation of the transferase on the basis of a chemical modification. Inactivation of the activated form of the galactosyltransferase can also be shown to be time-dependent and exhibits a pH curve opposite to that of activation. Many examples in recent years for this type of regulation of enzyme activity have been described (51) for bacterial and animal metabolic enzymes. In these systems, chemical modifications occur by attachment and removal of modifying groups by means of special modifying enzymes. The most prominent modifying groups which are reported to occur, e.g., in glycogen phosphorylase, glycogen synthase, and glutamine synthetase, are phosphate and adenylate (51). On the basis of the above observations, it is very probable that auxiliary enzymes may lead in a similar way to an activation-inactivation of the IFP-synthase involved in osmotic regulation in *Ochromonas.* Future studies may indicate which modifying group is involved and how the modifying enzymes are themselves modulated. One of the signals might be a change in cellular pH. In addition, participation of amplifying substances such as cAMP or Ca^{2+}, as in the case of animal systems (51), can be envisaged as a working hypothesis although no experimental evidence is available as yet in *Ochromonas*.

Dunaliella

Species of the green wall-less flagellate *Dunaliella* are rather salt-tolerant and are found in a broad range of salinities from fresh water up to almost complete NaCl saturation. At first the ecological adaptation to high salt concentrations was thought to be based on the fact that the membranes of *Dunaliella* are highly permeable to NaCl and thus the intracellular concentration of the salt may be the same as the extracellular one (52,53). However, such a hypothesis now appears unlikely because of the sensitivity of various enzymes and metabolic reactions in *Dunaliella* to salts (54–56). In addition, cells of different species of *Dunaliella* are able to form glycerol in amounts that increase with increasing external salt concentrations (13,14,57). The maximum intracellular concentrations reported, calculated on the basis of total cellular water, are about 2.0 M for *D. parva* cultured (13) in 1.5 M NaCl, 1.4 M for *D. tertiolecta* cultured in 1.36 M NaCl, and 4.4 M for *D. viridis* grown (56) in 4.25 M NaCl. When NaCl concentration was decreased, the glycerol content also decreased rapidly (13). The fate of the disappearing glycerol is not known exactly, although part of it might be excreted from the cells. Only a minor portion of the ^{14}C from [^{14}C]glycerol, produced either inside or fed to the algae at lowered π_0, appears to be transferred to other products (58).

These observations indicate sufficient glycerol concentrations to balance more than half of the external osmotic pressure and indicate that glycerol is a major agent for osmotic regulation in *Dunaliella*. However, the experimental data available do not as yet allow for a calculation of the exact balance. Future studies which will have to be done should include determinations of nonosmotically bound water, and should take into consideration the fact that glycerol is partly leaking out of the cells (14,57), preferentially (in reference 13, but not in reference 58) at lower π_0. It has also been observed that glucose concentration in *Dunaliella* is inversely related (56) to extracellular salt concentration, that active transport of K^+ might occur (56), and that appreciable amounts of sucrose can be formed as an end product of photosynthesis, especially at higher temperatures (K. Wegman, personal communication). It appears possible, therefore, that all these substances may play an additional role in the osmotic regulation of *Dunaliella*, in a way similar to that in *Ochromonas* (see under "Ochromonas").

A few biochemical details of glycerol metabolism in *Dunaliella* are known. The accumulation of glycerol is strongly inhibited by the uncoupler FCCP (FCCP: carbonyl cyanide *p*-trifluoromethoxyphenylhydrazone) (13) and an inhibitor of photosynthesis, dichlorophenyldimethylurea (DCMU) (58), but not by antimycin A (58), an inhibitor of respiratory electron transport. Although these findings might be taken to indicate a requirement for energy in the form of ATP, they are to some extent contradictory. The experiments should be repeated with simultaneous determinations of the extent of inhibition of respiration and photosynthesis. No final conclusion on the light requirement can be drawn from the literature, because some investigators could not find any formation of glycerol at all in the dark (58), whereas others state that the same accumulation of glycerol was observed both in light and in the dark (59).

One of the enzymes involved in glycerol metabolism appears to be the NADP-specific dihydroxyacetone reductase, as demonstrated in *D. parva* (60) and *D. tertiolecta* (56). This enzyme has some unusual properties. The apparent K_m value for dihydroxyacetone is in the mM range, whereas that for glycerol is rather high and varies between 1.4 M and 3.4 M, depending on the pH value (56,60). Although the enzyme was first described as a dihydroxyacetone reductase (60), it may well also operate in the opposite direction in vivo, when high concentrations of glycerol are accumulated. A pH value of 9.0 is optimal for dehydrogenation, whereas pH 7.5 is optimal for hydrogenation (56,60). In addition to the NADP-dependent glycerol dehydrogenase described, an NAD-dependent glycerol-3-phosphate dehydrogenase which produces dihydroxyacetone phosphate has been demonstrated recently in *D. tertiolecta* (K. Wegman, personal communication). Although both enzymes are likely to be involved in the glycerol metabolism related to osmotic regulation in *Dunaliella,* their cooperation and regulation is unknown.

Other Organisms

A few other lower organisms that employ special substances for osmotic regulation have been described. Mainly to document the diversity, they will be briefly mentioned here, although very little is known of the biochemical details of the respective regulation mechanisms.

In many cases (Table 1), sugar alcohols play a role as osmotic agents in cells. These organic molecules appear to be particularly suitable for these purposes. They are less inhibitory than NaCl or glucose when enzymes (glucose-6-phosphate dehydrogenase (56), NADP-specific isocitrate dehydrogenase (23)) are assayed in their presence. The term “compatible solute” has been coined to designate this property (23).

Although the observed changes in internal concentrations appear to be high enough to consider them as having a function in osmotic regulation, it is not known whether a change in other solutes also contributes to the osmotic process. The fact that certain osmotic solutes can be replaced to some extent by others is best illustrated with the marine hyphomycete *Dendryphiella.* During growth the hyphae maintain a constant solute concentration; the predominant soluble carbohydrates are mannitol and arabitol (21). When the nonmetabolizable 3-*0*-methylglucose is offered to the cells, the sugar analogue is actively transported into the cells by the hexose transport system and accumulates. Simultaneously, the metabolizable constituents mannitol and arabitol are converted to insoluble reserve material to an extent just sufficient to maintain the total solute concentration at a constant level. Several different types of osmotically active substances have been reported for various species or strains of *Saccharomyces* (22–24). These include glycerol, arabitol, mannitol, trehalose, K^+, and various organic acids, and it appears that these substances can also replace each other to some extent under different experimental conditions.

Several cases are known in which alterations in pool size of the free amino acids appear to be of significance for osmotic regulation. Two ciliates may be listed here although these organisms are not plants. In *Tetrahymena pyriformis* (61), osmotic regulation is mainly achieved by changes in Na^+, glutamic acid, glycine, and alanine. In contrast, in *Miamensis avidus* (62) almost any of the many amino acids present appear to increase in concentration, although changes in alanine, glycine, and proline are the most prominent ones. These data suggest a similarity to the situation reported for diatoms. In *Phaeodactylum tricornutum* (15) proline is the only amino acid which shows changes in concentration which parallel the external salt concentration changes, whereas in *Cyclotella meneghiniana* (63), both proline and its biosynthetic precursor Δ'-pyrroline-5-carbonic acid behave similarly. However, from the published data, exact calculations of the contribution of proline to the internal osmotic value are not possible, although the observed changes (based on fresh weight (15)) indicate that proline could at least partially affect osmotic regulation in diatoms (see under "Role of Proline" for higher plants). In contrast, proline accumulation alone may fully compensate the osmotic pressures of the growth medium in several taxonomically unrelated bacteria (6). In addition to proline, other bacterial species accumulate glutamic acid and γ-aminobutyric acid; such an accumulation of acids is accompanied by a strongly increased intracellular level of K^+ which is necessary in order to maintain a constant internal pH value. Potassium accumulation may interfere with several metabolic enzyme systems, including glutamate synthesis (6), in the cytoplasm of bacteria. The utilization of proline as an osmotic agent may, therefore, represent a more refined and more evolved mechanism, because it does not require the simultaneous accumulation of a neutralizing ion.

OSMOTIC REGULATION IN HIGHER PLANTS

It was pointed out in the introductory paragraph that studies on the biochemistry of osmotic regulation in higher plants are complicated either by problems related to the determination of the exact volumes or the cellular water content in situ. On the other hand, the regulation mechanisms are often obscured by the fact that several physiological parameters influence them simultaneously. The numerous physiological effects caused by small-to-moderate changes in osmotic pressure (2) (see also under "Feedback Loop for Osmotic Agents") must surely operate on a biochemical basis and must, therefore, be clarified by a combination of both physiological and biochemical methods. For example, the synthesis of α-amylase in the aleurone cells of barley seeds is regarded in textbooks as a case of pure "molecular biology." In whole seeds, however, a further regulation is superimposed by the osmotic pressure caused by the build-up in sugars produced in turn by the action of the amylase (26). The demonstration that this involves a more general inhibition of protein synthesis (27) is in accordance with the observation that the number of polyribosomes is reduced, but the ribosome activity is unimpaired (28). A biochemical signal must now be found to provide an explanation of feedback regulation (see Figure 1) of protein biosynthesis by use of changes in cellular osmotic pressure. In a similar way, other cases of osmotic influences in higher plants should be elucidated, such as the modulation of crassulacean acid metabolism (64) or the turgor regulation during extension growth (see under "Expanding Cells").

Expanding Cells

In elongating cells, the cell volume increases, and consequently solute concentration must be kept constant in order to maintain the more or less constant turgor required as a driving force (65). Changes in turgor which were produced experimentally to manipulate growth rates are accompanied by various changes in physiological and metabolic activities (2,65), including protein synthesis (66). The results from these and other physiological experiments 67) imply that turgor is important not only as a driving force but also as a trigger of some of the physiological effects mentioned above. In the light of the role of hydrostatic pressure in sensing turgor changes in algae (see, under "Pressure-sensing Mechanism"), it appears possible that pressure is a feature which might coordinate the complex events combining to allow cell extension. No experimental evidence is available with regard to the nature of the biochemical signals involved in such an integrated regulation system; however, hormones such as IAA may play a role as external signals (68) (see Figure 1).

The main solutes responsible for the maintenance of osmotic pressure in standard systems used for growth measurement appear to be K^+, Na^+, Cl^-, and organic acids and sugars such as glucose and fructose (25). Exact measurements of all of these in the same experiment before and after defined increases in cell volume appear to be unavailable. It is possible to speculate in the light of the results using lower plant systems (see under "Osmotic Regulation with Organic Compounds in Lower Plants") that the relative contribution of the above substances may vary with the tissue used and may depend on the experimental conditions (e.g., potassium phosphate buffer, sugar availability). A stoichiometric correlation of K^+ uptake and malate accumulation (almost 1 μmol/μEq) has been found recently for IAA-induced extension growth in *Avena* coleoptiles (30). Because malate formation is accompanied by dark fixation of CO_2, the reported effects of CO_2 on growth rates of coleoptiles (68) appear in a new light. The accumulation of K^+ and malate during cell extension has also been shown for cotton fibers growing in vivo and in vitro (29). Only about one μmol of K^+/μmol of malate was found when the ovules were grown in vivo, whereas amounts closer to stoichiometry were observed in ovules cultured in vitro. The authors state that the fluctuating component of the osmotic potential of the cells in the latter case might be fully accounted for by K^+-malate. The importance of these two substances for extension of the fiber cells is emphasized by the observation that fiber growth in cultures in vitro is only possible when optimum concentrations of K^+ and atmospheric CO_2 are supplied.

Stomata and Pulvini

Potassium also plays an important role in osmotic regulation mechanisms in specialized plant cells in which changes in turgor are required for the performance of work. The physiological and ecological problems encountered in the opening and closing reactions of stomata have recently been expertly reviewed (4). Several lines of evidence show clearly that the opening of stomata is accompanied by the movement of K^+ from surrounding cells into stomatal cells, whereas, when the stomata shut, K^+ ions move out of the stomatal cells. The amounts measured are sufficient, at least in order of magnitude, to explain the necessary and observed alterations in turgor (4). It is not clear, however, to what extent the malate is transported into the cells and to what extent it is synthesized de novo in the stomata by dark fixation of CO_2 (4,68). Investigations into the biochemistry of stomatal functions are complicated by at least two factors; the CO_2 concentration is an alternative stimulus to trigger the mechanism and, also, only small amounts of cell material are available for experimental manipulation because there is only a scattered occurrence of stomata in the epidermis (see also under "Feedback Loop for Osmotic Regulation").

Leaflets of several plants such as *Albizzia* (5) or *Samanea* (70, 71) can open and close. These movements are either induced by external stimuli or, alternatively, are under the control of circadian rhythms and result from differential changes in turgor and shape of motor cells on opposite sides of the pulvinus. The turgor changes are related to changes in the K^+ content of the motor cells; the counterions for the K^+ are unknown. The differences in K^+ in the cells of opposite sides of the pulvinus are brought about by movement of the ions from one side to the other, the total amount being constant. The underlying mechanisms for this kind of osmotic regulation must include coordinated changes in passive and active transport mechanisms, but the biochemistry and regulation of such processes is as yet a very complex subject (see under "Systems Employing Ion Transport in Algae").

Role of Proline

There are numerous reports (2) that proline content (and to a lesser degree some other free amino acids) in either whole plants or tissues can alter dramatically. This phenomenon can be demonstrated by exposing plants to water stress (2) or by exposure of halophytes to elevated salt concentrations (72–74). The observations are generally discussed as examples of osmotic regulation. Proline accumulation is thought to provide the cells with a higher osmotic potential to compete for water. Unfortunately, the data are often collected with consideration of only ecological aspects in mind, and consequently are given on a dry weight or fresh weight basis; such data make a determination of the quantitative contribution of proline to the osmotic potential difficult. In only a few halophytes (72) do the amounts appear to be high enough to cause a considerable increase in the osmotic values of the total cells. Thus, it is argued that compartmentation may restrict the location of proline to the cytoplasmic part of the cells, whereas osmotic compensation in the vacuoles is brought about instead by salts such as NaCl. This possibility is not unlikely, but supporting experimental evidence is still very vague. Another problem is that significant accumulation of proline due to water stress (75) does not occur under moderate, but only under *severe,* conditions, under which even wilting is already evident. This observation may mean that in addition to a role as an osmotic solute, which appears to be partly established for some bacteria and lower plants (see under "Other Organisms"), proline may also have additional functions in the prevention of membrane damage, as discussed in connection with frost resistance (76).

ACKNOWLEDGMENTS

I am greatly indebted to my colleagues D. J. Bowles, H. Cruse, W. Lang, and K. Thomson for discussions and help with the manuscript and to numerous other colleagues for access to manuscripts prior to publication.

REFERENCES

1. Cram, W. J. (1976). *In* A. Pirson and M. Zimmerman (eds.), Encyclopedia of Plant Physiology, New Series, 2A:284–316. Springer-Verlag, Berlin.
2. Hsiao, T. C. (1973). Annu. Rev. Plant. Physiol. 24:519.
3. Milsum, J. M. (1966). Biological Control Systems Analysis. McGraw-Hill, New York.
4. Raschke, K. (1975). Annu. Rev. Plant Physiol. 26:309.
5. Satter, R. L., Applewhite, P. B., Kreis, D. J., Jr., and Galston, A. W. (1973). Plant Physiol. 52:202.
6. Measures, J. C. (1975). Nature 257:398.
7. Kesseler, H. W. (1964). Helgol. Wiss. Meeresunters. 10:73.
8. Zimmermann, U., and Steudle, E. (1971). Marine Biol. 11:132.
9. Gutknecht, J. (1968) Science 160:68.
10. Zimmermann, U., and Steudle, E. (1974). J. Membr. Biol. 16:331.
11. Gutknecht, J., and Dainty, J. (1968). Oceanogr. Marine Biol. Annu. Rev. 6:163.
12. Bisson, M. A., and Gutknecht, J. (1975). J. Membr. Biol. 24:183.
13. Ben-Amotz, A., and Avron, M. (1972). Plant Physiol. 51:875.
14. Wegmann, K. (1971). Biochim. Biophys. Acta 234:317.
15. Besnier, V., Bazin, M., Marchelidon, J., and Genevet, M. (1969). Bull. Soc. Chim. Biol. 51:1255.
16. Hellebust, J. A. (1973) Plant Physiol. (Suppl.) 51:20.
17. Kirst, G. O. (1975). Z. Pflanzenphysiol. 76:316.
18. Craigie, J. S. (1969). J. Fish. Res. Bd. Canada 26:2959.
19. Kauss, H. (1967). Z. Pflanzenphysiol. 56:453.
20. Kauss, H. (1974). *In* U. Zimmermann and J. Dainty (eds.), Membrane Transport in Plants, pp. 90–94. (Springer, Berlin.)
21. Jennings, D. H., and Austin, S. (1973). J. Gen. Microbiol. 75:287.
22. Peña, A. (1975). Arch. Biochem. Biophys. 167:397.
23. Brown, A. D., and Simpson, J. R. (1972). J. Gen. Microbiol. 72:589.

24. Ikeda, K., and Ottolenghi, P. (1970). Abstract 8/Ab-2, X, International Meeting of Microbiology, August, Mexico City.
25. McNeil, D. L. (1976). Austr. J. Plant Physiol. 3:311.
26. Jones, R. L., and Armstrong, J. E. (1971). Plant Physiol. 48:137.
27. Chrispeels, M. J. (1973). Biochem. Biophys. Res. Commun. 53:99.
28. Armstrong, J. E., and Jones, R. L. (1973) J. Cell Biol. 59:444.
29. Dhindsa, R. S., Baesley, C. A., and Ting, I. P. (1975). Plant Physiol. 56:394.
30. Haschke, H. P., and Lüttge, U. (1975). Plant Physiol. 56:696.
31. Hastings, D. F., and Gutknecht, J. (1976). J. Membr. Biol. 28:263.
32. Kregenow, F. M. (1971). J. Gen. Physiol. 58:372.
33. Kregenow, F. M. (1971). J. Gen. Physiol. 58:396.
34. Gardner, J. D., Klaeveman, H. L., Bilezikian, J. P., and Aurbach, G. D. (1973). J. Biol. Chem. 249:516.
35. Parker, J. C. (1974). Fed. Proc. 33:266.
36. Kregenow, F. M. (1974). J. Gen. Physiol. 64:393.
37. McRobbie, E. A. C. (1975). *In* F. Bronner and A. Kleinseller (eds.), Current Topics in Membranes and Transport, Vol. 7, pp. 1–48. Academic Press, New York.
38. Jeschke, W. K. (1973). *In* W. P. Anderson (ed.), Ion Transport in Plants, pp. 285–296. Academic Press, New York.
39. Whittam, R., and Chipperfield, A. R. (1975). Biochim. Biophys. Acta 415:149.
40. Glynn, I. M., and Karlish, S. J. D. (1975). Annu. Rev. Physiol. 37:13.
41. Hastings, D. F., and Gutknecht, J. (1974). *In* U. Zimmermann and J. Dainty (eds.), Membrane Transport in Plants, pp. 79–83. Springer, Berlin.
42. Coster, H. G. L., Steudle, E., and Zimmermann, U. Plant Physiol., in press.
43. Schobert, B., Untner, E., and Kauss, H. (1972). Z. Pflanzenphysiol. 67:385.
44. Kauss, H., Lüttge, U., and Krichbaum, R. M. (1975). Pflanzenphysiol. 76:109.
45. Kauss, H. (1973). Plant Physiol. 52:613.
46. Kauss, H., and Schobert, B. (1971). FEBS Lett. 19:131.
47. Albrecht, G. J., and Kauss, H. (1971). Phytochemistry 10:1293.
48. Quader, H., and Kauss, H. (1975). Planta 124:61.
49. Quader, H. (1974). Ph.D. thesis, University of Kaiserslautern.
50. Kauss, H., and Quader, H. (1976). Plant Physiol. 58:295.
51. Holzer, H., and Duntze, W. (1971). Annu. Rev. Biochem. 40:345.
52. Marrè, E., and Servattaz, O. (1959). Atti Accad. Naz. Lincei Red. Cl. Sci. Fis. Mat. Natur. 26:272.
53. Ginzburg, M. (1969). Biochim. Biophys. Acta 173:370.
54. Johnson, M. K., Johnson, E. J., Macelory, R. D., Speer, H. L., and Bruff, B. S. (1968). J. Bacteriol. 95:1461.
55. Ben-Amotz, A., and Avron, M. (1972). Plant Physiol. 49:240.
56. Borowitzka, L. J., and Brown, A. D. (1974). Arch. Microbiol. 96:37.
57. Craigie, J. S., and McLachlan, J. (1964). Can. J. Botany 42:777.
58. Frank, G., and Wegmann, K. (1974). Biol. Zbl. 93:707.
59. Ben-Amotz, A. (1974). *In* U. Zimmermann and J. Dainty (eds.), Membrane Transport in Plants, pp. 95–100. Springer, Berlin.
60. Ben-Amotz, A., and Avron, M. (1973). FEBS Lett. 29:153.
61. Stoner, L. C., and Dunham, P. B. (1970). J. Exp. Biol. 53:391.
62. Kaneshiro, E. S., Holz, G. G., Jr., and Dunham, P. B. (1969). Biol. Bull. 137:161.
63. Schobert, B. (1974). Z. Pflanzenphysiol. 74:106.
64. Lüttge, U., Kluge, M., and Ball, E. (1975). Plant Physiol. 56:613.
65. Cleland, R. E. (1971). Annu. Rev. Plant Physiol. 22:197.
66. Dhindsa, R. S., and Cleland, R. E. (1975). Plant Physiol. 55:778.
67. Green, P. B., and Cummins, W. R. (1974). Plant Physiol. 54:863.
68. Evans, M. L. (1974). Annu. Rev. Plant Physiol. 25:195.
69. Pallaghy, C. K., and Fischer, R. A. (1974). Z. Pflanzenphysiol. 71:332.
70. Satter, R. L., Geballe, G. T., Applewhite, P. B., and Galston, A. W. (1974). J. Gen. Physiol. 64:413.
71. Satter, R. L., Geballe, G. T., and Galston, A. W. (1974). J. Gen. Physiol. 64:431.

72. Stewart, G. R., and Lee, J. A. (1974). Planta 120:279.
73. Moore, P. D. (1975). Nature 253:399.
74. Treichel, S. (1975). Z. Pflanzenphysiol. 76:56.
75. Waldren, R. P., and Teare, I. D. (1974). Plant Soil 40:689.
76. Heber, U., Tyankova, L., and Santarius, K. A. (1971). Biochim. Biophys. Acta 241:578.

International Review of Biochemistry
Plant Biochemistry II, Volume 13
Edited by D. H. Northcote

5
Biochemistry of Plant Pathogens

J. FRIEND
Department of Plant Biology, The University of Hull, HU6 7RX, England

---

BIOCHEMICAL CONTROL OF GERMINATION OF FUNGAL SPORES AND SCLEROTIA 142
Spore Germination Inhibitors and Stimulators 143
Control of Sclerotial Germination by Host Plants 144

CELL WALL DEGRADATION AND RELATED PHENOMENA 144
Cell Wall-degrading Enzymes 145
Pectic Enzymes 145
Hemicellulases and Cellulases 147
Function of Cell Wall-degrading Enzymes in Pathogenesis 147
Maceration of Plant Tissue 150
Protoplast Death 150
Cell Wall Degradation During Entry of Biotrophic Pathogens in Host Cells 151

EXTRACELLULAR TOXINS PRODUCED BY PATHOGENS 151
Host-specific Toxins 152
Nonspecific Toxins 153

EFFECTS OF PATHOGEN ON METABOLISM IN HOST PLANTS DURING ESTABLISHMENT OF BIOTROPHIC INTERACTIONS 155
Alteration in Photosynthesis and Respiration of Diseased Plants 155
Effects of Infection on Carbohydrate Translocation 156
Carbohydrate Metabolism in Infection Site 157
Effects of Infection on Nitrogen Metabolism 158
Nucleic Acid Metabolism of Infected Host 158

RESISTANCE OF PLANT TO PATHOGENIC ORGANISMS 159
Role of Inhibitors Present in Plants Before Infection 159
Antimicrobial Compounds Found in Specific Plants 159
Antimicrobial Compounds of Wide Distribution in Plant Kingdom 162
Inhibition of Cell Wall Degradation by Extracellular Enzymes of Pathogens 163
Metabolic Responses of Infected Plants Which are Involved in Inhibition of Pathogens 164
Production of Specific Antimicrobial Compounds or Phytoalexins 164
Increased Biosynthesis of Common Phenolic Compounds After Infection; Relationship to Resistance 171
Relationship of Hypersensitive Cell Death to Biochemical Reactions Involved in Resistance 174
Control of Resistance Reactions 175

CONCLUSIONS 175

A recent textbook (1) defines a plant pathogen as a living organism, especially a bacterium or a fungus, which attacks plants and causes disease. Pathogens are separated into necrotrophs and biotrophs; necrotrophs are those parasites which grow on dead plant tissue, which they kill in advance, whereas biotrophs are those which are able to colonize only on living tissue.

These definitions are only related to the host-pathogen interaction and not to the ability of the pathogen to grow in culture, which is implicit in the older terms "obligate" and "facultative" parasites. The difficulties in defining pathogens according to host-parasite relationships have recently been considered in detail by Lewis (2).

Biochemically there is now information about many aspects of necrotrophic and biotrophic pathogens. Books with extensive reviews of many aspects are in publication (3, 4), and this review will, therefore, concentrate on the control of germination and infection, cell wall degradation, production of disease symptoms, alteration of host metabolism, and the mechanisms by which hosts may control infection by pathogens.

BIOCHEMICAL CONTROL OF GERMINATION OF FUNGAL SPORES AND SCLEROTIA

Many fungal infections of plants, especially of leaves, are caused by spores which may be dispersed in a variety of ways. Sclerotia, as well as spores, are involved in infections of root systems.

Spore Germination Inhibitors and Stimulators

There are a number of factors which govern the germination of spores either in vitro or on the host plant. For example, many spores contain germination inhibitors whose function is presumably to stop their germination before they alight on their respective hosts. The self-inhibitors in rust uredospores can be removed by leaching in water; they have recently been identified (5) as methyl-ferulate in uredospores of *Puccinia graminis* and methyl-3,4-dimethoxycin-namate in uredospores of *Uromyces phaseoli, Puccinia helianthi, Puccinia antirrhini* and *Puccinia sorghi.* The *cis*-isomers of these compounds which are the naturally occurring forms are far more potent inhibitors of spore germination than the *trans*-isomers.

An inhibitor of spore germination of *Peronospora tabacina* occurs in tobacco leaves after infection. Named "quiesone," it has been tentatively identified as 5-isobutyroxy-β-ionone, and it is suggested that it might also be present in conidia, in which case it would also be a self-inhibitor (6). Some of the biochemical events occurring during spore germination have been reported (7). There is catabolism of neutral lipid (8), carbohydrate (9), and metabolism of phospholipids (10); these biochemical changes seem to provide starting materials for cell wall formation and the production of new cell membranes (11, 12).

There is evidence from *U. phaseoli* (13) and *P. tabacina* (14) that protein synthesis can occur at different stages of germination. However, it has also been shown that, whereas the ribosomes of spores of *P. graminis tritici, Melampsora lini,* and *Cronatium fusiforme* remained active during germination, those of *U. phaseoli* and *P. helianthi* declined (15). Coffey (16) has accordingly concluded that uredospores fail to carry out appreciable synthesis of protein.

Inhibition of protein synthesis has been suggested by Sussman and Douthit (7) as a general mechanism of action for self-inhibitors of fungal spore germination. This has so far only been demonstrated for quiesone. Quiesone inhibits the accelerated protein synthesis measured as the incorporation of ^{14}C-amino acids into protein that occurs early in the germination of conidia of *P. tabacina.* However, its effect is probably an induced one because it does not inhibit in vitro protein synthesis by cell-free systems prepared from conidia (6).

Germination stimulators may also help in the establishment of infection, and the best known, but so far unexplained, stimulatory effect is that of water-soluble extracts of pollen grains on the early stages of infection of broad bean leaves by *Botrytis* spp. (17). The active principle is water-soluble, dialyzable, heat stable, and appears to be present in pollen of all species of plants tested (18). In addition to increasing the rate of spore germination of germ tube growth, the effects of pollen on *Botrytis* spp. include restoring the infectivity of old spores and increasing the virulence and the severity of infection. Similar effects of pollen are shown for *Alternaria brassicicola* on detached cabbage cotyledons (19), for *Helminthosporium sativum* on rye leaves (20), and for *Phoma betae* on sugar beet leaves (21). The mechanism by which pollen extracts enhance the virulence of *Botrytis cinerea* on broad bean leaves has been examined by Mansfield and Deverall (22). They found that pollen extract made the fungus less sensitive to wyerone acid, the phytoalexin of broad bean.

Extracts of wheat anthers have a specific stimulatory effect in vitro on *Fusarium graminearum* (23). This fungus only infects wheat to cause headblight after the anthers have emerged; the fungus grows profusely on anthers in vivo, but can hardly be seen macroscopically on other organs. The stimulant which was dialyzable was resolved into a basic and a neutral component which were identified as choline and betaine (24). The interaction between *F. graminearum* and wheat plants illustrates a plant disease of which susceptibility seems to be an attribute of the host and of which the host possesses compounds which promote the growth of the parasite in vivo.

Control of Sclerotial Germination by Host Plants

Several cases exist in which fungal propagules such as sclerotia will germinate only in the presence of host plants, and in the case of the soil-borne fungus *Sclerotium cepivorum,* which causes white rot of onion and other species of *Allium,* the chemistry of the interaction has been defined (25). *S. cepivorum* forms sclerotia, or resting bodies, which produce hyphae on germination; these are long-lived, but because of the inhibitory effects of the microbial population they do not normally germinate in natural soil (26, 27). However, if a host plant is present, the inhibitory effects are overcome and the sclerotia germinate (28).

The roots of host plants exude alkyl cysteine sulfoxides (29), which are then metabolized by the soil bacteria to yield a mixture of corresponding alkyl thiols and sulfides (30) which are volatile; all of the volatile propyl compounds derived from onion and the allyl compounds from garlic are active stimulators of germination.

A similar response of host-stimulated germination occurs with sclerotia of *Stromatinia gladioli,* which cause dry rot of *Gladiolus* and some other members of the Iridaceae (31). However, the specific germination stimulator in the host plants has not yet been identified.

CELL WALL DEGRADATION AND RELATED PHENOMENA

At some stage of all plant disease, either at entry into or during development within the host, the pathogen comes into contact with host cell walls and generally causes some chemical alteration.

In biotrophic interactions, many of the hyphae are intercellular. This means that they grow between cells but they do not seem, from microscopic observations, to cause marked alterations in cell wall structure apart from those resulting from actual penetration of host cells by hyphae or haustoria. On the other hand, there are diseases known as soft rots, in which parenchyma tissue is colonized, usually by a necrotrophic pathogen, and the affected tissues become soft and waterlogged. Under these conditions, cells become separated from one another, by a process known as *maceration,* which is assumed to involve degradation of the wall material between cells. In the so-called dry rots, which are also caused by necrotrophic pathogens, the lesion has a dry texture, and there is degradation of cell walls although there is no maceration of the tissue.

The majority of the published information on cell wall degradation during the development of plant diseases has been obtained from studies on organisms which produce soft rots and dry rots.

Cell Wall-degrading Enzymes

These are enzymes which degrade carbohydrate polymers of the cell wall and which are classified according to the nature of polymeric carbohydrate substrate. The traditional division of plant cell wall polysaccharides has been based on their solubilities after the application of a series of extractants. The pectic fraction is soluble in hot water or chelating agents and the hemicellulose in alkali; α-cellulose is the insoluble residue left after hemicellulose extraction. The structure of each of the substrate polysaccharides is discussed together with the type of enzyme which catalyzes its degradation.

The majority of published papers deal with enzymes produced by pathogens grown in culture rather than in vivo. In addition, the substrates used for determination of enzyme activity have often been chosen for their availability rather than for their exact similarity in structure to the polymers which can be obtained from the wall of the natural host plant. This approach has sometimes caused difficulties in applying the results from in vitro experiments to in vivo situations.

Pectic Enzymes It has generally been assumed that the pectic fraction of all cell walls consists of a linear polymer based on 1,4-linked α-galacturonic acid units (32). A polymer of this type is known as a pectic acid; one in which the carboxyl groups are completely methylated is known as a pectin, and one with only partial methylation of the carboxyl groups as a pectinic acid. However, it is now realized that the majority of pectic polysaccharides rich in polygalacturonic acid contain significant amounts of neutral sugar substituents and that pure galacturonans are of infrequent occurrence (33). L-Rhamnose occurs in the interior of galacturonan chains forming rhamnogalacturonans, whereas the other neutral sugars, the most important of which are galactose and arabinose, occur as arabinan and galactan side chains (33). The isolation of acidic and neutral polymers from pectins may well represent random degradation of pectins which contain an uneven distribution of neutral sugars in their structure. The acidic pectic polymer of the primary cell walls of sycamore cells grown in suspension cultures is a rhamnogalacturonan in which 1,2-linked rhamnose occurs as rhamnosyl-1,4-galacturonosyl-1,2-rhamnosyl units alternating in the polymer with between 6–12 residues of α-1,4-linked D-galacturonosyl units (34). It has been suggested that the galactosyl residues are present as a linear chain which is attached at its reducing end to the rhamnosyl residues of the rhamnogalacturonan main chain; the arabinosyl residues seem to be in the form of a branched chain (34).

For most of the studies on pectic enzyme production by plant pathogens, the substrates traditionally used have been high galacturonic acid pectins such as citrus pectin and its derivatives, which still seem to be called pectinic acids, pectic acids, and sodium polygalacturonate. Pectic enzymes which act on these substrates may be divided primarily into those which de-esterify pectin or pectinic acids to give either pectic or pectinic acids of lower methoxyl content (the pectin methyl esterases, PME) and those which catalyze the cleavage of the polygalacturonide chain (hydrolases and lyases).

A pectin methyl esterase has been purified and characterized from *Fusarium oxysporum* f.sp. *vasinfectum,* and it seems that about half the pectinesterase activity occurs at the reducing ends of pectin chains, whereas the rest of the activity occurs at some other locus or loci on the pectic chain (35).

The enzymes which catalyze the cleavage of polygalacturonide chains have a variety of names in the older literature, but the generally accepted classification is that of Bateman and Millar (36), who divided the enzymes into two groups according to the mechanism of cleavage of the α-1,4-glycosidic bond between the uronide residues. Hydrolases are those enzymes which cleave the bonds hydrolytically. Lyases (sometimes called *trans*-eliminases) catalyze a *trans*-eliminative cleavage. These two groups can be further subdivided into exo- and endoenzymes, depending upon the site of cleavage of the chain, i.e., whether terminal or random cleavage occurs. It should be pointed out that, whereas exohydrolases release monomeric galacturonic acid, the exolyases are unusual in that they attack at the reducing end of the molecule to release unsaturated dimers (37). The final subdivision depends upon the relative rates of cleavage of methylated polymers (pectins and pectinic acids) and methylated polymers. In order for pectin cleavage to be catalyzed by an enzyme specific for pectic acid, it is first necessary for the pectin to be demethylated by a pectin methyl esterase; the correlation between pectinesterase and exolyase attack on the reducing end of chains is therefore noteworthy.

The galactan component of pectins can also be hydrolyzed by fungal enzymes. These enzymes may be the major pectic enzymes of organisms pathogenic on plants which contain pectins with a high proportion of galactose. Chain-splitting activity with the use of apple pectin as a substrate could not be demonstrated in culture filtrates of *Phytophthora infestans,* the potato blight fungus (38). However, following the later demonstration that potato pectin contains more galactose than galacturonic acid (39, 40), culture filtrates of *P. infestans* were shown to contain an enzyme which would catalyze the hydrolysis of the β-1,4-galactans of both potato and white lupin pectins (39, 41, 42). Although the culture filtrates of *P. infestans* were found to have hydrolytic activity against citrus polygalacturonate, they did not release products containing galacturonic acid from potato cell walls (42). The *P. infestans* enzyme is an endogalactanase (41); enzymes with similar activity have been found in potato tubers infected by *Fusarium caeruleum* and by *Phytophthora erythroseptica* (43); exogalactanase activity is also present in *P. erythroseptica* rots (43).

An arabanase from *Schlerotinia fructigena* has been extensively purified (44). It acts on a pectic araban which is composed of both α-1,3- and α-1,5-linked arabinofuranose units (45) to produce free arabinose; because it has not been possible to determine its linkage specificity, it is called an α-L-arabinofuranosidase.

Hemicellulases and Cellulases Hemicelluloses are polymers which contain xylose and mannose as major sugar components, often combined with glucose or galactose. There are xylans that have chains of β-1,4-linked D-xylanopyranose units which may have short side chains of α-1,2-linked 4-*O*-methyl-D-glucuronic acid or β-1,3-linked arabinofuranose. Glucomannans are mixed polymers of randomly arranged β-1,4-linked D-glucopyranose and D-mannopyranose units; the polymers which also have 1,6-linked D-galactopyranose side branches are the galactoglucomannans. Similarly, the mannans are β-1,4-linked D-mannopyranose chains, and the galactomannans have 1,6-linked D-galactopyranose side branches (46). In the primary cell wall of the sycamore cells grown in suspension culture, the major hemicellulose is a xyloglucan which consists of a chain of β-1,4-linked glucopyranose with terminal branches of 1,6-linked D-xylanopyranose (47).

Enzymes which hydrolyze β-1,4-xylans to xylose are produced by a range of organisms including *Diplodia viticola* (48), *Sclerotium rolfsii* (49), and *Fusarium roseum* "Avenaceum" (50). *S. rolfsii* and other fungi also produce enzymes which hydrolyze β-1,4-mannans (49, 51).

The xyloglucan of cultured sycamore cell walls can be hydrolyzed by an endoglucanase from *Colletotrichum lindemuthianium* (47).

Cellulose, which consists of long chains of β-1,4-linked D-glucopyranose residues containing as many as 8,000–12,000 residues, occurs mainly as microfibrils in plant cell walls. Cellulose microfibrils consist of both crystalline and amorphous regions. The crystalline regions are postulated to arise from intra- and intermolecular hydrogen bonding of cellulose chains which are arranged in an antiparallel manner (52).

Cellulolytic enzymes or cellulases are enzyme complexes. One enzyme, called C_1, converts native, insoluble cellulose to linear chains which are then susceptible to hydrolysis catalyzed by a second enzyme, C_x, which is also able to catalyze the hydrolysis of various modified celluloses. C_x enzyme is apparently an endo-β-1,4-glucanase which yields cellobiose; this is hydrolyzed in turn by cellobiose to glucose (51, 53, 54).

Function of Cell Wall-degrading Enzymes in Pathogenesis

As has been indicated earlier, the bulk of the published information on cell wall-degrading enzymes has been obtained from pathogenic organisms grown in culture. In order to demonstrate that these enzymes are involved in the invasion and colonization of the host tissue by the pathogen, it is important that the enzymes be isolated from infected tissue although difficulties may arise if the host tissue produces similar enzymes. A number of such isolations which have been made are listed in Table 1.

In a few cases, the properties of the enzyme isolated from infected tissue are different from similar enzymes produced by that pathogen in vitro (57). It is also noteworthy that enzymes of more than one type may be found in the infected tissue, for example, in potato infected by the dry rot pathogen *F. roseum* "Avenaceum" (50). This fungus produces a similar range of enzymes when it is cultured on isolated potato cell walls.

Recent studies on isolated cell walls have emphasized that degradation of the rhamnogalacturonan moiety of the pectic fraction is important in the first stage of wall degradation. Karr and Albersheim (58) found that treatment with a "wall-modifying enzyme," which appeared to have polygalacturonate hydrolase activity, was necessary before many cell wall polysaccharide hydrolases could catalyze hydrolytic reactions on cell walls isolated from *Phaseolus vulgaris.* Treatment of isolated potato cell walls (which have a pectic fraction with a high galactose to galacturonic acid ratio) with a bacterial endopolygalacturonate lyase solubilized more than 50% of the galactose, rhamnose, and galacturonic acid, as well as 20 and 24% of the xylose and arabinose, respectively (59). When isolated potato cell walls were treated with a culture filtrate of *F. roseum* "Avenaceum" which contained in addition to endopolygalacturonate lyase, two arabanases,

Table 1. Cell wall-degrading enzymes found in infected plant tissue

Plant	Pathogen	Enzymes
Bean, *P. vulgaris* var red kidney (hypocotyls)	*R. solani*	Polygalacturonase Polygalacturonate lyase Xylanase Galactomannanase Arabanase Cellulose (55)
Bean, *P. vulgaris* var red kidney (hypocotyls)	*S. rolfsii*	Polygalacturonase (56) Galactanase (49) Xylanase Mannanase
Potato tubers	*P. infestans*	Galactanase (41) Polygalacturonase (42)
Potato tubers	*F. roseum* "Avenaceum"	Exopolygalacturonase (50) Endopolygalacturonate lyase Endo-β-1,4-galactanase Endoxylanase 2 Arabanases Endocellulase (C_x)

endo-β-1,4-galactanase, and xylanase, solubilized more than 65% of the galactose and galacturonic acid and 50% of the xylose (60). Thus, it seems that enzymes other than the endolyases or endohydrolases are not of great importance in the initial solubilization of wall polysaccharides.

The necessity for at least partial degradation of the rhamnogalacturonan fraction of the wall before enzymes which hydrolyze other polysaccharides can act on wall fragments is further established by several reports on the sequence of production of polysaccharide-degrading enzymes by pathogens. In all the cases so far reported, in which pathogens have been cultured on isolated plant cell walls, the first enzymes produced are those that degrade the rhamnogalacturonan polymers, followed by enzymes which hydrolyze the hemicelluloses; cellulase is the last enzyme to be produced (50, 61–63).

These findings emphasize the key role of the rhamnogalacturonan chains in the structure of the plant cell wall; this is particularly evident in the model structure proposed for the cell wall of sycamore cells grown in suspension culture (64, 65). Degradation products of the wall were obtained after partial hydrolysis by polysaccharide-degrading enzymes. Enzymatic degradation of the rhamnogalacturonan fraction had to be performed before the other hydrolases could hydrolyze their substrates. The wall and wall fragments were methylated, hydrolyzed, and converted to methylated alditol acetates, which were then analyzed by combined gas-liquid chromatography and mass spectrometry (34, 47, 64). In the model structure, it is proposed that the pectic and hemicellulose components are covalently linked to each other and that the cellular microfibrils are hydrogen-bonded to the xyloglucan hemicellulose. The model also contains the glycoprotein which is now generally accepted as a structural component of cell walls. The glycoprotein contains a high content of hydroxyproline and its sugars are arabinose and galactose; it is assumed that the arabinose is linked to hydroxyproline and the galactose to serine (66, 52).

Production of cell wall-degrading enzymes is inducible, usually by the monosaccharide which forms the basis of the polymer degraded by the enzyme. For example, polygalacturonase and polygalacturonate lyase are induced by galacturonic acid, and arabanase, galactanase, and xylanase are induced by arabinose, galactose, and xylose, respectively, in cultures of *Verticillium albo-atrum* and *Fusarium oxysporum* f.sp. *lycopersici*. In the case of cellulase (C_x), cellobiose acted as an inducer for both fungi (63). These experiments were carried out with the use of a system in which the rate of supply of the sugars to the medium could be varied. It was also found that if the rate of supply of inducer exceeded rather low values, catabolite repression of the rate of enzyme synthesis occurred.

The rate of production of cell wall-degrading enzymes appears to be significant in the development of pathogens in their hosts. Thus, the establishment of infection of *Rhizoctonia solani* on *P. vulgaris* can be varied by altering the supply of sugars which seems to regulate the production of cell wall-degrading enzymes (67).

Glucose repression of polysaccharidases appears to be common in many plant pathogens and has been demonstrated by in vitro studies (68, 69).

Host plants may also contain stimulatory factors which increase the production of cell wall-degrading enzymes. Polygalacturonase production by *F. oxysporum* f.sp. *lycopersici* is greatly stimulated by a low molecular weight carbohydrate fraction from tomato cell wall which has a high content of anhydrogalacturonic acid (61), and a heat-labile factor from potato enhances polygalacturonate lyase production in the bacterium *Erwinia carotovora* (70).

Maceration of Plant Tissue

Maceration of plant tissue is often caused by organisms producing soft rots, and it was suggested by de Bary as long ago as 1886 (71) that extracellular enzymes were involved in the process. The first demonstration of the role of polygalacturonate hydrolase enzymes in maceration coincided with the demonstraton of the pectic nature of the middle lamella (72–74). The demonstration that, in culture filtrates from *S. fructigena,* macerating activity against potato tuber discs could be separated chromatographically from polygalacturonase activity (75) and its possible identity with α-L-arabinofuranosidase (76) was resolved when it was found that in this case maceration was caused by pectin methyltranseliminase (77); macerating activity of the bacterium *E. carotovora* is caused by pectate transeliminase (78). Purified endopectin methyltranseliminase from culture filtrates of either *Penicillium italicum* or *Penicillium digitatum,* which are both pathogens of oranges, causes maceration of orange rind tissue (79), and two purified endopolygalacturonate hydrolase enzymes isolated from onion tissue infected with *S. cepivorum* macerate onion epidermal tissue (80).

Macerating enzymes, which also cause cell death and which are produced by the potato pathogens *F. caeruleum* and *P. erythroseptica,* have been isolated and purified from the dry rot and pink rot lesion tissue caused by these organisms. Although the most active macerating enzymes isolated were endopolygalacturonate lyases and hydrolases, maceration was also caused by the purified galactanases; the separate galactanases from *F. caeruleum* caused maceration, whereas those from *P. erythroseptica* would only cause maceration in combination (43).

Protoplast Death

De Bary (71) suggested that, in addition to causing maceration of tissue, soluble substances secreted in advance of the hyphae of pathogens also killed cell protoplasts. Brown (81) was unable to separate the factors responsible for cell death and maceration of plant tissue in an enzyme extract from spores of *B. cinerea.* Tribe (82) also found that culture filtrates of *B. cinerea* and *Bacterium aeroideae* would both macerate and kill plant tissues; however, protoplast death could be retarded if tissues were macerated in hypertonic solutions. Although many hypotheses have been proposed to explain cell death (51, 83), it is clear from recent experiments using highly purified pectate-degrading enzymes that endopolygalacturonate lyase from *E. carotovora* (84) and *Erwinia chrysanthemi* (59) or endopolygalacturonate hydrolase from onion tissue infected with *S. cepivorum* (80) cause cell death merely by degrading the cell wall of the test tissue to such an extent that the turgor pressure of the protoplasts causes them to rupture. However, pectic enzymes do not cause death of isolated protoplasts; proteinases and phospholipases may be involved (83, 84).

Cell Wall Degradation During Entry of Biotrophic Pathogens in Host Cells

Apart from the investigations on *P. infestans* which have shown that there is a loss of galactan from the pectic fraction during the early stages of pathogenesis (40, 85), which is presumably due to the galactanase produced by the fungus (39, 41, 42), there is relatively little information on cell wall degradation by biotrophic pathogens. Spores of *P. graminis tritici* can secrete endopolygalacturonase, hemicellulase, and cellulase during germination (86). Pectinase activity seems to be present in the bark of *Pinus monticola* infected by *Cronartium ribicola,* because there is a marked reduction in the amount of extractable pectic material in infected bark (87).

Electron microscopic examination of the early stages of penetration of lettuce by *Bremia lactucae* indicate that during penetration there is only a limited degradation of host cell wall material, just sufficient to allow the entry of the fungal penetration peg (88). Recent unpublished experiments by Shimony and Friend show that *P. infestans* seems to penetrate potato tuber cells in a similar manner.

EXTRACELLULAR TOXINS PRODUCED BY PATHOGENS

Several plant pathogens are known to produce extracellular toxins in infected plants which, when applied to uninfected host plants, will produce some or all of the characteristic symptoms of the disease. In a few cases, there is evidence that the ability to produce toxin is the major factor determining virulence toward specific hosts. Toxins are usually classified as host-specific and nonspecific. A host-specific toxin is defined as the metabolic product of a pathogenic microorganism which is toxic only to the host susceptible to that pathogen (89). The specificity of host-specific toxins is such that they usually display their toxic effects on susceptible, but not resistant, cultivars of the same species. Nonspecific toxins are those which will produce disease symptoms in a range of nonhost plants, even though they may be produced by a pathogen which is specific for its host (90).

Chemical identification of many toxins has been made in recent years. They include carbohydrates such as polysaccharides, oligosaccharides, lipomucopolysaccharides, glycopeptides, and glycosides; additional classes are terpenoids and amino acid-derived toxins (91). The mechanism of action of some host-specific and nonspecific toxins has been actively investigated in the past few years.

Host-specific Toxins

Three important host-specific toxins whose effects have been examined in detail are those from *Helminthosporium victoriae, Periconia circinata,* and *Helminthosporium carbonum,* parasites of oats, sorghum, and maize, respectively (92). The toxins are designated HV, PC, and HC toxins, respectively.

The initial biochemical lesion in the case of each of these host-specific toxins appears to be in the plasma membrane of the susceptible host cultivars even though the expression of the plasma membrane damage may manifest itself in a different manner for each disease. For example, HV toxin induces loss of electrolytes within 2 min (93), and PC toxin induces detectable leakage within 15 min (94). HC toxin has its earliest known effect within 2 hr, and an increase in uptake of nitrate and other solute losses of materials from tissues occur 12–24 hr later (95). The effects of these three toxins on electropotentials of their respective host plant cells have been measured. HV and PC toxins both caused gradual decreases in the negative electropotentials of single cells, whereas HC toxin caused a rapid but transient increase in the negative electropotential. It was concluded, therefore, that HV and PC toxins affect passive efflux of ions, or electrically neutral ion exchange systems, across the plasma membrane; HC toxin possibly has a direct effect on the electrogenic pumps (96).

Scheffer and Yoder (92) suggested that resistant cultivars of the host plants of these three fungi either lack a receptor or sensitive site for the toxin or have a receptor with a considerably lower affinity for toxin. No specific receptor for HV, PC, or HC toxins has yet been found.

Recent proposals that a specific receptor exists for a host-specific toxin in sugar cane plants susceptible to eyespot disease caused by *Helminthosporium sacchari* have been made by Strobel, who reviewed his findings recently (91, 97, 98). The toxin helminthosporoside, which is probably 2-hydroxycyclopropyl-α-D-galactopyranoside, was found to bind to a receptor protein in plasmalemma of susceptible cells. Resistant sugar cane plants have an immunologically identical protein which does not, however, bind helminthosporoside. These two proteins have the same molecular weight and both have 4 subunits; however, they differ in 4 amino acid residues and in their mobility in gel electrophoresis. It appears that the protein functions in α-galactoside transport into the cells of susceptible cultivars. Because there is no α-galactoside transport into resistant cells, interference with this function is not the reason for the toxicity of helminthosporoside. Strobel suggests that by binding to the receptor there is a perturbation of the lipid layer of the membrane which in turn activates the KMg-ATPase. Such an activation would upset the ion balance of the cell.

A host-specific toxin is produced by race T of *Helminthosporium maydis,* which specifically causes southern corn leaf blight on male sterile maize plants that contain the Texas cytoplasm male sterility gene (T-cms plants (99–101)). The pathotoxin specifically affected several reactions of mitochondria isolated from etiolated shoots of T-cms maize, but it had no effect on mitochondria isolated from maize plants with normal cytoplasm (102). The effects included inhibition or stimulation of respiration depending upon the substrate, reduction of respiratory control, P:O ratios, and irreversible swelling of mitochondria in KCl medium.

A more detailed examination of these effects shows that uncoupling of oxidative phosphorylation occurs earlier than inhibition of malate oxidation (103). The site of the inhibition is in the first complex of the mitochondrial electron chain, and it seems likely that there is a specific toxin-binding site on the inner mitochondrial membrane (104). It has been suggested that other membrane components of T-cms cells, especially the plasma membrane, may also be affected by the toxin (105, 106); however, it seems that the primary effect is on the mitochondria, which seem to be more sensitive than the plasma membrane functions to low concentrations of the toxin. In addition, there is evidence that, although the *H. maydis* race T toxin increases the efflux of ^{86}Rb from maize leaves, the effect is not host-specific (107), and it has recently been shown that there is no difference in the effect of the toxin on purified K^+-ATPase isolated from microsomes of roots of maize plants resistant or susceptible to *H. maydis* race T (108). Mitochondrial inhibition by *H. maydis* race T toxin occurs so rapidly that it can be used as a simple biochemical assay for T-cms cytoplasm in maize (109).

Phyllosticta maydis, which causes yellow leaf blight of maize, produces a toxin in culture which is also specific for T-cms maize plants (110, 111). Its effects on mitochondria from etiolated T-cms maize shoots were identical with those of the *H. maydis* toxins, namely, uncoupling of oxidative phosphorylation, stimulation or inhibition of oxygen uptake depending upon the substrate, and irreversible swelling (110).

Nonspecific Toxins

Rhizobium japonicum, which is the symbiont of root nodules of soybean plants, produces a toxin, rhizobitoxine, which can cause chlorosis in soybeans (112). Its structure has been determined as 1-amino-4-(2-amino-3-hydroxypropoxy)-*trans*-but-3-enoic acid (113).

Rhizobitoxine was found to inhibit the growth of *Salmonella typhimurium;* the inhibition could be overcome by methionine and homocysteine but not by cystathionine and homoserine. Rhizobitoxine inhibits the β-cystathionase of *S. typhimurium,* the transsulfurase which catalyzes the cleavage of cystathionine to homocysteine, pyruvate, and ammonia (114); it also inhibits β-cystathionase isolated from spinach leaves (115). Treatment of spinach or corn seedlings with rhizobitoxine reduced the specific activity of β-cystathionase to about 30–40% of that in control seedlings during the first few hours after administration of the toxin. There was an accompanying accumulation of cystathionine. It was not clear that there was sufficient deficiency of methionine to account for the chlorosis although this might have been caused by a localized deficiency (116). Other pathways of methionine metabolism may be inhibited because rhizobitoxine also inhibits ethylene production and particularly the incorporation of ^{14}C from [U-^{14}C]methionine into ethylene in apple tissue (117). This latter inhibition may be related to a pyridoxal phosphate requirement in ethylene biosynthesis from methionine because rhizobitoxine inhibition of spinach β-cystathionase is related to pyridoxal phosphate binding by the enzyme (118).

Phaseotoxin is the extracellular toxin produced by *Pseudomonas phaseolicola,* the bacterium which produces halo blight of beans (*P. vulgaris*) (90) and which will produce the chlorotic symptoms characteristic of the disease. The toxin contains L-serine, but several details of its structure are unclear (90). There is considerable evidence that phaseotoxin acts as a reversible specific inhibitor of the enzyme ornithine carbamoyl transferase (carbamoyl phosphate: L-ornithine carbamoyl transferase, EC 2.1.3.3) (OCT), the enzyme that catalyzes the carbamylation of ornithine into citrulline.

There is both accumulation of ornithine and a reduction of the level of arginine in plants infected by *P. phaseolicola* (118, 119); similar effects, as well as chlorosis, are found in toxin-treated plants (120). Partially purified phaseotoxin inhibited ornithine carbamoyl transferase; citrulline pretreatment of plants inhibited toxin-induced chlorosis (121). By using more highly purified toxin, it was found that OCT was the only enzyme of the ornithine cycle inhibited, that both citrulline and arginine would relieve chlorosis, and that inhibition of OCT was by allosteric competitive inhibition in relation to carbamoyl phosphate and noncompetitive inhibition in relation to ornithine (122). Plants treated with purified toxin do not accumulate ornithine but do show chlorosis (123); it is likely, therefore, that ornithine can be metabolized by other pathways. However, it is believed by Patil et al. (123) that chlorosis is related to inhibition of citrulline synthesis, which in turn is necessary for chlorophyll synthesis, perhaps by being required for the chlorophyll-synthesizing enzymes.

One feature of the bacterial toxins such as rhizobitoxine and phaseotoxin is that although the toxins themselves are not host-specific, the bacteria themselves are host-specific in the sense that there are host plants which are either resistant or susceptible to the bacteria. This is in contrast to those toxin-producing fungi which are host-specific and whose toxins are also host-specific. Because the host-specificity of the diseases caused by these bacteria cannot be explained in terms of the host-specificity of the toxin, other explanations have been offered.

In the case of phaseotoxin, for example, pretreatment of leaves of a resistant cultivar of the host plant (*P. vulgaris*) with phaseotoxin before inoculation with *P. phaseolicola* seems to overcome the resistance because bacteria multiply to the same extent as in leaves of a genetically susceptible plant (124). The mechanism for the suppression of resistance by pretreatment with phaseotoxin is explained by Patil and Gnanamanickam (125) as a suppression of production of the phytoalexins, the antimicrobial compounds which are synthesized by *P. vulgaris* after infection. They were also unable to detect phaseotoxin in inoculated resistant tissues of bean and, therefore, suggested that the factor which determines the resistance of bean tissues to *P. phaseolicola* is related to suppression of the production of phaseotoxin. Patil (90) suggested that bacterial toxin production in host tissues could be under regulatory control, and he gave as an example the finding that rhizobitoxine is produced in greater concentrations in nodules of susceptible plants than in those of resistant plants, even though growth of the bacterium in the nodules is the same (112).

EFFECTS OF PATHOGEN ON METABOLISM IN HOST PLANTS DURING ESTABLISHMENT OF BIOTROPHIC INTERACTIONS

Infection by biotrophs causes marked changes in the metabolism of the host plant. Many of these changes are necessary for the diversion of nutrients to the pathogen.

Carbohydrate metabolism has been extensively investigated and attention has been particularly devoted to changes in photosynthesis and respiration, carbohydrate movement, particularly from host to parasite, and the form in which carbohydrate is accumulated in the parasite.

Alteration in Photosynthesis and Respiration of Diseased Plants

Infection of host leaf tissues by a biotrophic fungus eventually results in the impairment of photosynthetic ability of the host tissue (54, 126); a wide range of pathogens causes an increase of respiration after infection (127). Although there are reports that the decline in photosynthesis may initially be preceded by a rise in the photosynthetic rate (128, 129, 130), the initial rise may have been an artifact caused by high $^{14}CO_2$ concentration used for measurement of photosynthesis (131). In barley leaves, infection by *Erysiphe graminis* (powdery mildew) causes a biphasic inhibition of photosynthesis at physiological concentration of $^{14}CO_2$; the suggested explanation for the first phase of inhibition is either by an inhibition of glycollate oxidation or partial closure of stomata (131).

However, other mechanisms may be involved, such as the inhibition of photophosphorylation found in chloroplasts isolated from leaves of broad bean (*Vicia faba*) infected with the rust *Uromyces fabae.* The primary inhibition is of electron transport from water to NADP or ferricyanide; there is no effect on either ATP:2*e* ratios or on cyclic photophosphorylation (132).

Increase in respiration is probably due to an alteration in pathways of carbohydrate metabolism because it is clear in many cases of rust infection and in powdery mildew infection that there is a decrease in $C_6:C_1$ ratios (133, 134) as well as an increase in the activities of glucose-6-phosphate dehydrogenase and 6-phosphogluconate dehydrogenase (130).

In the case of barley leaves infected with powdery mildew, it has been suggested that the increased rate of respiration is linked to the decline in photosynthesis because the increase in respiration and in the activities of the pentose phosphate pathway dehydrogenases following infection only occur in green plants and not in susceptible etiolated plants (130). In infected leaves, there is some chloroplast breakdown, and $NADP^+$ moves from inside the chloroplasts to the cytoplasm; it is postulated that the increased concentration of $NADP^+$ in the cytoplasm could be responsible for the increased activity of the pentose phosphate pathway (135).

Infection of young rice plants with the necrotrophic fungus *Pyricularia oryzae* decreased the net rate of photosynthesis. In addition, there was no effect upon either the rate of respiration of leaves or the pattern of $^{14}CO_2$ production from 1-[^{14}C]-, 2-[^{14}C]-,3,4-[^{14}C]-, and 6-[^{14}C]glucose supplied to leaf segments. It was, therefore, concluded that *P. oryzae* infection did not lead to any marked change in the carbohydrate metabolism of the leaf as a whole (136).

Effects of Infection on Carbohydrate Translocation

There is considerable evidence that fungal infection of single leaves, especially with rusts and powdery mildews, alters the patterns of translocation of newly photosynthesized assimilates (137). Translocation after $^{14}CO_2$ feeding has been measured in wheat plants infected with *Puccinia striiformis* (129), bean infected with *U. phaseoli* (138), and barley infected with *E. graminis* (139). Infected leaves export less ^{14}C-labeled assimilate than healthy ones. Whereas in wheat infected leaves could not attract assimilate from other healthy leaves, the situation was markedly different in bean when a marked reversal of the polarity of transport occurred in older infected leaves which received assimilate at the expense of the youngest leaves. The differences between wheat and bean were explained by Thrower and Thrower (140) on the basis of the possible routes of translocation in mono- and dicotyledons. The results with barley confirm this hypothesis (139). Translocation in the mature primary leaf of barley is strongly basipetal (from tip to base). A mildew colony in the middle 30-cm section of the primary leaf can stop translocation of ^{14}C fixed only in the infected area and also in uninfected parts which are acropetal (nearer to the tip than the infection). There is no import of assimilate from the second leaf. However, when there is general infection of wheat plants by *P. striiformis* (141), there is an alteration of translocation from single leaves; the proportion of translocate to the roots is lowered and that going to the leaves—particularly from the second, sixth, and seventh leaves—is raised.

The infection of coltsfoot (*Tussilago farfara*) by *Puccinia poarum* causes alterations of translocation similar to those of rusted bean, namely, a decrease in carbohydrate export from infected mature leaves and, when there is heavy infection of these leaves, an enhanced capacity to import carbohydrate (142). After a 15-min exposure to $^{14}CO_2$ in the light, followed by 24 hr in the dark, there was marked accumulation of ^{14}C in the region of the pustule, part of which is due to degradation in the dark of ^{14}C-labeled starch. In contrast, there was no evidence in potato leaves infected by *P. infestans* that movement of carbohydrate from uninfected to infected tissue was a significant factor in the accumulation of ^{14}C-labeled assimilate in the colonized tissue surrounding the necrotic regions (143). The enhanced ^{14}C fixation in infected tissue was associated with the permanently wide stomata at the infection sites; accumulation resulted from retention of assimilates at the infection site. Farrell (143) suggests that, whereas a rust or powdery mildew, which has a relatively long symbiotic relationship with its host, may need a continual supply of photosynthate, in the case of more transient infections caused by *P. infestans* the carbohydrate requirements would be smaller.

Although it has been claimed that alteration of cytokinin levels in diseased plants alter patterns of translocation to the advantage of the fungus (2), Sequeira (144) has pointed out that many of the assay methods used are not specific for cytokinins and are, therefore, not entirely reliable.

Carbohydrate Metabolism in Infection Site

In many infected plants, the formation of the characteristic carbohydrates found in most fungi (except phycomycetes), trehalose and the polyols, especially mannitol and arabitol (137), is accompanied by an accumulation of host carbohydrates (145). When *P. vulgaris* leaves infected by the bean rust fungus *U. fabae* incorporate $^{14}CO_2$, almost three-quarters of the radioactivity in the soluble fraction was found in the trehalose, mannitol, and arabitol, with the remainder in sucrose, glucose, and fructose. These latter three sugars accounted for four-fifths of the radioactivity in the soluble fraction in healthy leaves (128). It appears that host sugars are taken up by the fungus and then converted to the fungal carbohydrates. In *Tussilago* leaves infected with *P. poarum,* the major soluble sugars formed during photosynthesis are sucrose and, especially in host tissues around the pustules, glucose and fructose (142). At the infection site, photosynthetic products are converted to specific fungal products including mannitol, arabitol, trehalose, glycogen, a glucomannan component of the cell wall, and lipids. Host fructan also accumulates in infected leaves (146). There is an accumulation of invertase in infected *Tussilago* leaves which is correlated with the high levels of reducing sugars also found (147).

Long et al. (147) propose that sucrose is translocated to the pustule and is then hydrolyzed by the invertase, which is probably located in cell walls of both host and fungus, before uptake as hexose into host and fungal cells. Whereas the fungus will convert the hexose into fungal sugars, in host cells the hexose will either stay as reducing sugar or be converted to fructan. A similar role for invertase is also proposed for the accumulation of trehalose in the phycomycetous pathogen *Albugo tragopogonis* on *Senecio squalidis* leaves. This host-parasite combination is another in which starch accumulates in photosynthetic cells around the pustule (148), and it is proposed that invertase is involved in the absorption of carbohydrate into these cells before conversion to starch (147). Starch accumulation in wheat leaves infected by *P. striiformis* is determined by the activity of ADP-glucose pyrophosphorylase, which seems to be controlled by the inorganic phosphate levels (149).

Effects of Infection on Nitrogen Metabolism

The accumulation of amides is another metabolic change in wheat leaves following infection with stem rust (*P. graminis*) (150, 151). Asparagine, glutamine, and glutamate increase, but aspartate decreases in barley leaves in the first 4 days after infection with *E. graminis* (152); there is an accompanying accumulation of ammonium ions and increased ammonia production. There are increases in both NAD^+- and $NADP^+$-dependent glutamate dehydrogenases and of glutamic and asparagine synthetases. Similar increases in glutamine synthetase occur in wheat leaves following infection with stem rust (150).

The incorporation of ammonium ions into amino acids by infected leaves is probably limited by the supply of α-ketoglutarate. Sadler and Scorr (152) suggest that the increased level of ammonium ions is caused by an increase in the level of amide hydrolysis; the concurrent accumulation of amides is explained on the basis of compartmentation. The accumulation of amides appears to require a source of "exogenous ammonium" which is different from the separate pool of ammonium used for glutamate synthesis.

Nucleic Acid Metabolism of Infected Host

The changes in nucleic acid metabolism which occur particularly in biotrophic interactions have recently been described in detail (153).

There is a significant increase in total RNA content of plants infected with either rust or mildew fungi compared with uninfected controls. A substantial portion of this RNA seems to be fungal in origin in mildewed wheat (154) and in bean plants infected with the rust *U. fabae* (155). However, the change in RNA content may be less important than alterations in the rate of RNA synthesis which increases significantly in both host and parasite in rusted wheat leaves 6 days after infection (156); in flax cotyledons, the increase in RNA synthesis which occurs within 48 hr of infection by the rust *Melampsora lini* appears to be in the host rather than in the fungus (157).

Chloroplast RNA synthesis seems to be particularly affected by powdery mildew infections. In cucumber leaves infected by *Erysiphe cichoracearum,* there is a retardation of synthesis of chloroplast ribosomal RNA and an acceleration of the rate of senescence (158). In susceptible barley leaves infected with *E. graminis,* there is a loss of both chloroplast RNA and chloroplast ribosomes (159); the polyribosome content is reduced as early as one day after inoculation (160). These changes could have been caused either by an impairment of photophosphorylation or a decreased availability of amino acids in addition to changes in ribosome metabolism or changes in the synthesis and degradation of chloroplast RNA.

Callow (153) summarizes evidence that there is also a small, early increase in the synthesis of RNA in susceptible cereal leaves infected by rust fungi. He also points out that it may not be indicative of synthesis of new protein required for the establishment of the biotrophic interaction, but could be a secondary consequence of infection resulting from changed hormone levels in diseased tissue.

Accompanying the increases in RNA synthesis, there are also increases in RNase, which have been analyzed in some detail (161). Two types of RNase are found at different stages of the infection sequence; they are named "early" and "late," respectively. The "early" RNase occurs in both compatible and incompatible interactions, whereas the "late" RNase appears to be specifically confined to compatible interactions. However, the role of these two enzymes has not yet been elucidated.

RESISTANCE OF PLANT TO PATHOGENIC ORGANISMS

Resistance of plants to pathogens is far more common than susceptibility, and it seems likely that all plants have resistance mechanisms which protect them from attack by many microorganisms. This is presumably the reason why most pathogens have a relatively restricted range of host plants.

Investigations into the mechanism of disease resistance have shown that chemical mechanisms of defense may be divided broadly into two categories. One includes antimicrobial substances and inhibitors of cell wall-degrading enzymes of pathogen that are present in plants before inoculation either in a free or combined state; the other depends upon the synthesis of antimicrobial substances after infection or the chemical modification of host cells so that the pathogens can either no longer attack the host or obtain nutrient from it.

Role of Inhibitors Present in Plants Before Infection

The term "inhibitors" is used to include antimicrobial substances present in plants before infection, such as compounds specific to a particular plant or group of plants (162) and compounds which have a wide distribution in the plant kingdom, yet which reach very high and potentially antimicrobial levels in some plants. In some cases, the presence of these antimicrobial compounds has been evoked as an explanation of disease resistance. Other inhibitors are not antimicrobial compounds but inhibit cell wall-degrading enzymes of pathogens; such inhibitions have recently been postulated to have a role in resistance.

Antimicrobial Compounds Found in Specific Plants Oats which are resistant to the fungus *Gaeummanomyces graminis* contain an antifungal substance, avenacin, which is detoxified by an enzyme avenacinase from culture filtrates of *G. graminis* var *avenae,* which can parasitize oats (163). Avenacin, which was isolated from oat roots (164) was found to be a potent inhibitor of *G. graminis* and 15 other microorganisms at a concentration of 3–50 μg/ml. It was determined that the structure (165) is:

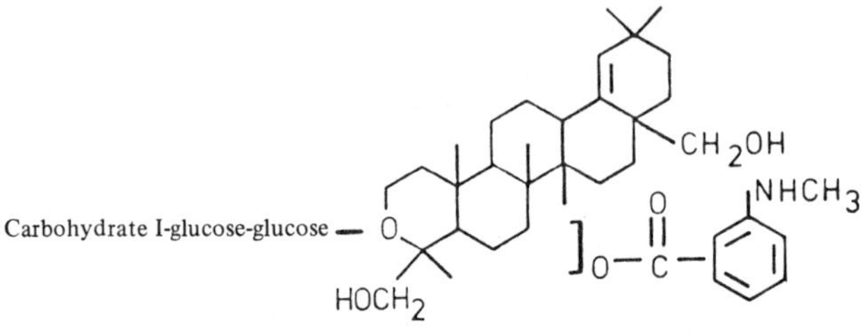

Carbohydrate I-glucose-glucose — O HOCH2 CH2OH NHCH3 O O C

Avenacin

and it was suggested that the detoxification of avenacin by avenacinase is by removal of the carbohydrate I portion from the remainder of the molecule. It has been proposed that terpene glycosides such as avenacin cause membrane disruption by binding to sterols; the difference between sensitive and insensitive fungi is related to the content of ergosterol in the fungal plasma membrane (166, 167).

Five-day-old barley coleoptiles contain a series of antifungal compounds, the hordatines (168), which were characterized as hordatine A, hordatine B, and the hordatines M (169).

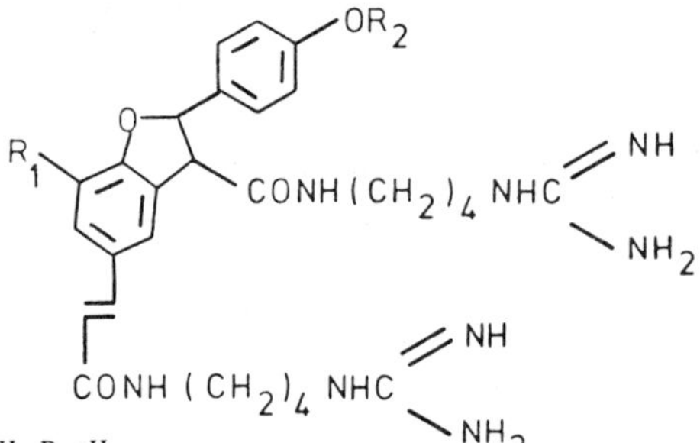

OR2 R1 O CONH(CH2)4 NHC NH NH2 CONH (CH2)4 NHC NH NH2

Hordatine A R_1=H R_2=H
B R_1=OMe R_2=H
Hordatines M R_1=H R_2=α-D-glucopyranosyl
R_1-OMe R_2=α-D-glucopyranosyl

The presence of hordatines explains the resistance of young barley seedlings to *H. sativum;* their antifungal activity is not expressed in older seedlings because of the presence of an inhibitor, which may be Ca^{2+} ions because the addition of a chelating agent can restore antifungal activity. The antifungal activity of the hordatines is an inhibition of spore germination rather than of mycelial growth (170). The chemically related *p*-coumarylagmatine has also been isolated from barley, but has only low antifungal activity (171).

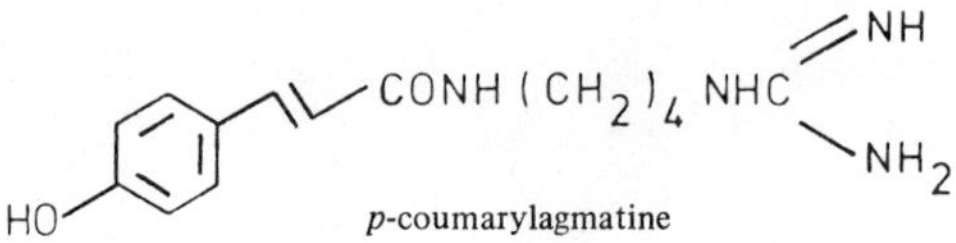

HO CONH (CH2)4 NHC NH NH2

p-coumarylagmatine

Antifungal compounds have been isolated from cereal and were identified as 2(3)-benzoxazolinone from rye and 6-methoxy-2(3)-benzoxazolinone from maize and wheat. However, it was later found that the benzoxazolinones were artifacts of the isolation procedure and that they exist in the plants as the 4-*O*-glucosides of 2,4-dihydroxy-1,4-benzoxazin-3-one and 2,4-dihydroxy-7-methoxy-1,4-benzoxazin-3-one in rye, maize, and wheat, respectively (172). When the plants are crushed, an enzyme, presumably a glucosidase, hydrolyzes the glucoside to its corresponding aglycone, which is unstable; it then breaks down to the corresponding benzoxazolinone.

enzymatic hydrolysis

R=H – Rye
R=OMe – Wheat and maize

It has been suggested that, in some varieties of wheat, resistance to stem rust (*P. graminis* f. sp *tritici*) may depend upon the breakdown of the glucoside, which has very little antifungal activity, to the more highly antifungal aglycone and benzoxazolinone (173, 174).

Resistance of tulip pistils to *B. cinerea* and susceptibility to *Botrytis tulipae* have been explained on the basis of hydrolysis of pre-existing glucosides known as tuliposides. *B. cinerea* releases a larger quantity of tuliposides because it causes a greater loss of permeability in the pistils. In addition *B. cinerea* converts tuliposides to the highly fungitoxic lactones, whereas *B. tulipae* only forms the less fungitoxic acids (175).

$HOCH_2CH(R).C(=CH_2)-C(=O)-CO.Gl$ —B. tulipae→ $HOCH_2CH(R).C(=CH_2)-C(=O)-COOH$

B. cinerea

Tuliposide A r=H
B R=OH

The diterpenes sclareol and episclareol have been isolated from leaves of *Nicotiana glutinosa* (176). Although a mixture of the two compounds inhibits the radial growth on agar colonies of several fungi, it does not inhibit spore germination in vitro. However, germination of spores of the rust fungi *Uromyces appendiculatur, Uromyces viciae-fabae,* and *Puccinia recondita* f.sp. *tritici* is inhibited by a sclareol-episclareol mixture; preliminary experiments suggest that spraying plants with such a mixture can be used to control rust on French bean, broad bean, and wheat (177).

Fungitoxic compounds which are present in wood include pinosylvin, the mansonones, and other sesquiterpenes. These have been reviewed recently (162).

Antimicrobial Compounds of Wide Distribution in Plant Kingdom The best known example of antimicrobial compounds which are not restricted to the host species is that of phenolic compounds in the onion, *Allium cepa* (178); phenolic compounds are present in a resistant host and, in that species, seem to be responsible for resistance to a pathogen. Onion varieties with colored outer scale leaves are resistant to attack by the fungus *Colletotrichum circinans,* which causes smudge disease on onion varieties with colorless scale leaves. Although the spores of *C. circinans* do not germinate in drops of water on dry scales of colored onions, they will germinate on the scales of colorless varieties. The colored scales contain catechol and protocatechuic acid, which seem to diffuse into the drops and prevent spore germination.

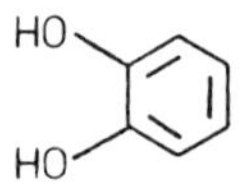

HO HO COOH

HO HO

Catechol 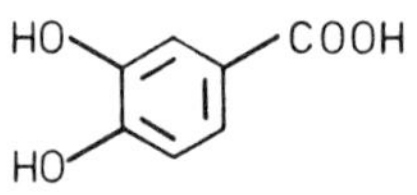 **Protocatechuic acid**

If the outer scales of the colored onions which contain these phenolics are removed, the inner fleshy scales can be infected as readily as the fleshy scales of colorless bulbs.

A closely related compound, 3,4-dihydroxybenzaldehyde, is found in skins of green Cavendish bananas; it inhibits the germ tube growth of *Gloeosporium musarum,* a fungus which causes a rot in the banana as it ripens and as the level of fungistatic activity declines (179).

Chlorogenic acid, a quinic acid ester of caffeic acid, occurs widely in plants; it has been proposed many times that it acts as an antifungal agent. It is the major phenolic compound in potato tubers (180), and a correlation has been obtained between the chlorogenic acid content of various potato organs and their resistance to such pathogens as *Streptomyces scabies* (181), *Verticillium albo-atrum* (182, 183), and *P. infestans* (184, 185). However, Kuć (186) has pointed out that the levels of chlorogenic and caffeic acids could not necessarily account for all the inhibitory activity of potato extracts against a range of fungi. In addition, it has been reported that growth of *P. infestans* is stimulated by 0.1–0.2% chlorogenic acid (187).

There is evidence that the oxidation state of phenolic compounds is important in their function as antimicrobial agents and that, after oxidation, they may inhibit enzymes produced by pathogens. Spore germination of *S. fructigena,* which causes brown rot of apple, was little inhibited by naturally occurring phenols at concentrations of the order of those encountered in fresh fruit (188). A similar range of phenols had little effect on either macerating activity or endopolygalacturonase of culture filtrates of *S. fructigena*. However, after oxidation of the phenols, either by autooxidation or by phenolase, they had marked inhibitory effects on both these enzyme activities. The most effective inhibitory compounds, after oxidation, were *d*-cathecin, *l*-epicatechin, and leucocyanidin (from cacao). Inhibition of the polygalacturonase activity of *Rhizoctonia solani* by oxidized catechin appears to be involved in the resistance of cotton plants to this pathogen (189). However, not all the catechin may have been present in the plant before infection; some of it appeared to have been synthesized after infection.

There is now evidence that in many higher plants phenols are stored in the reduced state either in the vacuoles of all cells or in specialized cells that are randomly scattered throughout the tissue (190, 191). Damage of the tonoplast membrane or plasmalemma by a pathogen will release phenols which can then be oxidized by phenolases in the host to form inhibitory quinones.

The implication of these reports of the inhibition of cell wall-degrading enzymes by oxidized polyphenols is that the inhibition is due to a "tanning" of the enzyme, that is, hydrogen bonding of the oxidized phenol to active sites on the enzyme. However, another possible mode of action is that polymerized phenolic oxidation products form a protective seal on the pectic substrates of the cell wall. Recent experiments have shown that artificial membranes of calcium pectate-pectin are disrupted by oxalate (a known synergist involved in cell wall degradation by *S. rolfsii* (56)), but after infusion with the oxidation products of 3-hydroxytryptamine they are resistant to oxalate degradation (192). Furthermore, infusion of membranes with phenolic oxidation products makes them far less susceptible to degradation by *F. oxysporum* f.sp. *cubense*. Infusion of natural perforation plates in banana root vessels with similar phenolic oxidation products also makes them resistant to oxalate degradation. It is postulated that vascular browning, following infusion of infection sites with oxidized phenolics, may both localize the infection by vascular wilt pathogens and insulate the infection from surrounding healthy tissue (192, 193).

Melanization of cell walls in potato tubers infected by potato virus X has been suggested as the reason for the greater resistance of these tubers to *F. roseum* "Avenaceum" than virus-free tubers. The virus-infected tissue contains more phenolase than virus-free tissue; melanization by phenol oxidation would make the cell walls more resistant to the polysaccharide-degrading enzymes of the pathogen (60).

Inhibition of Cell Wall Degradation by Extracellular Enzymes of Pathogens It has been postulated by Albersheim et al. (194) that varietal resistance to pathogens might be determined by variations in the chemistry of cell walls which would in turn cause differential induction of the polysaccharide-degrading enzymes produced by the pathogen. However, it is now apparent that there is not sufficient variation between the cell wall structure of different plant species to support this hypothesis; apart from differences between monocotyledonous and dicotyledonous plants, all primary cell walls are very similar (195). In addition, cell walls from stems of tomato cultivars which are either susceptible or resistant to *F. oxysporum* f.sp. *lycopersici* are equally efficient as inducers of cell wall-degrading enzymes by the pathogen. There is no difference in the sequence of production of enzymes between cell walls from either cultivar (61).

Albersheim and Anderson-Prouty (195) still consider that control of cell wall degradation may be part of a general mechanism of resistance and suggest that proteins which have been isolated from a range of host plants and which specifically inhibit endopolygalacturonases would be involved. These protein inhibitors have been isolated from cell walls of bean hypocotyls, tomato stems, and suspension-cultured sycamore cells (196). The purified protein from red kidney beans was a far more effective inhibitor of the endopolygalacturonase of *Colletotrichum lindemuthianum* than of *F. oxysporum* and had no effect against the enzyme from *S. rolfsii,* although the crude ammonium sulfate fraction was active. The kidney bean protein also inhibits the endopolygalacturonase of *Aspergillus niger* (197), and the possibility that such proteins could both have a wide spectrum of inhibitory activity and be widely distributed in plants has been raised (195). Their importance in inhibition of pathogenesis is related to the findings that rhamnogalacturonan degradation is the first stage in cell wall degradation (58–60), and endopolygalacturonase is the first cell wall-degrading enzyme to be produced by several plant pathogens (61–63). In addition, the inhibitor from sycamore cells almost completely inhibited the solubilization of tomato stem cell walls by a mixture of *F. oxysporum* extracellular enzymes (61).

Presumably these inhibitors will not be effective against those pathogens in which either exopolygalacturonase or endopolygalacturonate lyase are the major rhamnogalacturonan-degrading enzymes.

Metabolic Responses of Infected Plants Which are Involved in Inhibition of Pathogens

Production of Specific Antimicrobial Compounds or Phytoalexins The term "phytoalexin," which was first used by Müller and Börger (198), is applied to an antifungal compound produced by the metabolic interaction of a fungal parasite and a host plant. Phytoalexins were postulated to be the substances responsible for the inhibition of fungal development in hypersensitive tissues. The original experiments of Müller and Börger were on potatoes and *P. infestans;* isolation and chemical identification of a potato phytoalexin were not made until 1968 (199, 200). However, Müller (201) was able to demonstrate that when pods of French beans (*P. vulgaris*) were infected with spore droplets of *Monilinia fructicola,* a fungus which is not a normal pathogen of beans, phytoalexin activity diffused into the infection droplets. The work was extended to pea pods by Cruickshank and Perrin (202), and the active principle was isolated, crystallized, and identified as pisatin (203); the substance produced by bean pods was then isolated as phaseollin and its structure determined (204, 205).

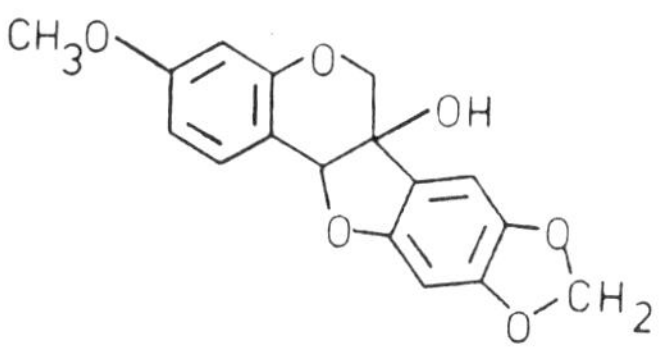

CH_3O OH O O CH$_2$ O

Pisatin

HO O O O CH$_3$ CH$_3$

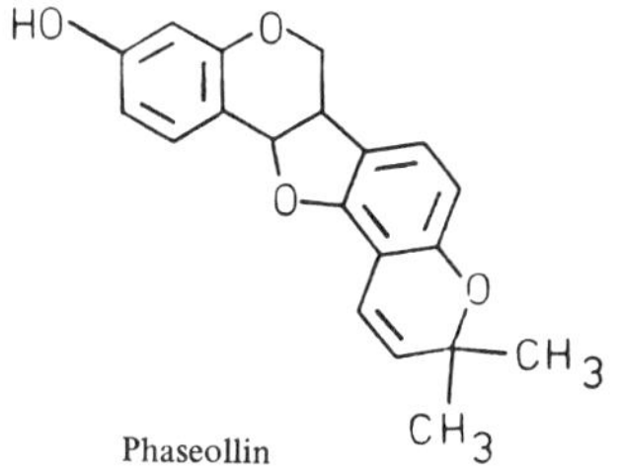

Phaseollin

Phytoalexin production in the Leguminosae has been intensively investigated and has proved fruitful, because it is now known that a wide range of phytoalexins is produced; these have been listed in a recent review (206). It is interesting that the majority are isoflavonoids and in particular many are pterocarpans (207); the only nonmembers of this class are wyerone and wyerone acid, which occur in *V. faba* after infection. A range of terpenoid phytoalexins occurs in the Solanaceae (208). The structures of those, other than pisatin and phaseollin—which have been found in the French bean, soybean, alfalfa, broad bean, and potato—are given in Table 2.

There have been two major technical methods which have helped in the identification of phytoalexins. One of these is the so-called "drop-diffusate" method used by Müller (201) for showing the production of phytoalexins; a spore suspension is applied to the host tissue and after incubation the liquid is collected, centrifuged, and tested for activity. In the case of pisatin, this can be isolated from the diffusate by extraction with light petroleum. Cruickshank and Perrin (220) point out that in addition to the simplicity of the technique, it uses the plant cell membrane as the primary "filter" in the extraction procedure and avoids the possibilities of the technique itself producing artifacts.

The second is the direct screening of chromatograms, particularly thin layer chromatographic plates, for antifungal activity by spraying them with fungal spore suspensions and then observing zones of inhibition on the chromatogram as the areas where the fungus has not grown. The usual test organism used is *Cladosporium cucumerinum* (221). This technique has been used particularly in the demonstration that phytoalexins are produced in leguminous plants as a response not only to fungal, but also to viral, infections (221, 222) and has also been used to screen plants for pre-existing antifungal compounds (176).

The implications of the phytoalexin hypothesis are that resistance to pathogens can be explained on the basis of high levels of phytoalexins after infection of resistant plants. Invasion of the plants by a virulent pathogen would occur only if that pathogen were insensitive to the phytoalexin, could degrade it to a less toxic metabolite, or could suppress phytoalexin synthesis. In addition,

Table 2. Phytoalexins found in French bean, soybean, alfalfa, broad bean, and potato

French bean (*P. vulgaris*)

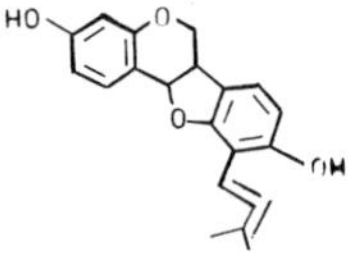

HO O O OH HO O HO O CH_3 CH_3

Phaseollidin (209) 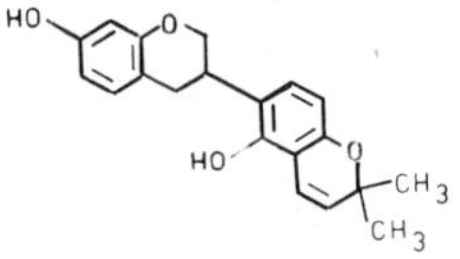 Phaseollinisoflavan (210)

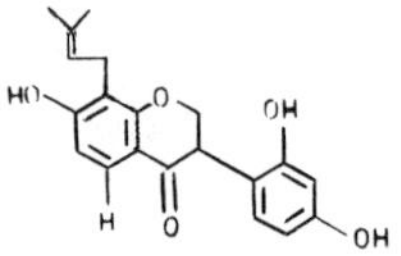

HO O OH H O OH HO O MeO O CH_3 CH_3

Kievitone (219) 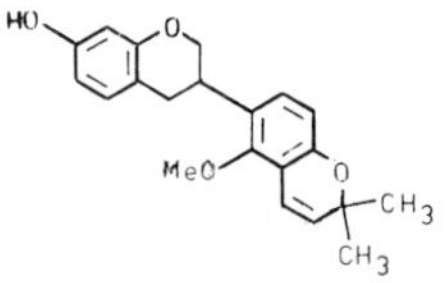 2′-Methoxyphaseollinisoflavan (211)

Phaseollin (structure given on p. 165)

Soybean (*Glycine max*) CH_3 H_3C O O OH O OH

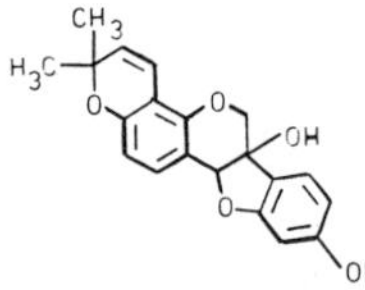

Hydroxyphaseollin (synonyms: glyceollin or glycinin (212)

Alfalfa (Lucerne) HO O O H OCH_3

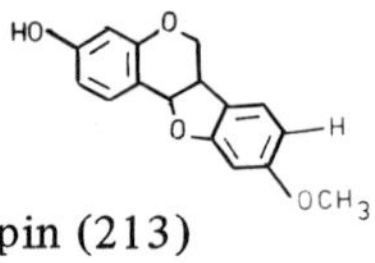

Medicarpin (213)

Medicago sativa HO O

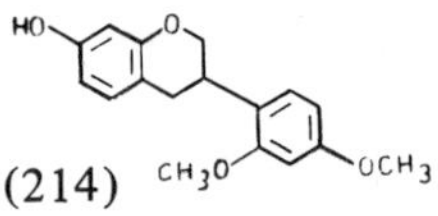

Sativan (214) CH_3O OCH_3

Broad bean (*V. faba*)

$CH_3CH_2CH = CH \cdot C \equiv C \cdot CO$–(furan, O)–$CH = C \cdot CO_2R$ (215)

R = H Wyerone acid R = CH_3 Wyerone

Potato (*Solanum tuberosum*)

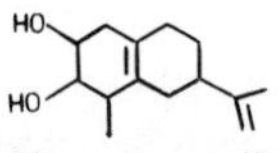

HO HO OH OH

Rishitin (200) 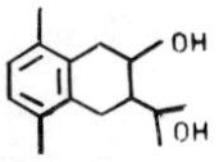 Rishitinol (217)

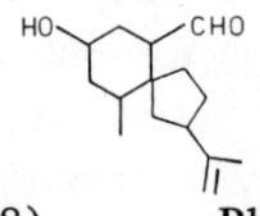

HO CHO O O OAc

Lubimin (218) 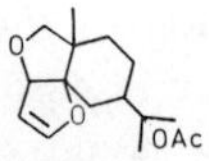 Phytuberin (219)

plants treated before inoculation to make them more resistant would be expected to increase phytoalexin production, and pretreatments which make them more susceptible would decrease phytoalexin production. There is evidence that all these phenomena can occur in a range of host-parasite combinations. Examples are taken from studies on a range of host plants in the Leguminosae and the Solanaceae.

Isoflavonoid Phytoalexins in Leguminosae Many experiments have been carried out with the use of French bean (*P. vulgaris*) as the host plant and a number of pathogens of different degrees of virulence.

The response of *P. vulgaris* when inoculated with different races of *C. lindemuthianum,* to which it is either susceptible or resistant, has been examined in some detail. In the first series of experiments, only phaseollin accumulation was determined (223). There was an accumulation of phaseollin at an early stage after inoculation with an incompatible race (race δ) of *C. lindemuthianum;* phaseollin increased soon after penetration when the infected cells turned brown. However, following inoculation with a race of the fungus giving a susceptible reaction, there was little or no phaseollin present at the stage when intracellular growth was taking place, but only after the stage of cellular collapse and browning. Moreover, phaseollin was localized in the smallest amounts of brown tissue that could be separated from the surrounding green tissue. It was later shown that, in addition to phaseollin, there was accumulation of phaseollidin, phaseollinisoflavan, and kievitone in the brown necrotic tissue (224). Phytoalexin accumulation continued after the host cell had died. From microscopic observations, it was concluded that cellular necrosis occurred at an early stage of the hypersensitive reaction, and hyphal growth continued very slowly in brown cells (225).

In bean hypocotyl tissue inoculated with *Rhizoctonia solani,* kievitone and phaseollin were the two major phytoalexins which accumulated in the lesions, but the rates of accumulation differed. Kievitone was present in steadily increasing amounts in young, intermediate, and mature lesions, whereas phaseollin was present in lower concentrations in young and intermediate lesions and then dramatically increased in mature lesions (226). The levels of the two phytoalexins, particularly of kievitone in the earlier stages of infection, were considered by comparison with in vitro assays to be sufficient to cause inhibition of the mycelial growth of *R. solani* associated with lesion limitation.

Phytoalexin accumulation has also been advanced as an explanation for the hypersensitive response of soybean leaves to the bacterial pathogen *Pseudomonas glycinea* (227). *P. glycinea*-soybean combinations which reacted hypersensitively contained much higher levels of hydroxyphaseollin (syn: glyceollin-glycinin) and the two constitutive isoflavonoids commestrol and diadzein than did compatible combinations. In the hypersensitive reactions, there was rapid accumulation of the three isoflavonoids between 20 and 40 hr after inoculation of the resistant leaves; necrosis of under and upper surfaces of leaves occurred at 24–30 hr and 36–40 hr, respectively, after inoculation. The bacterial populations remained constant. In the compatible combinations, hydroxyphaseollin was first detected about 60 hr after inoculation when the symptoms of water congestion also appeared. The three isoflavonoids then increased and followed necrosis and chlorosis, but the levels attained only 20% of those in the hypersensitively reacting leaves.

Attempts to correlate the susceptibility or resistance of plants to fungal infection with the sensitivity of the pathogens to the phytoalexins do not seem to have given clear-cut results.

Pisatin accumulates in pea seedlings infected with *F. solani* f.sp. *pisi, Aphanomyces euteiches,* and *R. solani.* Although very high concentrations of pisatin were found in *F. solani* f.sp. *pisi* lesions, the pathogen was insensitive to pisatin in vitro. In *Rhizoctonia*-infected tissue, lesion restriction and attainment of maximum pisatin concentration in the lesions occurred coincidentally, but pisatin had earlier accumulated to a concentration which should have been inhibitory to *R. solani.* The results with *A. euteiches* are definitely not compatible with the concept that susceptibility is based on tolerance of the pathogen to pisatin or failure of the pathogen to induce pisatin. *A. euteiches* is very sensitive to pisatin in vitro, yet its hyphae produced larger lesions and colonized pea tissue much more rapidly than those of the insensitive pathogen *F. solani* f.sp. *pisi.* By 36 hr after inoculation, pisatin concentration in the lesions had reached 8 times that which entirely prevented growth of *A. euteiches* in vitro. Although the pisatin concentration decreased after 36 hr, it still remained higher than that which completely inhibited growth in vitro; the lesion continued to expand (228).

Metabolism of phytoalexins seems to occur in susceptible reactions. In the case of French bean hypocotyls infected with *F. solani* f.sp. *phaseoli* (229), a pathogen of bean, kievitone was not detected; there were only traces of phaseollidin and 2′-methoxyphaseollinisoflavan, a larger amount of phaseollinisoflavan, and a very large accumulation of both phaseollin and 1-α-hydroxyphaseollone,

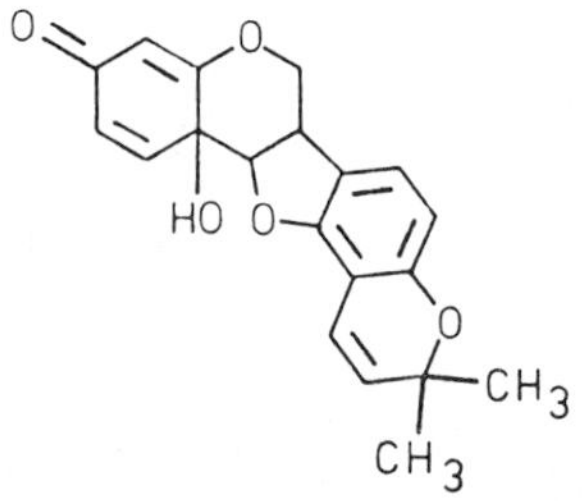

a fungal metabolite of phaseollin (230). It is suggested that conversion of phaseollin to 1-α-hydroxyphaseollone is a detoxication mechanism; the latter compound is much less inhibitory to radial growth of a number of fungi than is phaseollin. Recent experiments by P. Kuhn and D. A. Smith (unpublished observations) have shown that *F. solani* f.sp. *phaseoli* can also metabolize kievitone. This may be the reason for the absence of this compound from bean hypocotyls infected by the fungus.

In the case of alfalfa leaves, the role of medicarpin in limiting fungal lesions produced by the leaf pathogens *Stemphyllium botryosum, Phoma herbarum* var *medicagensis,* and *Leptosphaerulina briosiana* seems less clear (231). Small amounts of medicarpin can be extracted from the leaves of infected plants; at these levels medicarpin did not inhibit mycelial growth. It is clear that all three pathogens will degrade medicarpin. However, determinations which might resolve the question of the amounts of medicarpin at the cellular level in diseased tissue have not yet been accomplished.

There is evidence that treatments of plants which make them more susceptible to pathogens reduce phytoalexin accumulation. Hydroxyphaseollin increases in soybeans which are resistant to *Phytophthora megasperma* var *sojae* 10–100 times faster than in susceptible plants. Heat treatment, which breaks the resistance conferred by a resistance allele, concomitantly decreases the rate of hydroxyphaseollin accumulation in inoculated hypocotyls by 90% (232).

Lowering the oxygen tensions of both pea and bean reduced the rate of pisatin and phaseollin biosynthesis, respectively, when pod tissues were inoculated with *M. fructicola.* At oxygen tensions of 1%, there was maximum fungal growth and a large inhibition of phytoalexin synthesis (233). Conversely, triggering phytoalexin production (hydroxyphaseollin) in soybeans by ultraviolet irradiation caused them to become more resistant to attack by *P. megasperma* var *sojae* (234).

The hypersensitive reaction which French beans show to bacterial infection with *P. phaseolicola* may be overcome by the application of the endotoxin phaseotoxin (125). Pretreatment of resistant plants with phaseotoxin prior to inoculation with *P. phaseolicola* led to increased numbers of bacteria and suppression of the typical symptoms of the hypersensitive response; in addition, the levels of phaseollin, phaseollidin, phaseollinisoflavan, and kievitone were lower than in inoculated control plants.

Isoflavonoid phytoalexin synthesis can be stimulated by a range of abiotic agents, such as salts of heavy metals, and by molecules isolated from fungal culture filtrates (195). *M. fructicola* produces a peptide, monilicolin A, which stimulates phaseollin production in *P. vulgaris* (235). An elicitor from culture filtrates of *P. megasperma* var *sojae* appears to be a glucan that contains 3-linked, 6-linked, 3,5-linked, and terminal glucose (236); it may resemble the mycolaminarins or β-1,3-glucans which are found in the cytoplasm of *P. megasperma* var *sojae* (237) and can act as nonspecific elicitors of phytoalexin production. A polysaccharide fraction, consisting predominantly of 3- and 4-linked glucosyl residues, isolated both from culture filtrates and cell walls of *C. lindemuthianum,* causes cell browning and accumulation of phytoalexins in *P. vulgaris* (238). There is some doubt about the host-specificity of these elicitor molecules.

Wyerone in Broad Bean (Vicia faba) The relationship of the synthesis of the acetylenic phytoalexin wyerone acid to the resistance of *V. faba* to *B. cinerea* and susceptibility to *B. fabae* after inoculation of detached leaves with spore droplets has been examined (239). There appeared to be an inhibitor of spore germination and germ tube growth of *B. cinerea* on the surface which was not related to wyerone acid. The acid was produced in response to infection by both fungi, and it rapidly increased in concentration at the stage at which there was visible host cell damage. The highest concentrations of wyerone acid were found in the sites bearing limited lesions. It appeared to be localized in the brown cells where the *B. cinerea* hyphae ceased to grow. *B. fabae* seemed to be a typical necrotroph, causing disorganization of tissue in advance of fungal hyphae and then growing into the dead tissue (240). Only low concentrations of wyerone acid were found in tissue completely colonized by *B. fabae;* reduced wyerone acid, a less toxic metabolic product of wyerone acid, was found (241), and it has been concluded that the ability to reduce wyerone to its less toxic metabolite is an extremely important factor in the virulence of *B. fabae* (206, 242). There is now evidence from fluorescence microspectrography that wyerone acid accumulates in living cells adjacent to the lesion (243), which suggests that phytoalexin production may well be in these living cells. It was not possible to determine whether or not wyerone acid was accumulating in necrotic cells because of the ultraviolet-absorbing material produced by necrotic browning.

Terpenoid Phytoalexins in Solanaceae Although potatoes can produce several phytoalexins (200, 217–219), in most of the experiments on the role of phytoalexins in disease resistance measurements seem to have been made only of rishitin levels.

Rishitin was first isolated from tuber slices of the potato variety Rishiri, which carries the R_1 gene and gives a resistance reaction with with race O of *P. infestans* (199). Rishitin was also produced after inoculation with *F. solani* f.sp. *phaseoli,* but there were only traces in sliced uninoculated tissue and in tissue infected by a compatible race of *P. infestans* (race 1) (199). Other cultivars were found to accumulate rishitin when inoculated with incompatible races of *P. infestans* (244); most of the rishitin in tuber tissue was present in infected brown and adjacent cells, as shown by cutting into thin slices (245). Cell death associated with the incompatible reaction seemed to be the trigger for rishitin synthesis which occurred about 8 or 9 hr after death of 20% of the tuber cells; this was 10 hr before the lesion stopped developing (246). The concentration of rishitin was calculated to be 100 μg/g of tissue, fresh weight, when development of the lesion was ceasing. This concentration would be sufficient to cause almost complete inhibition of fungal growth in vitro. Although the highest levels of rishitin occur in the first 0.5-mm-thick layer of cut tubers inoculated with an incompatible race of *P. infestans,* the next 0.5-mm-thick layer is far more active in rishitin synthesis, as determined by the incorporation of 2-[^{14}C]acetate into rishitin. It is, therefore, suggested that rishitin is synthesized in healthy tissue and is transported to the brown cells, where it accumulates (247).

Rishitin and phytuberin accumulate in potato tubers, which are susceptible to all races of *P. infestans,* after inoculation with the bacterium *E. carotovora* (248). Accumulation of these terpenoids after infection may be related to varietal resistance to the bacterium (249).

It is clear that the ability to accumulate rishitin is not solely confined to potato varieties which contain major genes for resistance to *P. infestans* because, with the appropriate treatment, rishitin can be produced in tubers with no major resistance genes. Such treatments are, in addition to bacterial inoculation, addition of either fresh or boiled cell-free sonicates of *P. infestans* to tuber slices (250–252). When incompatible races of the fungus or fungal sonicates are used to trigger rishitin production, there is an accompanying necrosis of the tissue. However, necrosis caused by chemicals does not cause rishitin accumulation (252). There must be a complex series of interactions which determines whether or not potato cells will accumulate rishitin in response to fungal or bacterial attack. In order to explain control of rishitin production in tuber slices, Kuć et al. (208) have accordingly suggested that all races of *P. infestans* contain a factor called RIA (rishitin-inducing activity) which may well be a saponin linked to cell walls. In all cases of infection with the fungus or treatment of tuber slices with hyphal sonicate, RIA shifts the acetate-mevalonate pathway from the synthesis of the steroid glycoalkaloids, which are normally produced as part of the wounding reaction, to rishitin synthesis. The fungus also contains a blocker which is numbered according to the race of the fungus. If the fungus is one which gives an incompatible reaction, its blocker has no effect on the receptor site in the host cell and rishitin synthesis proceeds. However, in a race of the fungus which gives a compatible reaction, the blocker reacts with a receptor site in the rishitin biosynthetic pathway; rishitin synthesis is suppressed and non-fungitoxic terpenoids accumulate.

Increased Biosynthesis of Common Phenolic Compounds After Infection: Relationship to Resistance Phenolic compounds are of widespread occurrence in the plant kingdom. In many plants there is an acceleration of phenolic biosynthesis after infection with plant pathogens (187, 253). In many cases this increased phenolic biosynthesis is greater in resistant than in susceptible hosts. However, the role of the additional phenolic synthesis in resistance is unclear. One possibility is that the increased levels of phenolics may lead to greater phenolic oxidation, which in turn might cause the type of inhibition of pathogenesis which has been described under "Antimicrobial Compounds of Wide Distribution in Plant Kingdom."

Chlorogenic acid increases in potato tubers, showing resistance reactions to *H. carbonum* (254). However, the previously recorded comments of Kuć (186) about the levels of chlorogenic acid being insufficiently fungistatic apply here also. Other phenolics may also increase, such as scopolin in potatoes infected with *P. infestans* (255). However, it now appears that the levels of scopolin increase correlate with the susceptibility of the tubers to *P. infestans* (256).

Benzoic acids, as well as hydroxycinnamic acids, are common plant constituents; in some cases these may reach sufficiently high concentrations after infection to play some role in either resistance or at least lesion restriction. Antifungal concentrations of *p*-hydroxybenzoic, salicylic, and vanillic acids were found in apples after infection with *S. fructigena* (257). This appears to be an interesting case in which inhibitory compounds are synthesized by the pathogen from host-produced precursors such as chlorogenic acid (258). The same three hydroxybenzoic acids are more closely related to a resistance reaction of suspension cell potato cultures to *P. infestans* (259). Filtrates from suspension cultures of a resistant cultivar of potato inoculated with *P. infestans* contained material which inhibited the germination of *P. infestans* zoospores in an in vitro test. There was 50% inhibition of zoospore germination by the filtrate 2 days after inoculation and complete inhibition after 7 days. The filtrate obtained from an inoculated culture of cells from a susceptible cultivar of potato contained no inhibitory activity (260). The active compounds in the toxic fraction were shown by thin layer and gas-liquid chromatography and mass spectrometry to be *p*-hydroxybenzoic, salicylic, and vanillic acids (259). However, it is not absolutely clear whether the acids are biosynthesized in response to the fungus, because they are also present in a bound form, possibly as glycosides, in resistant cells before inoculation (261). It is interesting to note that these hydroxybenzoic acids are not produced after inoculation of potato tuber tissue (262).

The antifungal compound found in arrested lesions formed by *Nectria galligena* on unripe Bramley's seedling apples is benzoic acid (263). It was also found that a protease which could be isolated either from infected tissue or from culture filtrates of the fungus induced the formation of benzoic acid in immature apples (264). *N. galligena* eventually grows out from the arrested lesions as the apples ripen; this regrowth is associated with degradation of benzoic acid to β-oxoadipic acid (264, 265).

There is an increase in a lignin-like polymer in potato tuber slices infected with *P. infestans* (40); the accumulation of this material is greater in tuber discs of a resistant than of a susceptible variety inoculated with the same race of *P. infestans* (262, 266, 267).

Tuber slices and leaves of potato cultivars with major resistance genes also accumulate this lignin-like material after inoculation with an incompatible race of *P. infestans,* but inoculation with a compatible race always gives lower levels of this material (268, 269). Lignification had been proposed as a general mechanism of resistance (270), and it has been postulated that lignification of potato tuber tissue occurring in incompatible reactions with *P. infestans* could be part of the resistance reaction (262, 266, 267). Such lignification need not be extensive; it could merely alter the β-1,4-galactan substrates of the potato cell wall which are the presumed substrates for the fungal galactanase involved in the limited cell wall degradation which occurs during pathogenesis (39–41, 85). More recently it has been found that the lignin-like material consists partly of insoluble esters of ferulic and *p*-coumaric acids (268, 269); it has accordingly been suggested that esterification of the galactan would be a sufficient modification of the galactan to hinder hydrolysis by the fungal enzymes (269). Synthesis and deposition of insoluble phenolic material appeared to be involved in the accumulation of suberin in potato tuber tissue (271) and lignification of wheat leaf tissue (272), which are associated with the resistance of these plants to bacterial and fungal pathogens. However, lignification after inoculation seems to be limited to one isoline of wheat showing a resistant reaction to *E. graminis* f.sp. *tritici*(273).

In view of the relatively widespread distribution of ferulic and *p*-coumaric esters of cell wall carbohydrates in plants (274), it is of interest to determine how widespread resistance mechanisms based on these compounds are. After feeding [*U*-^{14}C]quinate and [*U*-^{14}C]shikimate, more radioactivity accumulated in the insoluble esters of ferulic and *p*-coumaric acid and in the insoluble, nonhydrolyzable residues of resistant than in susceptible wheat leaves inoculated with *P. graminis* f.sp. *tritici.* The *N-p*-coumaryl and *N*-ferulyl amides of hydroxyputrescine formed are not related to resistance (275, 276). The results obtained from the same host-pathogen combinations after feeding [*U*-^{14}C]phenylalanine and [*U*-^{14}C]ferulate (277) may not be related to reactions at the infection site because it has recently been shown that after infusing [*U*-^{14}C]phenylalanine or [3-^{14}C]cinnamic acid into the bases of detached wheat leaves, the bulk of the radioactivity remains near the cut basal end of the leaves (278).

Care may need to be taken in the interpretation of results of experiments on the incorporation of radioactive precursors of phenolic compounds into infected tissue. Colorimetric analysis shows that there is a greater increase in phenolic compounds in leaves of resistant than in susceptible leaves of rice plants after inoculation with *P. oryzae* (279), whereas when [*U*-^{14}C]phenylalanine and [*U*-^{14}C]tyrosine were incorporated into segments of rice leaves infected with *P. oryzae* there was no difference in the labeling of phenolic fractions between uninfected and inoculated leaves of the resistant variety. However, there was an increase in the labeling from [*U*-^{14}C]phenylalanine of soluble phenolics in susceptible leaves after infection (280). There are also discrepancies between measurements of phenolic metabolism in potato tissue infected by *P. infestans* determined by spectrophotometric methods and measured as the incorporation of radioactivity into phenolic compounds from [3-^{14}C]cinnamate (268).

The activities of enzymes associated with phenolic biosynthesis and metabolism increase in infected plant tissue. Phenylalanine ammonia lyase (PAL) activity increases after infection of sweet potato roots with *Ceratocystis fimbriata* (281), and in a range of plant tissues the rise is correlated with the production of specific phenolic compounds. In pea and French bean pods, PAL increases correlate with pisatin and phaseollin synthesis (282, 283) rather than with a general increase in phenolic biosynthesis (284). In tubers and leaves of potato inoculated with an incompatible race of *P. infestans,* the increases in PAL are associated with the increase of insoluble phenolics in the resistance reaction (266–268); in apples the rise in PAL correlates with benzoic acid production after infection with *N. galligena* or treatment with a fungal protease (285). The

PAL increase in the Pm 2 isoline of wheat is associated with lignification and resistance to *E. graminis* (273). In potato tuber tissue resistant to *P. infestans,* it is likely that deposition of phenolic material is controlled more closely by the levels of caffeic acid-*O*-methyltransferase activity than by PAL (286). In rice plants infected by *P. oryzae,* the levels of PAL were always higher in leaves of susceptible than resistant cultivars. Inoculation doubled the activity in both susceptible and resistant leaves within 24 hr. The high activity was maintained in susceptible but not in resistant leaves (280).

Increases in phenolase and peroxidase activity have been reported in many infected plants (187, 253). Presumably, the increases in phenolase could be related to the type of resistance reactions involving oxidized phenolics, which have been discussed under "Antimicrobial Compounds of Wide Distribution in Plant Kingdom." Although there is evidence that peroxidase increases may be involved in the control of protein and histone (287), there is also evidence that increased peroxidase activity is not causally related to resistance of wheat to *P. graminis* (288). Both phenolase and peroxidase activities increase in potato half-tubers only after inoculation with incompatible and not compatible races of *P. infestans* (289). These increases in activities may be related to the greater degree of cell death found in the hypersensitive resistance reaction, because there are not marked differences between the increased activities of each enzyme found in tuber slices of two varieties, one giving a compatible and the other an incompatible reaction with *P. infestans* (286). In the slices, there is an additional reaction associated with wound healing.

Relationship of Hypersensitive Cell Death to Biochemical Reactions Involved in Resistance

It has been generally accepted that the hypersensitive reaction of resistant plants to fungal attack is caused initially by the necrosis of a small group of cells around the point of penetration, which is followed by the death of the fungus; fungal death would then be caused by restriction of the nutrient supply and also by the production of fungitoxic metabolites (54, 290). There is evidence that hypersensitive cell death precedes rishitin synthesis and phenolic accumulation in potatoes inoculated with an incompatible race of *P. infestans* (246, 291). Nevertheless, on the basis of experiments in which chloramphenicol was applied to tubers of a potato variety having no major gene resistance to *P. infestans,* it was concluded that fungal death occurred before host cell death and rishitin accumulation (292). Electron microscopic examination of both leaves and tubers of major gene-resistant potato cultivars has shown that after inoculation with an incompatible race of *P. infestans* hypersensitive cell death precedes fungal death by at least 12 hr (293, 294). Similar electron microscopic results had been reported earlier for a resistant variety of lettuce inoculated with *B. lactucae* (295) and more recently for sweet pepper fruits giving an incompatible reaction with *P. infestans* (296). It seems likely that Kiraly et al. (292) were misled by the use of chloramphenicol, which inhibits the light-induced rise in activity of

PAL in potato tuber slices (297). As in the accumulation of phenolics and rishitin occurring after inoculation of potato with incompatible races of *P. infestans,* underlying healthy cells play a role in hypersensitive cell death (298). The interaction between the invaded and adjacent cells in infected plants seems to be complex and is probably related to the fact that phytoalexins and pathotoxic phenolic compounds seem to accumulate initially in the uninvaded cells adjacent to lesions (243, 247, 299).

Control of Resistance Reactions

Because hypersensitive cell death associated with resistance reactions starts very rapidly after penetration (293, 296), the recognition of the type of pathogen by the host cells must be a very early event and may possibly involve recognition of the cell wall or cell membrane of the pathogen by the host. However, there is, as yet, little understanding of the biochemical mechanisms involved in recognition. A number of reports, reviewed by De Vay et al. (300), show that bacterial and fungal pathogens share common antigens with their susceptible, but not their resistant, hosts. It is not clear how such common antigens would control the interaction between host and pathogen. There is evidence that determination of hypersensitivity of wheat to stem rust (*P. graminis* f.sp. *tritici*) may involve RNA (301). Extracts from leaves of wheat plants containing a gene for resistance to *P. graminis* inoculated with an avirulent race of the fungus were found to induce the formation of hypersensitive necrotic lesions when injected with leaves of similar plants inoculated with a virulent race of the fungus. The extracts had little effect on a susceptible cultivar inoculated with the same virulent race of the fungus. The active principle appeared to be RNA; it is not clear whether it represents RNA from the pathogen or a new RNA complex produced by the interaction, because control experiments with RNA from uninoculated leaves were not reported.

Recently Yamomoto and Matsuo (302) have reported that painting DNA isolated from a major gene-resistant potato cultivar onto cut surfaces of a susceptible tuber slice makes it resistant to a compatible race of *P. infestans.* Rather more surprising is their finding that when tuber slices from a major gene-resistant cultivar were painted with DNA from a susceptible cultivar, inoculation with an incompatible race gave rise to lesion development typical of a susceptible reaction.

CONCLUSIONS

There are now acceptable biochemical explanations of many metabolic events occurring before, during, and after the infection of plants by pathogenic fungi and bacteria. Yet there are still many features of plant-pathogen interactions for which biochemical explanations are either lacking or are insufficient. It has already been indicated that there is no biochemical explanation of the genetic control of specificity of resistance and susceptibility. In addition, explanations of the mechanism of action of germination stimulation and inhibitors and of the possible control of the cell wall-degrading enzymes of pathogens by cell wall proteins of the host are still lacking. There is also a need for information on the mechanism of action of phytoalexins.

Note added in proof: Phaseotoxin has been found to consist of four biologically active fractions, the first of which appears to be *N*-phosphoglutamic acid (303).

ACKNOWLEDGMENTS

I wish to thank my colleagues Dr. D. A. Smith and Dr. J. R. Coley-Smith for helpful discussions.

REFERENCES

1. Tarr, S. A. J. (1972). The Principles of Plant Pathology, p. 632. Macmillan, London.
2. Lewis, D. H. (1973). Biol. Rev. 48:261.
3. Friend, J., and Threlfall, D. R. Biochemical Aspects of Plant Parasite Relationships, in press. Academic Press, New York.
4. Williams, P., and Heitefuss, R. Encyclopedia of Plant Physiology, Vol. I, in press. Springer-Verlag, Berlin.
5. Macko, V., Staples, R. C., Renwick, J. A. A., and Pirone, J. (1972). Physiol. Plant Pathol, 2:347.
6. Leppik, R. A., Hollomon, D. W., and Bottomley, W. (1972). Phytochemistry 11:2055.
7. Sussman, A. S., and Douthit, H. A. (1973). Annu. Rev. Plant Physiol. 24:311.
8. Caltrider, P. G., Ramachandran, S., and Gottlieb, D. (1963). Phytopathology 53:86.
9. Daly, J. M., Knoche, H. W., and Wiese, M. V. (1967). Plant Physiol. 42:1633.
10. Langenbach, R. J., and Knoche, H. W. (1971). Plant Physiol. 48:735.
11. Trocha, P., Daly, J. M., and Langenbach, R. J. (1974). Plant Physiol. 53:519.
12. Trocha, P., and Daly, J. M. (1974). Plant Physiol. 53:527.
13. Trocha, P., and Daly, J. M. (1970). Plant Physiol. 46:520.
14. Hollomon, D. W. (1969). J. Gen. Microbiol. 55:267.
15. Staples, R. C., and Yaniv, Z. (1973). Physiol. Plant Pathol. 3:137.
16. Coffey, M. D. (1975). *In* D. H. Jennings and D. L. Lee (eds), Symbiosis, S. E. B. Symposium No. 25, pp. 297–323. Cambridge University Press, Cambridge.
17. Chou, M. C., and Preece, T. F. (1968). Ann. Appl. Biol. 62:11.
18. Preece, T. F. *In* J. Friend and D. R. Threlfall (eds.), Biochemical Aspects of Plant-Parasite Relationships, in press. Academic Press, New York.
19. Channon, A. G. (1970). Ann. Appl. Biol. 65:481.
20. Fokkema, N. J. (1971). *In* T. F. Preece and C. H. Dickinson (eds.), Ecology of Leaf Surface Microorganisms, pp. 278–282. Academic Press, New York.
21. Warren, R. C. (1972). Netherlands J. Plant Pathol. 78:89.
22. Mansfield, J. W., and Deverall, B. J. (1971). Nature 232:339.
23. Strange, R. N., and Smith, H. (1971). Physiol. Plant Pathol. 1:141.
24. Strange, R. N., Majer, J. R., and Smith, H. (1974). Physiol. Plant Pathol. 4:277.
25. Coley-Smith, J. R. *In* J. Friend and D. R. Threlfall (eds.), Biochemical Aspects of Plant Parasite Relationships, in press. Academic Press, New York.
26. Coley-Smith, J. R. (1959). Ann. App. Biol. 47:511.
27. Coley-Smith, J. R., King, J. E., Dickinson, D. J., and Holt, R. W. (1967). Ann. Appl. Biol. 60:109.
28. Coley-Smith, J. R., and Holt, R. W. (1966). Ann. Appl. Biol. 58:273.
29. Coley-Smith, J. R., and King, J. E. (1969). Ann. Appl. Biol. 64:289.
30. King, J. E. and Coley-Smith, J. R. (1969). Ann. Appl. Biol. 64:303.

31. Jeves, T. M. (1974). Ph. D. thesis, University of Hull.
32. Kertesz, Z. I. (1951). The Pectic Substances, p. 628. Wiley/Interscience, New York.
33. Aspinall, G. O. (1973). *In* F. Loewus (ed.), Biogenesis of Plant Cell Wall Polysaccharides, pp. 95–115. Academic Press, New York.
34. Talmadge, K. W., Keegstra, K., Bauer, W. D., and Albersheim, P. (1973). Plant Physiol. 51:158.
35. Miller, L., and MacMillan, J. D. (1971). Biochemistry 10:570.
36. Bateman, D. F., and Millar, R. L. (1966). Annu. Rev. Phytopathol. 4:119.
37. MacMillan, J. D., Phaff, H. J., and Vaughn, R. H. (1964). Biochemistry 3:572.
38. Clarke, D. D. (1966). Nature 211:649.
39. Knee, M., and Friend, J. (1968). Phytochemistry, 7:1289.
40. Friend, J., and Knee, M. (1969). J. Exp. Botany 20:763.
41. Knee, M., and Friend, J. (1970). J. Gen. Microbiol. 60:23.
42. Cole, A. L. J. (1970). Phytochemistry 9:337.
43. Sturdy, M. L. (1973). Ph. D. thesis, University of Hull.
44. Laborda, F., Archer, S. A., Fielding, A. H., and Byrde, R. J. W. (1974). J. Gen. Microbiol. 81:151.
45. Rees, D. A., and Richardson, N. G. (1966). Biochemistry 5:3099.
46. Northcote, D. H. (1972). Annu. Rev. Plant Physiol. 23:113.
47. Bauer, W. D., Talmadge, K. W., Keegstra, K., and Albersheim, P. (1973). Plant Physiol. 51:174.
48. Strobel, G. A. (1963). Phytopathology 53:592.
49. Van Etten, H. D., and Bateman, D. F. (1969). Phytopathology 59:968.
50. Mullen, J. M., and Bateman, D. F. (1975). Physiol. Plant Pathol. 6:233.
51. Bateman, D. F. *In* J. Friend and D. R. Threlfall (eds.), Biochemical Aspects of Plant–Parasite Relationships, in press. Academic Press, New York.
52. Preston, R. D. (1974). The Physical Biology of Plant Cell Walls, p. 491. Chapman and Hall, London.
53. Reese, E. T. (ed.). (1963). Advances in Enzymic Hydrolysis of Cellulose and Related Materials, p. 290. Macmillan Company, New York.
54. Wood, R. K. S. (1967). Physiological Plant Pathology, p. 570. Blackwell Scientific Publications, Oxford.
55. Bateman, D. F., VanEtten, H. D., English, P. D., Nevins, D. J., and Albersheim, P. (1969). Plant Physiol. 44:641.
56. Bateman, D. F., and Beer, S. V. (1965). Phytopathology 55:204.
57. Bateman, D. F. (1963). Phytopathology 53:197.
58. Karr, A. L., Jr., and Albersheim P. (1970). Plant Physiol. 46:69.
59. Basham, H. G., and Bateman, D. F. (1975). Physiol. Plant Pathol. 5:249.
60. Mullen, J. M., and Bateman, D. F. (1975). Phytopathology 65:797.
61. Jones, T. M., Anderson, A. J., and Albersheim, P. (1972). Physiol. Plant Pathol. 2:153.
62. English, P. D., Jurale, J. B., and Albersheim P. (1971). Plant Physiol. 47:1.
63. Cooper, R. M., and Wood, R. K. S. (1975). Physiol. Plant Pathol. 5:135.
64. Keegstra, K., Talmadge, K. W., Bauer, W. D., and Albersheim, P. (1973). Plant Physiol. 51:188.
65. Albersheim, P. (1975). Sci. Am. 232 (4):80.
66. Lamport, D. T. A. (1970). Annu. Rev. Plant Physiol. 21:235.
67. Weinhold, A. R., and Bowman, T. (1974). Phytopathology 64:985.
68. Horton, J. C., and Keen, N. T. (1966). Phytopathology 56:908.
69. Mullen, J. M., and Bateman, D. F. (1971). Physiol. Plant Pathol. 1:363.
70. Zucker, M., and Hankin, L. (1970). J. Bacteriol. 104:13.
71. de Bary, A. (1886). Bot. Z. 44:377.
72. Bateman, D. F. (1963). Phytopathology 53:1178.
73. McClendon, J. H. (1964). Am. J. Botany 51:628.
74. Zaitlin, M., and Coltrin, D. (1964). Plant Physiol. 39:91.
75. Byrde, R. J. W., and Fielding, A. H. (1962). Nature 196:1227.
76. Byrde, R. J. W., and Fielding, A. H. (1965). Nature 205:390.
77. Byrde, R. J. W., and Fielding, A. H. (1968). J. Gen. Microbiol. 52:287.
78. Dean, M., and Wood, R. K. S. (1967). Nature 214:409.

79. Bush, D. A., and Codner, R. C. (1970). Phytochemistry 9:87,
80. Beck, M. J. (1975). Ph. D. thesis, University of Hull.
81. Brown, W. (1915). Ann. Botany 29:313.
82. Tribe, H. T. (1955). Ann. Botany 19:351.
83. Wood, R. K. S. *In* J. Friend and D. R. Threlfall (eds.), Biochemical Aspects of Plant-Parasite Relationships, in press. Academic Press, New York.
84. Stephens, G. J., and Wood, R. K. S. (1975). Physiol. Plant Pathol. 5:165.
85. Knee, M. (1970). Phytochemistry 9:2075.
86. Van Sumere, C. F., Van Sumere-de-Preter, C., and Ledingham, G. A. (1957). Can. J. Microbiol. 3:761.
87. Welch, B. L., and Martin, N. E. (1974). Phytopathology 64:1287.
88. Sargent, J. A., Tommerup, I. C., and Ingram, D. S. (1973). Physiol. Plant Pathol. 3:231.
89. Pringle, R. B., and Scheffer, R. P. (1964). Annu. Rev. Phytopathol. 2:133.
90. Patil, S. S. (1974). Annu. Rev. Phytopathol. 12:259.
91. Strobel, G. A. (1974). Annu. Rev. Plant Physiol. 25:541.
92. Scheffer, R. P., and Yoder, O. C. (1972). *In* R. K. S. Wood, A. Ballio, and A. Graniti (eds.), Phytotoxins in Plant Diseases, pp. 251–272. Academic Press, New York.
93. Sammadar, K. R., and Scheffer, R. P. (1971). Physiol. Plant Pathol. 1:319.
94. Gardner, J. M.. Mansour, I. S., and Scheffer, R. P. (1972). Physiol. Plant Pathol. 2:197.
95. Yoder, O. C., and Scheffer, R. P. (1973). Plant Physiol. 52:518.
96. Gardner, J. M., Scheffer, R. P., and Higinbotham, N. (1974). Plant Physiol. 54:246.
97. Strobel, G. A. (1975). Sci. Am. 232 (1):80.
98. Strobel, G. A. *In* J. Friend and D. R. Threlfall (eds.), Biochemical Aspects of Plant Parasite Relationships, in press. Academic Press, New York.
99. Hooker, A. L., Smith, D. R., Lim, S. M., and Beckett, J. B. (1970). Plant Disease Reporter 54:708.
100. Lim, S. M., and Hooker, A. L. (1972). Phytopathology 62:968.
101. Ullstrup, A. J. (1972). Annu. Rev. Phytopathol. 10:37.
102. Miller, R. J., and Koeppe, D. E. (1971). Science 173:67.
103. Peterson, P. A., Flavell, R. B., and Barratt, D. H. P. (1975). Theor. Appl. Genet. 45:309.
104. Flavell, R. (1975). Physiol. Plant Pathol. 6:107.
105. Gengenbach, B. G., Miller, R. J., Koeppe, D. E., and Arntzen, C. J. (1973). Can. J. Botany 51:2119.
106. Tipton, C. L., Mondal, M. H., and Uhlig, J. (1973). Biochem. Biophys. Res. Commun. 51:725.
107. Keck, R. W., and Hodges, T. K. (1973). Phytopathology 63:226.
108. Tipton, C. L., Mondal, M. H., and Benson, M. J. (1975). Physiol. Plant Pathol. 7:277.
109. Peterson, P. A., Flavell, R. B., and Barratt, D. H. P. (1974). Plant Disease Reporter 58:777.
110. Comstock, J. C., Martinson, C. A., and Gengenbach, B. G. (1973). Phytopathology 63:1357.
111. Yoder, O. C. (1973). Phytopathology 63:1361.
112. Owens, L. D., and Wright, D. A. (1965). Plant Physiol. 40:927.
113. Owens, L. D., Thompson, J. F., Pitcher, R. G., and Williams, T. (1972). Chem. Commun. 714.
114. Owens, L. D., Guggenheim, S., and Hilton, J. L. (1968). Biochim. Biophys. Acta 158:219.
115. Giovanelli, J., Owens, L. D., and Mudd, S. H. (1971). Biochim. Biophys. Acta 227:671.
116. Giovanelli, J., Owens, L. D., and Mudd, S. H. (1973). Plant Physiol. 51:492.
117. Owens, L. D., Lieberman, M., and Kunishi, A. (1971). Plant Physiol. 48:1.
118. Patel, P. N., and Walker, J. C. (1963). Phytopathology 53:522.
119. Patel, P. N., and Walker, J. C. (1963). Phytopathology 53:855.
120. Rudolph, K., and Stahmann, M. A. (1966). Phytopath. Z. 57:29.

121. Patil, S. S., Kolattukudy, P. E., and Dimond, A. E. (1970). Plant Physiol. 46:752.
122. Tam. L. Q., and Patil, S. S. (1972). Plant Physiol. 49:808.
123. Patil, S. S., Tam, L. Q., and Sakai, W. S. (1972). Plant Physiol. 49:803.
124. Rudolph, K. (1972). *In* R. K. S. Wood, A. Ballio, and A. Graniti (eds.), Phytotoxins in Plant Disease pp. 373–375. Academic Press, New York.
125. Patil, S. S., and Gnanamanickam, S. S. (1976). Nature 259:486.
126. Goodman, R. N., Kiraly, Z., and Zaitlin, M. (1967). The Biochemistry and Physiology of Infectious Plant Disease, p. 354. Van Nostrand Co., Inc., Princeton.
127. Millerd, A., and Scott, K. J. (1962). Annu. Rev. Plant Physiol. 13:559.
128. Livne, A. (1964). Plant Physiol. 39:614.
129. Doodson, J. K., Manners, J. G., and Myers, A. (1965). J. Exp. Botany 16:304.
130. Scott, K. J., and Smillie, R. M. (1966). Plant Physiol. 41:289.
131. Edwards, H. H. (1970). Plant Physiol. 45:594.
132. Montalbini, P., and Buchanan, B. B. (1974). Physiol. Plant Pathol. 4:191.
133. Daly, J. M., Sayre, R. M., and Pazur, J. H. (1957). Plant Physiol. 32:44.
134. Shaw, M., and Samborski, D. J. (1957). Can. J. Bot. 35:389.
135. Ryrie, I. J., and Scott, J. J. (1968). Plant Physiol. 43:687.
136. Burrell, M. M., and Ap Rees, T. (1974). Physiol. Plant Pathol. 4:489.
137. Smith, D., Muscatine, L., and Lewis, D. (1969). Biol. Rev. 44:17.
138. Livne, A., and Daly, J. M. (1966). Phytopathology 56:170.
139. Edwards, H. H. (1971). Plant Physiol. 47:324.
140. Thrower, L. B., and Thrower, S. L. (1966). Phytopathol. Z. 57:267.
141. Siddiqui, M. Q., and Manners, J. G. (1971). J. Exp. Botany 22:792.
142. Holligan, P. M., Chen, C., McGee, E. E. M., and Lewis, D. H. (1974). New Phytol. 73:881.
143. Farrell, G. M. (1971). Physiol. Plant Pathol. 1:457.
144. Sequeira, L. (1973). Annu. Rev. Plant Physiol. 24:353.
145. Scott, K. J. (1972). Biol. Rev. 47:537.
146. Holligan, P. M., Chen, C., and Lewis, D. H. (1973). New Phytol. 72:947.
147. Long, D. E., Fung, A. K., McGee, E. E. M., Cooke, R. C., and Lewis, D. H. (1975). New Phytol. 74:173.
148. Long, D. E., and Cooke, R. C. (1974). New Phytol. 73:889.
149. MacDonald, P. W., and Strobel, G. A. (1970). Plant Physiol. 46:126.
150. Farkas, G. L., and Kiraly, Z. (1961). Physiol. Plant 14:344.
151. Blumenbach, D. (1968). Phytopathol. Z. 56:238.
152. Sadler, R., and Scott, K. J. (1974). Physiol. Plant Pathol. 4:235.
153. Callow, J. A. *In* J. Friend and D. R. Threlfall (eds.), Biochemical Aspects of Plant-Parasite Relationships, in press. Academic Press, New York.
154. Plumb, R., Manners, J. G., and Myers, A. (1968). Trans. Brit. Mycol. Soc. 51:563.
155. Heitefuss, R. (1968). Netherlands J. Plant Pathol. (Suppl. 1) 74:9.
156. Bhattacharya, P. K., and Shaw, M. (1967). Can. J. Botany 45:555.
157. Chakravorty, A. K., and Shaw, M. (1971). Biochem. J. 123:551.
158. Callow, J. A. (1973). Physiol. Plant Pathol. 3:249.
159. Bennett, J., and Scott, K. J. (1971). FEBS Lett. 16:93.
160. Dyer, T. A., and Scott, K. J. (1972). Nature 236:237.
161. Chakravorty, A. K., Shaw, M., and Scrubb, L. A. (1974). Nature 247:577.
162. Overeem, J. C. (1976). *In* J. Friend and D. R. Threlfall (eds.), Biochemical Aspects of Plant-Parasite Relationships, in press. Academic Press, New York.
163. Turner, E. M. C. (1961). J. Exp. Botany 12:169.
164. Maizel, J. V., Burkhardt, H. J., and Mitchell, H. K. (1964). Biochemistry 3:424.
165. Burkhardt, H. J., Maizel, J. V., and Mitchell, H. K. (1964). Biochemistry 3:426.
166. Olsen, R. A. (1973). Physiol. Plant 28:507.
167. Olsen, R. A. (1973). Physiol. Plant 29:145.
168. Ludwig, R. A., Spencer, E. Y., and Unwin, C. H. (1960). Can. J. Botany 38:21.
169. Stoessl, A. (1967). Can. J. Chem. 45:1745.
170. Stoessl, A., and Unwin, C. H. (1970). Can. J. Botany 48:465.
171. Stoessl, A. (1965). Phytochemistry 4:973.

172. Wahlroos, O., and Virtanen, A. I. (1959). Acta Chem. Scand. 13:1906.
173. El Naghy, M. A., and Linko, P. (1962). Physiol. Plant 15:764.
174. Knott, D. R., and Kumar, J. (1972). Physiol. Plant Pathol. 2:393.
175. Schönbeck, F., and Schroder, C. (1972). Physiol. Plant Pathol. 2:91.
176. Bailey, J. A., Vincent, G. G., and Burden, R. S. (1974). J. Gen. Microbiol. 85:57.
177. Bailey, J. A., Carter, G. A., Burden, R, S., and Wain, R. L. (1975). Nature 255:328.
178. Walker, J. C., and Stahmann, M. A. (1955). Annu. Rev. Plant Physiol. 6:351.
179. Mulvena, D., Webb, E. C., and Zerner, B. (1969). Phytochemistry 8:393.
180. Baruah, P., and Swain, T. (1959). J. Sci. Food. Agric. 10:125.
181. Johnson, G., and Schaal, L. A. (1952). Science 115:627.
182. Lee, S., and Le Tourneau, D. J. (1958). Phytopathology 48:268.
183. McClean, J. G., Le Tourneau, D. J., and Guthrie, J. W. (1961). Phytopathology 51:84.
184. Valle, G. (1957). Acta Chem. Scand. 11:395.
185. Rubin, B., and Aksenova, V. (1957). Biokhimiya (Eng. Trans.) 22:191.
186. Kuć, J. (1957). Phytopathology 47:676.
187. Farkas, G. L., and Kiraly, Z. (1962). Phytopathol. Z. 44:105.
188. Byrde, R. J. W., Fielding, A. H., and Williams, A. H. (1960). *In* J. B. Pridham (ed.), Phenolics in Plants in Health and Disease, pp. 95–99. Pergamon Press, Oxford.
189. Hunter, R. E. (1974). Physiol. Plant Pathol. 4:151.
190. Mace, M. E. (1963). Physiol. Plant 16:915.
191. Beckman, C. H., and Mueller, W. C. (1970). Phytopathology 60:79.
192. Beckman, C. H., Mueller, W. C., and Mace, M. E. (1974). Phytopathology 64:1214.
193. Butler, E. J. and Jones, S. G. (1955). Plant Pathology, p. 979. Macmillan, London.
194. Albersheim, P., Jones, T. M., and English, P. D. (1969). Annu. Rev. Phytopathol. 7:171.
195. Albersheim, P., and Anderson-Prouty, A. J. (1975). Annu. Rev. Plant Physiol. 26:31.
196. Albersheim, P., and Anderson, A. J. (1971). Proc. Natl. Acad. Sci. U.S.A. 68:1815.
197. Fisher, M. L., Anderson, A. J., and Albersheim, P. (1973). Plant Physiol. 51:489.
198. Müller, K. O., and Börger, H. (1940). Reichsanstalt. Landw. Forstw. Berlin 23:189.
199. Tomiyama, K., Sakuma, T., Ishizaka, N., Sato, N., Katsui, N., Takasugi, M., and Masamune, T. (1968). Phytopathology 58:115.
200. Katsui, N., Murai, A., Takasugi, M., Imaizumi, K., Masamune, T., and Tomiyama, K. (1968). Chem. Commun. 43.
201. Müller, K. O. (1958). Aust. J. Biol. Sci. 11:275.
202. Cruickshank, I. A. M., and Perrin, D. R. (1960). Nature 187:799.
203. Perrin, D. R., and Bottomley, W. (1962). J. Am. Chem. Soc. 84:1919.
204. Cruickshank, I. A. M., and Perrin, D. R. (1963). Life Sci. 2:680.
205. Perrin, D. R. (1964). Tetrahedron Lett. 29.
206. Deverall, B. J. *In* J. Friend and D. R. Threlfall (eds.), Biochemical Aspects of Plant-Parasite Relationships, in press. Academic Press, New York.
207. VanEtten, H. D., and Pueppke, S. *In* J. Friend and D. R. Threlfall (eds.), Biochemical Aspects of Plant-Parasite Relationships, in press. Academic Press, New York.
208. Kuć, J., Currier, W. W., and Shih, M. J. *In* J. Friend and D. R. Threlfall (eds.), Biochemical Aspects of Plant-Parasite Relationships, in press. Academic Press, New York.
209. Perrin, D. R., Whittle, C. P., and Batterham, T. J. (1972). Tetrahedron Lett. 1673.
210. Burden, R. S., Bailey, J. A., and Dawson, G. W. (1972). Tetrahedron Lett. 4175.
211. VanEtten, H. D. (1973). Phytochemistry 12:1791.
212. Burden, R. S., and Bailey, J. A. (1975). Phytochemistry 14:1389.
213. Smith, D. B., McInnes, A. G., Higgins, V. J., and Millar, R. L. (1971). Physiol. Plant Pathol. 1:41.
214. Ingham, J. L., and Millar, R. L. (1973). Nature 242:125.
215. Letcher, R. M., Widdowson, D. A., Deverall, B. J., and Mansfield, J. W. (1970). Phytochemistry 9:249.
216. Fawcett, C. H., Spencer, D. M., Wain, R. L., Fallis, A. G., Jones, E. R. H., Le Quan, M., Page, C. B., Thaller, V., Shubrook, D. C., and Whitman, P. M. (1968). J. Chem. Soc. C2455.

217. Katsui, N., Matsunaga, A., Imaizumi, K., Masamune, T., and Tomiyama, K. (1971). Tetrahedron Lett. 843.
218. Metlitskii, L. B., Ozeretskovskaya, O. L., Vulfson, N. S., and Chalova, L. I. (1971). Mikol. Fitopatol. 5:439.
219. Coxon, D. T., Curtis, R. F., Price, K. R., and Howard, B. (1974). Tetrahedron Lett. 2363.
220. Cruickshank, I. A. M., and Perrin, D. R. (1961). Aust. J. Biol. Sci. 14:336.
221. Bailey, J. A. (1973). J. Gen. Microbiol. 75:119.
222. Bailey, J. A., and Burden, R. S. (1973). Physiol. Plant Pathol. 3:171.
223. Bailey, J. A., and Deverall, B. J. (1971). Physiol. Plant Pathol. 1:435.
224. Bailey, J. A. (1974). Physiol. Plant Pathol. 4:477.
225. Skipp, R. A., and Deverall, B. J. (1972) Physiol. Plant Pathol. 2:357.
226. Smith, D. A., VanEtten, H. D., and Bateman, D. F. (1975). Physiol. Plant Pathol. 5:51.
227. Keen, N. T., and Kennedy, B. W. (1974). Physiol. Plant Pathol. 4:173.
228. Pueppke, S. G., and VanEtten, H. D. (1974). Phytopathology 64:1433.
229. VanEtten, H. D., and Smith, D. A. (1975). Physiol. Plant Pathol. 5:225.
230. Heuvel, J. V. D., VanEtten, H. D., Serum, J. W., Coffen, D. L., and Williams, T. H. (1974). Phytochemistry 13:1129.
231. Higgins, V. J. (1972). Physiol. Plant Pathol. 2:289.
232. Keen, N. T. (1971). Physiol. Plant Pathol. 1:265.
233. Cruickshank, I. A. M., and Perrin, D. R. (1967). Phytopathol. Z. 60:335.
234. Bridge, M. A., and Klarman, W. L. (1973). Phytopathology 63:606.
235. Cruickshank, I. A. M., and Perrin, D. R. (1968). Life Sci. 7:449.
236. Ayers, A., Ebel, J., and Albersheim, P. (1974). Proc. Am. Phytopathol. Soc. 1:23.
237. Keen, N. T., Wang, M. C., Bartnicki-Garcia, S., and Zentmeyer, G. A. (1975). Physiol. Plant Pathol. 7:91.
238. Anderson-Prouty, A. J., and Albersheim, P. (1975). Plant Physiol. 56:286.
239. Mansfield, J. W., and Deverall, B. J. (1974). Ann. Appl. Biol. 77:227.
240. Mansfield, J. W., and Deverall, B. J. (1974). Ann. Appl. Biol. 76:77.
241. Mansfield, J. W., and Widdowson, D. A. (1973). Physiol. Plant Pathol. 3:393.
242. Deverall, B. J. (1972). *In* J. B. Harborne (ed.), Phytochemical Ecology, pp. 217–233.
243. Mansfield, J. W., Hargreaves, J. A., and Boyle, F. C. (1974). Nature 252:316.
244. Sato, N., Tomiyama, K., Katsui, N., and Masamune, T. (1968). Ann. Phytopathol. Soc. Jap. 34:140.
245. Sato, N., and Tomiyama, K. (1969). Ann. Phytopathol. Soc. Jap. 35:202.
246. Sato, N., Kitazawa, K., and Tomiyama, K. (1971). Physiol. Plant Pathol. 1:289.
247. Nakajima, T., Tomiyama, K., and Kinukaw, M. (1975). Ann. Phytopathol. Soc. Jap. 41:49.
248. Lyon, G. D. (1972). Physiol. Plant Pathol. 2:411.
249. Lyon, G. D., Lund, B. M., Bayliss, C. E., and Wyatt, G. M. (1975). Physiol. Plant Pathol. 6:43.
250. Varns, J. L., Kuć, J., and Williams, E. B. (1971). Phytopathology 61:174.
251. Varns, J. L., and Kuć, J. (1971). Phytopathology 61:178.
252. Varns, J. L., Currier, W. W., and Kuć, J. (1971). Phytopathology 61:968.
253. Kosuge, T. (1969). Annu. Rev. Phytopathol. 7:195.
254. Kuć, J., Henze, R. E., Ullstrup, A. J., and Quackenbush, F. W. (1956). J. Am. Chem. Soc. 78:3123.
255. Hughes, J. C., and Swain, T. (1960). Phytopathology 50:398.
256. Clarke, D. D. (1973). Physiol. Plant Pathol. 3:347.
257. Fawcett, C. H., and Spencer, D. M. (1967). Ann. Appl. Biol. 60:87.
258. Fawcett, C. H., and Spencer, D. M. (1968). Ann. Appl. Biol. 61:245.
259. Robertson, N. F., Friend, J., Aveyard M. A., Brown, J., Huffee, M., and Homans, A. L. (1968). J. Gen. Microbiol. 54:261.
260. Ingram, D. S. (1967). J. Gen. Microbiol. 49:99.
261. Robertson, N. F., Friend, J., and Aveyard, M. A. (1969). Phytochemistry 8:7.
262. Friend, J. (1973). *In* R. J. W. Byrde and C. V. Cutting (eds.) Fungal Pathogenicity and the Plant's Response. pp. 383–396. Academic Press, London.

263. Brown, A. E., and Swinburne, T. R. (1971). Physiol. Plant Pathol. 1:469.
264. Swinburne, T. R. (1973). *In* R. J. W. Byrde and C. V. Cutting (eds.), Fungal Pathogenicity and the Plant's Response, pp. 365–382. Academic Press, London.
265. Brown, A. E., and Swinburne, T. R. (1973). Physiol. Plant Pathol. 3:453.
266. Friend, J., Reynolds, S. B., and Aveyard, M. A. (1971). Biochem. J. 124:29P.
267. Friend, J., Reynolds, S. B., and Aveyard, M. A. (1973). Physiol. Plant Pathol. 3:495.
268. Henderson, S. J. (1975). Ph.D. thesis, University of Hull.
269. Friend, J. *In* J. Friend and D. R. Threlfall (eds.), Biochemical Aspects of Plant-Parasite Relationships, in press. Academic Press, New York.
270. Hijwegen, T. (1963). Netherlands J. Plant Pathol. 69:314.
271. Zucker, M., and Hankin, L. (1970). Ann. Botany 34:1047.
272. Ride, J. P. (1975). Physiol. Plant Pathol. 5:125.
273. Green, N. E., Hadwiger, L. A., and Graham, S. O. (1975). Phytopathology 65:1071.
274. Hartley, R. D., and Harris, P. J. (1975). Nutritive Value of Plant Fibre, Agriculture Gp., Sc. I A. G. M.
275. Rohringer, R., Fuchs, A., Lunderstädt, J., and Samborski, D. J. (1967). Can. J. Botany 45:863.
276. Samborski, D. J., and Rohringer, R. (1970). Phytochemistry 9:1939.
277. Fuchs, A., Rohringer, R., and Samborski, D. J. (1967). Can. J. Botany 45:2137.
278. Barnes, C., and Friend, J. (1975). Phytochemistry 14:139.
279. Sridhar, R., and Ou, S. H. (1974). Phytopathol. Z. 79:222.
280. Burrell, M. M., and Ap Rees, T. (1974). Physiol. Plant Pathol. 4:497.
281. Minamikawa, T., and Uritani, I. (1964). Arch. Biochem. Biophys. 108:573.
282. Hadwiger, L. A., Hess, S. L., and Von Broembsens. (1970). Phytopathology 60:332.
283. Rathmell, W. G. (1973). Physiol. Plant Pathol. 3:259.
284. Rathmell, W. G., and Bendall, D. S. (1971). Physiol. Plant Pathol. 1:351.
285. Swinburne, T. R., and Brown, A. E. (1975). Physiol. Plant Pathol. 6:259.
286. Friend, J., and Thornton, J. D. (1974). Phytopath. Z. 81:56.
287. Stahmann, M. A., and Demorest, D. M. (1973). *In* R. J. W. Byrde and C. V. Cutting (eds.), Fungal Pathogenicity and the Plant's Response, pp. 405–420. Academic Press, London.
288. Daly, J. M., and Knoche, H. W. *In* J. Friend and D. R. Threlfall (eds.), Biochemical Aspects of Plant-Parasite Relationships, in press. Academic Press, New York.
289. Tomiyama, K., and Stahmann, M. A. (1964). Plant Physiol. 39:483.
290. Müller, K. O. (1960). *In* J. G. Horsfall and A. E. Dimond (eds.), Plant Pathology, pp. 470–519. Academic Press, New York.
291. Tomiyama, K., Takakuwa, M., and Takase, N. (1958). Phytopathol. Z. 31:237.
292. Kiraly, Z., Barna, B., and Ersek, T. (1972). Nature 239:456.
293. Shimony, C., and Friend, J. (1975). New Phytol. 74:59.
294. Shimony, C., and Friend, J. Israel J. Botany, in press.
295. Maclean, D. J., Sargent, J. A., Tommerup, I. C., and Ingram, D. S. (1974). Nature 249:186.
296. Jones, D. R., Graham, W. G., and Ward, E. W. B. (1975). Phytopathology 65:1274.
297. Reynolds, S. B. (1971). Ph.D. thesis, University of Hull.
298. Kitazawa, K., and Tomiyama, K. (1973). Phytopathol. Soc. Jap. 39:85.
299. Paxton, J., Goodchild, D. J., and Cruickshank, I. A. M. (1974). Physiol. Plant Pathol. 4:167.
300. De Vay, J. E., Charudattan, R., and Wimalajeewa, D. L. S. (1972). Am. Natl. 106:185.
301. Rohringer, R., Howes, N. K., Kim, W. K., and Samborski, D. J. (1974). Nature 249:585.
302. Yamomoto, M., and Matsuo, K. (1976). Nature 259:63.
303. Patil, S. S., Youngblood, P., Christiansen, P., and Moore, R. E. (1976). Biochem. Biophys. Res. Commun. 69, No. 4: 1019.

International Review of Biochemistry
Plant Biochemistry II, Volume 13
Edited by D. H. Northcote

6 Glycoproteins

R. G. BROWN AND W. C. KIMMINS
Department of Biology
Dalhousie University, Halifax, Nova Scotia, Canada

EXTRACTION AND PURIFICATION 185
Chemical Extractions 185
Alkali 185
Hydrazinolysis 186
Acetic Acid–Sodium Chlorite 188
Miscellaneous 188
Enzymic Extraction 189
Determination of Purity 189

STRUCTURE 190
Amino Acid Constituents 190
Carbohydrate Constituents 192
Arabinose 192
Xylose 193
Galactose 193
Glucose 193
Mannose 194
Glucosamine 194
Miscellaneous Sugars 194
Protein-Carbohydrate Linkages 194
Hydroxyproline 194
Serine 195
Asparagine 195
Miscellaneous Linkages 195

BIOSYNTHESIS 195
Protein 195
Hydroxylation 196
Glycosylation 196
Cellular Location 197

BIOLOGICAL SIGNIFICANCE 197
General 197
Cell Wall Structure 198
Cell Growth 200
Polysaccharide Synthesis 200
Pathogenesis 201
Cell Surface Interactions 202
Symbiosis 202
Reproductive Physiology 202
Frost Resistance 203
Germination 205
Enzymes 205

Proteins can be conveniently divided into two groups on the basis of their composition: simple proteins, which contain only amino acids, and conjugated proteins, containing nonamino acid components (the prosthetic group). A widely distributed group of conjugated proteins contain covalently linked carbohydrate and are variously referred to as glycoproteins, mucins, mucoproteins, proteoglycans, and peptidoglycans. There is no systematic distinction between these terms, but it is a common practice to use glycoprotein when the carbohydrate has a comparatively low degree of polymerization. The remaining terms are usually applied when the polysaccharide makes up most of the weight, as in heparin and chondroitin.

Glycoproteins are distributed throughout the biological kingdom, the majority being extracellular or associated with cell surface structures. Advances in polysaccharide chemistry during the 1960's (reviewed by Lindberg (1) and Aspinall and Stephen (2)) have been applied to animal glycoproteins with considerable success (3, 4). Unfortunately, with few exceptions, these techniques have not been applied in studies of plant glycoproteins. The difficulty of determining the sequence and anomeric configuration of the carbohydrate moiety may also be resolved in future studies by application of new specific degradation techniques. These have been reviewed by Lindberg et al. (5). Significant progress is also being made toward describing the biosynthesis of glycoproteins, but, again, very little of this work is being done with plants. Finally, the roles of the carbohydrate moieties are still not completely understood, and their biological significance remains partly speculative. However, the evidence increasingly supports functions related to molecular and cellular recognition, specificity, and structural matrices.

EXTRACTION AND PURIFICATION

Many methods have been used for extracting glycoproteins from plant tissue. The extraction method often determines the type of glycoprotein obtained and may alter structure by a variety of reactions, including base elimination or hydrolysis of acid-labile arabinofuranose units. Complexes which are artifacts of extraction technique may also be obtained (6, 7). Both chemical and enzymic methods have been used, either separately or in combination. It is becoming apparent that the structure of some plant glycoproteins can only be ascertained by using a variety of preparative methods, particularly if the glycoproteins are associated with the cell wall.

Chemical Extractions

The list of chemical extractants which have been used is long and varied. The methods which are discussed have been selected either because of their prevalent use or, as in the case of cell wall glycoproteins, because the extraction method bears on the structural studies.

Alkali Solutions of barium, sodium, or potassium hydroxide at a variety of temperatures have been widely used (6–12). Alkali extraction at high temperatures (90°C) has allowed the demonstration of hydroxyproline arabinosides in cell walls from a wide variety of plants (13). A trend exists in the more advanced groups toward longer arabinose oligosaccharide substituents with a maximum of 4 arabinose residues. Related material obtained by other methods (14–16) contained galactose in addition to arabinose. An "alkali-labile" linkage to the terminal arabinose of the hydroxyproline arabinosides has been proposed to explain this discrepancy (13). However, this proposal was made untenable by the isolation of hydroxyproline-rich galactopeptides which contain no arabinose but have single galactose residues linked glycosidically to serine (17, 18). A related cyclic glycopeptide with hydroxyproline as the only amino acid constituent and both arabinose and galactose, the latter presumably attached to arabinose, has been isolated (19). Thus, galactose can be linked to hydroxyproline through arabinose, as has been found in algal cell walls (20, 21). Furthermore, hydroxyproline-rich glycoproteins which contain arabinose and galactose as the main carbohydrate components have been obtained from cell walls (22, 23). The quantity of galactose present is too high to be accounted for by single galactose residues attached to serine. With some exceptions, the presence of carbohydrate having a molecular weight higher than hydroxyproline tetraarabinoside has not been reported in hot barium hydroxide extracts of plant cell walls. A polysaccharide from tomato (*Lycopersicum* L.) cell walls which contains arabinose and galactose with lower amounts of uronic acid is released by a β elimination reaction, indicating that this polysaccharide is linked to serine (24). Presumably, degradation from the reducing end of this polysaccharide could take place during hot barium hydroxide extraction of hydroxyproline arabinosides, although extraction of lupin (*Lupinus*) hypocotyl cell walls by this method released the bulk of cell wall galactose as a high molecular weight nondialyzable polymer which also contained arabinose (25).

Extraction of cell walls from lupin hypocotyls with alkali (10% KOH) at 2°C rapidly removed most of the hemicellulose A fraction (75% xylose) but failed to extract hydroxyproline-rich glycoproteins until after 4 hr, when a slow release of hydroxyproline-containing material occurred (12). Increasing the temperature to 18–22°C caused an immediate increase in the extraction of hydroxyproline-rich glycoproteins. The glycoproteins contained mainly arabinose and galactose, but the ratio of these two sugars changed with extraction time. Extraction of glycoproteins rich in hydroxyproline by this method appears to be dependent on the use of lupin hypocotyl cell walls (11), although similar results were obtained with mung bean (*Phaseolus aureus* L.) hypocotyl segments (26), and extraction of oat (*Avena sativa* L.) coleoptile walls with "cold" alkali (25°C) removed almost all of the hydroxyproline proteins (27).

When abraded primary leaves of French bean (*Phaseolus vulgaris* L.) were extracted with sodium hydroxide (1 M) at 2°C, glycoproteins with low hydroxyproline content were obtained (8, 9). Simple abrasion of the leaves resulted in extraction of glycoproteins rich in arabinose and xylose, whereas, if tobacco necrosis virus was present, the glycoproteins obtained were enriched in glucose. Hydroxyproline-poor glycoprotein preparations with similar amino acid compositions have been obtained from cell walls of runner bean (*Phaseolus coccineus*) where this component is associated with hemicellulose A (23).

One disadvantage of alkali extraction is base elimination reactions involving uronic acid methyl esters or serine-*O*-glycosidic linkages, the latter being more important for glycoproteins. A study of this reaction indicated that base elimination in 0.2 M NaOH was not significant at 4°C, but room temperature caused 60% destruction of serine in 18 hr. At 50°C, the reaction was complete within 5 hr (18).

Hydrazinolysis Hydrazine is a poor base but a good nucleophile. When glycoproteins are treated (100°C) with this reagent and hydrazine sulfate to act as a catalyst, depolymerization of the protein moiety occurs. The first report on this procedure indicated that the reaction was essentially complete; however, depolymerization was measured by the disappearance of rapidly sedimenting material during ultracentrifugation (28). Because this would have detected only very high molecular weight fragments, the claim for the reaction proceeding to completion was not justified. A later study demonstrated complete fission of peptides at temperatures as low as 60°C with little decomposition of susceptible amino acids by hydrazine (29). Furthermore, a 1,000-fold range of rate constants was demonstrated in catalyzed hydrazinolysis of peptides and attributed to steric effects (30). This has important consequences because hydrazine has been used to extract the carbohydrate moieties of plant cell wall glycoproteins (7, 14, 19). Two adjacent hydroxyproline residues yield a cyclic product which retains carbohydrate side chains (19). However, carbohydrate which is attached to serine is cleaved by a β elimination reaction with destruction of the substituted serine residues (18). The stability to hydrazine of the cyclic dipeptide which arises from adjacent hydroxyprolines was attributed to the lactam linkages. If the carbonyl group of these linkages is stable to hydrazine then perhaps other amino acids could form stable combinations with protected carbonyl groups. This might explain the high molecular weight of the hydrazinolysis products of plant glycopeptides from French bean when measured by gel filtration versus the low value obtained by methylation analysis (9). The failure to detect hydroxyproline hydrazide or the cyclic hydroxyproline dipeptide in the hydrazinolysate also suggests the presence of stable amino acid combinations. Amino acid analyses of the high molecular weight compound in hydrazinolysates of other plant glycoproteins have demonstrated enrichment of hydroxyproline but also the presence of most of the original amino acids (Table 1). Presumably, the presence of these amino acids indicates incomplete depolymerization rather than rearrangement of amino acid hydrazides through transamination reactions to give peptides which are stable to hydrazine. Clearly, the full use of hydrazine for the study of plant glycoproteins is predicated on a better understanding of the reactions involved.

Table 1. Amino acid composition of high molecular weight component in hydrazinolysates of plant glycoproteins

	Source of glycoprotein		
Amino acid	Sycamore cell wall[a]	Wheat endosperm (31)	*Phaseolus aureus*[b]
Lys	9.0	0	10
His	1.4	0	
Arg	1.8	0	
Hyp	36.1	35.0	
Asx	2.8	3.2	15
Thr	3.0	3.8	7
Ser	9.4	11.2	10
Glx	4.8	11.2	13
Pro	8.4	2.8	
Gly	3.7	9.9	16
Ala	4.3	8.1	10
Val	7.5	9.1	2
Met		0.2	
Ilc	2.5	1.6	6
Leu	3.9	2.1	10
Tyr	1.4	1.0	
Phe	Trace	0.9	

[a]Calculated from the data of Heath and Northcote (14).
[b]Franz, personal communication.

Material solubilized by hydrazinolysis of apple (*Malus domestica*) fruit cell walls was partly separated into a high molecular weight galactose-rich fraction and a low molecular weight hydroxyproline-arabinose-rich fraction (7). Degradation of the latter with arabinosidase suggested that it contained hydrazides of hydroxyproline arabinosides. The high molecular weight galactose-containing fraction may be similar to the galactose-rich polysaccharide released by hot barium hydroxide extraction of lupin hypocotyl cell walls (25).

Acetic Acid–Sodium Chlorite Attempts to extract intact glycoproteins from cell walls by chemical means have generally involved initial extraction of pectin, then "delignification" of plant tissue. Some tissue should be delignified with caution as certain procedures can cause significant losses of material (25). Parenchyma tissue and cell suspension cultures are free of lignin, thus the latter step is unnecessary when these materials are used as sources of glycoproteins (23). A novel use of chlorite-HOAc, a delignifying agent, for the isolation of hydroxyproline-rich glycoproteins from depectinated cell wall material has been reported (22, 23). The procedure can be used for isolating cell wall proteins from even heavily lignified tissues. Its primary disadvantage is that some of the constituent amino acids are either destroyed or modified. The main amino acids affected are tyrosine, cystine, methionine, and lysine. Cystine, methionine, and lysine give rise to cysteic acid, methionine sulfoxide, and α-aminoadipic acid, respectively. Fractionation of parenchyma cell wall material from runner bean showed that at least two main types of wall glycoproteins were present. One, relatively rich in hydroxyproline, is associated with α-cellulose, whereas the other is relatively poor in hydroxyproline and is associated with hemicellulose A. Only the former is extracted by the chlorite-HOAc treatment.

Miscellaneous Hot water has been used to extract proteoglycans from rice bran (*Oryza satira* L.) (32). Hydroxyproline-containing material can be extracted from apple fruit tissue by the following agents: water, 5 mM phosphate (pH 7), M KCl (pH 7.5), 0.2 M Tris-phosphate containing 0.01 M DIECA (pH 8), and 0.2 M Tris (pH 10) (7). Water extracts of wheat (*Triticum aestivum* L.) flour that had first been extracted with hot 80% ethanol contained only 2% protein, all of which was firmly bound to polysaccharide. If the ethanol treatment was omitted, up to 20% of the extracted polymeric fraction was protein which was not associated with polysaccharide (33). Generally, alcohol treatment is to be avoided because cytoplasmic proteins, nucleic acids, starch, and polyphenols are precipitated and solubilization of glycoproteins can be made difficult (6, 34). Rigorous extraction with the use of alkali, alkali with reducing agents, or hot trichloroacetic acid (TCA) was required to obtain hydroxyproline-containing material from corn (*Zea mays* L.) pericarp (35). Hot TCA was the most effective, but probably caused some degradation. About 30% of the hydroxyproline content of leaves of broad bean (*Vicia faba* L.) was extracted by cold, aqueous TCA (36). Hydroxyproline-containing glycoproteins in membranous organelles from carrot (*Daucus carota* L.) phloem parenchyma have also been extracted with cold TCA, whereas glycoproteins from cell walls were extracted with 0.2 M $CaCl_2$ at 2°C (37). The most effective agent for solubilizing glycoproteins from mung bean hypocotyl segments at room temperature was the powerful chaotropic salt, guanidinium thiocyanate (6 M).

Enzymic Extraction

Enzymic solubilization of intact plant glycoproteins has not been particularly successful although its potential usefulness, particularly for the isolation of cell wall glycoproteins, is obvious. A crude cellulase preparation which contained both protease and carbohydrase activity released glycopeptides from walls of tomato cells grown in suspension culture (16). Carbohydrase activity was not essential because pronase also released glycopeptides. The source of cell walls affects the effectiveness of enzymic treatment because only 10–20% of the sycamore wall hydroxyproline was released by the crude cellulase preparation, in contrast to 70–75% from tomato or tobacco (*Nicotiana*) walls. The major glycopeptides were large, containing from 34–45 sugar and amino acid residues. Pectinase, hemicellulase, chitinase, takadiatase, trypsin, chymotrypsin, pronase, subtilisin, pepsin, elastase, and collagenase released only negligible amounts of hydroxyproline-rich material from sycamore walls.

A combination of mild acid hydrolysis to cleave arabinofuranose residues, followed by treatment with trypsin, released glycopeptides containing single galactose residues linked *O*-glycosidically to serine from tomato cell walls (18). Application of this technique to carrot cell walls resulted in the production of similar glycopeptides, but the galactose side chains contained either 1 or 2 residues (17). Mild acid hydrolysis of a hydroxyproline-containing protein obtained from leaves of sandal (*Santalum album* L.) rendered the protein susceptible to hydrolysis by pronase, but the protein remained resistant to collagenase (38). Glycoproteins extracted from primary leaves of French bean did not need pretreatment to make them susceptible to hydrolysis by pronase (8, 9).

Treatment of sycamore (*Acer pseudoplatanus* L.) cell walls with endopolygalacturonase and endoglucanase, followed by protease, released glycopeptides which contained pectic fragments, thus providing evidence for a covalent connection between pectic polysaccharides and cell wall protein (39). Pretreatment of apple fruit walls with polygalacturonase increased the release of hydroxyprolyl residues by protease, although protease treatment alone could release hydroxyproline-containing material (7). No evidence was found to support covalent attachment of polyuronide to apple fruit wall protein, although evidence was obtained which indicated that polyuronide and wall protein are physically associated in the wall so that removal of polyuronide facilitates protease attack.

Determination of Purity

Protein homogeneity implies the presence of only one polypeptide chain of determinant molecular weight, whereas purity of a polysaccharide preparation means the presence of a family of closely related carbohydrate chains of indeterminant molecular weight. Any method of determining purity must take cognizance of both the protein and polysaccharide moieties of glycoproteins. Acrylamide gel electrophoresis and isopycnic centrifugation are perhaps the most widely used methods of determining homogeneity. Glycoproteins often give diffuse bands on acrylamide gel electrophoresis as a result of the microheterogeneity of the carbohydrate portion of the glycoprotein (40). Despite this shortcoming, gel electrophoresis followed by staining for both protein and polysaccharide components in split or parallel gels has been used extensively (8, 9, 32, 38, 40).

As polysaccharides have buoyant densities in the range 1.6–2.0 g/ml and proteins have buoyant densities of less than 1.4 g/ml, glycoproteins would be expected to be in the range 1.4–2.0 g/m (33). Isopycnic centrifugation of hydroxyproline-rich cell wall glycoprotein revealed a density of 1.436 g/cm^3 (37), whereas, the buoyant density of an arabinogalactan peptide from wheat endosperm was 1.65 (33). The profile of the arabinogalactan peptide was diffuse, which probably indicated heterogeneity in composition and hence buoyant density.

STRUCTURE

No plant glycoprotein has been studied thoroughly enough to reveal a complete structure. The amino acid and sugar composition of a large number of glycoprotein preparations has been determined, but only fragmentary knowledge about amino acid sequence is available. Likewise, knowledge of the sequence, linkage, and configuration of carbohydrate constituents is incomplete and in some cases contradictory. The nature of the covalent linkage between protein and carbohydrate moieties has received the most attention, and consequently, more is known about this particular aspect of glycoprotein structure.

Amino Acid Constituents

Plant glycoproteins contain all the amino acids normally found in proteins (9, 14, 21, 23, 32, 35, 36, 38, 41–48). The content of methionine and cysteine is generally low, as is the content of the basic amino acids, histidine and arginine. Ornithine has been tentatively identified in two plant glycoproteins (41, 46). Proline appears to be an almost universal component of plant glycoproteins, even those which possess enzymic activity (42, 45). The hydroxyproline content can vary from very high (18, 36, 41, 46, 49) to complete absence (42, 45). In this connection, furfural, but not 5-hydroxymethylfurfural, reacts with ninhydrin to give a series of three "yellow peaks," all eluted before aspartic acid during amino acid analysis (50). The likelihood of furfural formation during acid hydrolysis of glycoproteins suggests that it may be responsible for unidentified peaks (9) unless care is taken to remove it by lyophilizing hydrolysates before amino acid analysis. Glycoproteins which have been exposed to alkali may contain *allo*-4-hydroxyproline in addition to *trans*-4-hydroxyproline (13, 31, 51). The hydroxyproline content of the primary cell wall of monocotyledons is low (less than 0.2%) in comparison to dicotyledons (1.28% in sycamore cell walls) (52, 53). Cell walls of dicotyledons appear to contain both hydroxyproline-rich and hydroxyproline-poor glycoproteins (23). *Pleurochrysis scherrfelii* cell walls are composed of scales in a gelatinous matrix (43). Purification yields scale material containing 32% protein, characterized by a high content of serine, glycine, and glutamic acid, but no hydroxyproline. *Chlamydomonas gymnogama* cell walls contain a hydroxyproline-rich glycoprotein whose amino acid composition is remarkably similar to preparations from higher plants, thus indicating the ubiquitous nature of this glycoprotein in green plant cell walls (21). A mucopolysaccharide from corn pericarp contained hydroxyproline, serine, and threonine in nearly equimolar amounts (35).

Information on the amino acid sequence of only two glycoproteins is available. Trypsin treatment of acid-stripped cell wall glycoprotein ("extensin") yields peptides having serine at the NH_2 terminus and lysine at the COOH terminus (Figure 1) (54). Thus, lysyl-serine, a linkage cleaved slowly by trypsin, occurs with high frequency. The small number of tryptic peptides means that there is probably only one hydroxyproline-rich protein in the wall. The peptides contain only 6 of the 18 amino acids known to occur in plant cell walls; therefore, only a small portion of cell wall glycoprotein is represented by these peptides. The peptides are rich in hydroxyproline, but sequences of more than 4 adjacent hydroxyproline residues do not occur. Lamport suggests that a limit of 4 contiguous hydroxyproline residues may be related to the observation of Okabayashi et al. that helical conformation in the series of proline oligopeptides

A. Ser-Hyp-Hyp-Hyp-Hyp-Ser-Hyp-Ser-Hyp-Hyp-Hyp-Hyp-("Tyr"-Tyr)-Lys
Exists with 3, 2, or 1 galactose residues.

B. Ser-Hyp-Hyp-Hyp-Hyp-Ser-Hyp-Lys
Exists with 2, 1, or 0 galactose residues.

C. Ser-Hyp-Hyp-Hyp-Hyp-Thr-Hyp-Val-Tyr-Lys
Exists with 1 or 0 galactose residues.

D. Ser-Hyp-Hyp-Hyp-Hyp-Lys
Exists with 1 or 0 galactose residues.

E. Ser-Hyp-Hyp-Hyp-Hyp-Val-"Tyr"-Lys-Lys
Exists with 1 or 0 galactose residues.

Figure 1. Peptides isolated from acid-stripped tomato cell walls by treatment with trypsin (102).

commences after the tetramer. Sequences of five or greater complete the helical structure; perhaps, the role of serine in the tryptic peptides is to prevent helix formation. Heath and Northcote, quoting Rippon et al., state that proline oligopeptides of as few as 3 residues exist in the conformation of polyproline II, i.e., a left-handed helix in aqueous solution (14). In molecular aggregates of poly(*O*-acetyl-4-hydroxy-L-proline), these helices are folded and stacked antiparallel and are much more stable to changes of pH and salt concentration than polyproline II. Because the presence or absence of helical conformation in plant cell wall glycoproteins may have important consequences related to function, this question needs further investigation.

The partial amino acid sequence of bromelian has recently been determined (55). Comparison with papain indicated that the carbohydrate moiety is attached to residue 118 (asparagine) (55).

Carbohydrate Constituents

Arabinose The most common carbohydrate constituent of plant glycoproteins is arabinose (7, 9–11, 12, 14, 16, 19–24, 32, 35, 36, 38, 41, 46, 48, 49, 56). It is generally assumed to be L-arabinose, but few studies of glycoproteins have confirmed this. Biosynthesis of L-arabinose proceeds by epimerization of uridine 5′-(α-D-xylopyranoxyl pyrophosphate) at C-4 to yield uridine 5′-(β-L-arabinopyranosyl pyrophosphate) (57). No satisfactory explanation for the origin of L-arabinofuranose residues in plants is available, but direct formation of uridine 5′-β-L-arabinofuranosyl pyrophosphate from uridine 5′-β-L-arabinopyranosyl pyrophosphate seems to be the simplest and the most likely possibility. Glycosidically linked arabinofuranose residues are acid-labile, and all glycoproteins tested have arabinose only in this ring form. Determinations of the configuration of arabinofuranosidic linkages are few. Optical rotation (31) and enzymic hydrolysis (58, 59) indicate that arabinose is α-linked, but the β configuration was proposed in view of the resistance of tetraarabinoside to emulsin, a preparation reported to have a α-L-arabinofuranosidic activity (54).

Most of the arabinose in extensin occurs as a short tetrasaccharide side chain containing (1→2) and (1→3) linkages. The sequence of these linkages is the subject of considerable confusion. Lamport gives the sequence $A_{(f)}1{\rightarrow}3\ A_{(f)}1{\rightarrow}2\ A_{(f)}1{\rightarrow}3\ A_{(f)}$ (in the text) and $A_{(f)}1{\rightarrow}3\ A_{(f)}1{\rightarrow}2\ A_{(f)}1{\rightarrow}2\ A_{(f)}$ (in a diagram) in the same article (54). Talmadge et al. (56) reported the sequence $A_{(f)}1{\rightarrow}2\ A_{(f)}1{\rightarrow}2\ A_{(f)}1{\rightarrow}3\ A_{(f)}$ and claim this is the same sequence reported by Karr (60), but the article quoted does not contain this information. Brown et al. (9) reported that hydroxyproline-poor glycoproteins of *P. vulgaris* contain *O*-2-linked terminal arabinose. If these glycoproteins contain tetraarabinoside, the sequence $A_{(f)}1{\rightarrow}2\ A_{(f)}1{\rightarrow}3\ A_{(f)}1{\rightarrow}3\ A_{(f)}$ is implied. The ratio of (1→3) to (1→2) linkages in hydroxyproline-poor glucoproteins is 2:1 (9), whereas hydroxyproline-rich glycoproteins appear to have the opposite ratio, 1:2 (56). On this basis, the sequence in the two types of glycoproteins may well be different.

Hydroxyproline oligosaccharides from *Chlamydomonas reinhardtii* contain glucose and galactose in addition to arabinose, and the sequences Hyp–Ara–Glc–Ara–Gal–Ara and Hyp–Ara–Ara–Gal–Ara have been proposed (20).

Xylose Only a limited number of plant glycoprotein preparations contain D-xylose (9, 22, 23, 32, 35, 48, 61, 62). Biosynthesis of D-xylose occurs by decarboxylation of UDP-D-glucuronic acid (57). The reaction is catalyzed by the enzyme uridine 5′-β-D-glucuronic acid pyrophosphate carboxylyase and probably proceeds through a 4-keto intermediate. Only two investigations have confirmed the presence of xylose in glycoproteins. Pineapple (*Ananas comosus*) bromelain contains xylose as a β-linked nonreducing terminal residue attached to glucosamine (62, 63). Hydroxyproline-poor glycoproteins from *P. vulgaris* contain β-(1→4)-linked D-xylopyranose residues which form chains having a molecular weight up to 34,490. Xylanase treatment hydrolyzed all the susceptible xylose residues, suggesting that these residues are on the periphery of the glycoprotein.

Galactose Many plant glycoproteins contain D-galactose (7, 11, 12, 14, 16, 18–24, 31, 32, 35, 36, 38, 46, 48, 64). Although this distribution testifies to the importance of this sugar constituent, few studies have investigated its linkage. Infrared spectroscopy and optical rotation studies demonstrated that the arabinogalactan peptide from wheat endosperm contains β-linked D-galactopyranosyl residues (31). The resistance of the arabinogalactan peptide to enzymic hydrolysis suggests that the galactose residues are not joined by β-(1→4) linkages. However, the specificity of the arabinogalactanase used was not completely defined, and the arabinofuranosyl residues in the polysaccharide could render the D-galactose resistant to enzymic attack.

There is increasing evidence that cell wall hydroxyproline-rich glycoproteins contain a polyuronide arabinogalactan attached via serine (7, 22–24). Single D-galactose residues have previously been shown to be *O*-glycosidically linked to serine in extensin from tomato (18), whereas extensin from carrot contains a short chain of galactose residues linked to serine (17).

Glucose The occurrence of glucose in plant glycoproteins is of considerable interest because of the potential role of such glycoproteins in starch and cellulose biosynthesis. D-Glucose has been found in a number of glycoproteins (8, 9, 17, 32, 35, 43, 48, 64, 65–68). The glycoprotein acceptor involved in initiation of starch biosynthesis may contain glucose linked in two ways (66). Glucose on the periphery of the glycoprotein is hydrolyzed by β-amylase, whereas glucosidic linkages closer to the protein are resistant. The glycoprotein inducer of sexuality in *Volvox* contains glucose and a smaller quantity of mannose (48). The binding of the sexual hormone to concanavalin A demonstrates that either glucose or mannose or both must be α-linked (69). Glycoproteins having β-(1→4)-linked glucose residues have been found in *P. scherrfelii* (43), *Z. mays* (35), *P. vulgaris* (8, 9), and *P. aureus* (70). The scales of *P. scherrfelii* contain a cellulose–protein complex having cellulose associated with the peptide moiety in a way which does not interfere with the cellulosic crystal order (43). Cellulase treatment of the hydroxyproline-containing mucopolysaccharide from corn pericarp removes most of the glucose, suggesting that the glucose-containing polysaccharide is peripheral (35). Glycoproteins from *P. vulgaris* contain short chains of β-(1→4)-linked glucose residues (8, 9). A particulate membrane-bound glucan synthetase from *P. aureus* catalyzes the formation of a glycoprotein containing β-(1→3)- and β-(1→4)-glucosidic linkages (70).

Mannose The distribution of D-mannose in plant glycoproteins is restricted, and a low content of this hexose generally occurs in glycoproteins which also contain glucosamine (17, 40, 47, 48, 61, 63, 70–76). The storage glycoprotein, vicilin, contains only one carbohydrate chain/mol, and it is postulated that the glycoprotein is a simple tetramer composed of 4 identical subunits with a molecular weight of 50,000 (71). The major glycoprotein of French bean seeds (47, 75) can form a tetramer with a sedimentation constant of 19 S. The mannose content of these glycoproteins is low, i.e., 1–4%. Soybean (*Glycine max*) hemagglutinin contains mannose α-linked to *N*-acetylglucosamine (74), whereas pineapple bromelain contains mannose β-linked to di-*N*-acetylchitobiose (73).

Glucosamine D-Glucosamine is a common constituent of animal glycoproteins, and there are many reports that plant glycoproteins contain this sugar (35, 36, 40, 42, 45, 47, 48, 60–63, 71–75, 77, 78). However, evidence for the occurrence of glucosamine in many plant glycoproteins is indirect. Generally the content of this sugar is low, i.e., 0.2–4%. Soybean hemagglutinin contains *N*-acetyl-D-glucosamine β-linked to asparagine (74), whereas pineapple bromelian has the disaccharide, di-*N*-acetylchitobiose β-linked to asparagine (73).

Miscellaneous Sugars Glycoprotein preparations from a variety of plant sources have been reported to contain the following sugars: L-rhamnose (22, 23, 32, 46), D-galacturonic acid (22, 23, 46, 49), D-galactosamine (48, 79), mannitol (64), and fucose (61, 62).

Protein-Carbohydrate Linkages

To verify the presence of a glycoprotein, the region containing the covalent linkage between carbohydrate and protein moieties must be isolated. Only three amino acids, namely hydroxyproline, serine, and asparagine, have been demonstrated to participate in such linkages in plant glycoproteins.

Hydroxyproline Perhaps the most common linkage found in plant glycoproteins is L-arabinose linked *O*-glycosidically to hydroxyproline (13, 20, 21, 51). Despite the isolation of this dimer by several workers (21, 51), the configuration of this glycosidic bond has not been unambiguously demonstrated. Treatment of the dimer with α-L-arabinofuranosidase (58) might be useful for this purpose. D-Galactose is also linked *O*-glycosidically to hydroxyproline (20, 31). Sugars linked to hydroxyproline are not removed by a β elimination reaction when treated with alkali. Thus, reduction of alkali-treated and untreated glycoproteins with $[^3H]$sodium borohydride resulted in the same incorporation of radioactivity in both cases (32). This demonstrated that no *O*-glycosidic linkages were cleaved during alkali treatment and suggested that sugar constituents were linked to hydroxyproline. Incorporation of $[^3H]$sodium borohydride in alkali used to extract glycoproteins might be useful in demonstrating the presence of alkali-sensitive bonds, that is, those involving serine, in glycoproteins extracted by this method.

Serine Galactosylserine, a linkage cleaved by alkali, has been demonstrated in glycopeptides from extensin (17, 18). If sodium borohydride or sodium sulfite is present during alkali treatment, serine is converted to alanine or cysteic acid, respectively. Radioactive sulfite (^{35}S) may be used to label the cysteic acid. No information on the anomeric nature of the galactosidic bond is available.

Asparagine The linkage 1-acetamido-1-*N*-(4′-L-aspartyl)-2-deoxy-β-D-glucopyranosylamide has been demonstrated in soybean hemagglutinin (74) and pineapple bromelain (73).

Miscellaneous Linkages Scales of *P. scherrfelii* contain a cellulose-glycoprotein rich in serine (43). Linkage between polysaccharide and protein is thought to occur via an ether bond between a sugar hydroxyl group and the hydroxyl group of serine or threonine. An alternative proposal is linkage through asparagine or glutamine, presumably involving a sugar other than *N*-acetylglucosamine.

BIOSYNTHESIS

The biosynthesis of plant glycoproteins has been studied by a small number of investigators who, with a few exceptions, have directed their attention to hydroxyproline-rich glycoproteins.

Protein

Protein synthesis seems to occur by similar pathways in all organisms studied. Much of our knowledge of this process in plants is fragmentary and relies heavily on information obtained from prokaryotic systems. However, there is sufficient information available to have warranted two recent reviews (80, 81).

Formation of polypeptides takes place on ribosomes which are free or membrane-bound in the cytoplasm, the mitochondrion, and the chromoplast. Evidence from a number of reports (82–84) shows that it would be imprudent to use the functional location as a basis to make assumptions on the site of synthesis of a particular protein. Observations on the incorporation of $[^{14}C]$proline into the cytoplasmic proteins of cultured cells (53, 85–87) suggests that the protein moiety of hydroxyproline glycoproteins is assembled in the cytoplasm. This location is also supported by the presence of radioactivity in polyribosome-associated proteins isolated from tissue incubated with $[^{14}C]$proline (88).

Hydroxylation

A number of reports indicate that following assembly the proline-rich polypeptide chain is modified by hydroxylation of some of the proline residues to form hydroxyproline (27, 85, 86, 89, 90); this has been confirmed through use of the chelator α,α′-dipyridyl, which inhibits hydroxylation of proline but not its incorporation (91). Only the 4-*trans*-hydrogen is displaced (92), and molecular oxygen is the source of oxygen (93–95). The reaction is catalyzed by a peptidyl hydroxylase from the cytoplasm and requires O_2, Fe^{2+}, ascorbate, and a keto acid. Ketoglutarate gives maximum activity, but pyruvate and oxaloacetate are also effective (96).

Kinetic data for incorporation of [^{14}C]proline (85, 96, 97) and analysis of Fe^{2+} reversal of α,α′-dipyridyl-inhibited hydroxylation in the presence of cyclohexamide, an inhibitor of protein synthesis (91), indicate that proline hydroxylation occurs in the cytoplasm after the polypeptide is released from the ribosome. The possibility of further hydroxylation catalyzed by peroxidases at the cell wall (98) has been re-investigated and concluded to be unlikely (99).

Glycosylation

A variable proportion of the hydroxyproline residues are covalently bonded via an *O*-glycosidic link to arabinose (100). Further glycosylation follows to form a series of hydroxyproline arabinosides up to 4 pentose units long (13). The substrate is uridine diphosphate arabinose, and the reaction is catalyzed by a group of particulate arabinosyl transferases (60) which may be located in the Golgi complex (101). Although mono-, di-, tri-, and tetraarabinosides are synthesized by this enzyme system (60), it is not known whether the linkages are identical with those in the cell wall glycoprotein. Because the hydroxyproline tetraarabinoside is linked by two different linkages, one of which occurs twice, it is possible that at least four enzymes are required.

Serine residues in the hydroxyproline protein are attached by an *O*-glycosidic link to galactose (17, 18), and it is possible that further glycosylation may occur through links to a polyuronide arabinogalactan (24). There is as yet no evidence for the location or other details of this reaction, but the extent of serine glycosylation may be related to the developmental stage of the tissue (102).

Examination of protein glycosylation in animal systems shows that for certain glycoproteins it involves the formation of lipid intermediates such as dolichol monophosphates which are polyisoprene derivatives.

$$(CH_2{=}\underset{}{\overset{CH_3}{\overset{|}{C}}}{-}CH{=}CH_2)_n - CH_2{-}\overset{CH_3}{\overset{|}{C}}H{-}CH_2{-}CH_2{-}O{-}PO_3^{2-}$$

Dolichol monophosphates

For example, mammalian mannosylphosphoryl polyisoprenol can serve as the precursor for $(Man)_n$-GlcNAc-lipid, which is a precursor for the carbohydrate moiety of the glycoprotein (103, 104). Related glycosylated lipids have now been detected in plants (105, 106) and fungi (107). In addition, Forsee and Elbein (108) found that particulate preparations from cotton (*Gossypium*) fibers catalyze the incorporation of mannose from GDP-[^{14}C]mannose into lipid-linked polysaccharides and thence to glycoprotein with manosylphosphoryl polyisoprenol being possible intermediates. The lipopolysaccharides are formed by the incorporation of GlcNAc and UDP-GlcNAc into lipid. These results are consistent with a glycosylation pathway for GlcNAc-containing glycoproteins that resembles the system described for animal cells (104).

Cellular Location

The newly synthesized hydroxyproline glycoproteins are partitioned between the cytoplasm and the cell wall with the distribution being dependent on the nature of the glycoprotein, cell type, and developmental stage. Most of the studies with callus cell cultures show the hydroxyproline-containing material becoming associated with the cell wall (86, 109) and, more specifically, for extensin, the α-cellulose fraction (14). Transport to the cell wall is mediated by smooth membranes (109, 110) which may be derived from the Golgi complex (101, 111). In contrast to these results, the elongating cells of oat coleoptile contain hydroxyproline-poor glycoproteins of which only a minor proportion are transported to the cell wall (27). Also, Steward et al. (112), in a useful elaboration of earlier studies (15, 113), suggest the hydroxyproline proteins are largely retained in the cytoplasm of actively growing callus culture cells and are transported to the wall only in differentiating, mature, or senescent cells. In interpreting these results, it seems reasonable to suggest that a fraction of the cytoplasmic material may be a precursor to cell wall glycoprotein (see Brysk and Chrispeels (37)), and the remainder could function in the multiple roles that are now becoming evident for plant glycoproteins.

BIOLOGICAL SIGNIFICANCE

General

The in vitro properties of proteins are modified by the presence of covalently bound carbohydrate so that they are less susceptible to denaturation (114, 115), to proteolysis (116), and are less affected by storage (117). In addition to these effects on the properties of glycoproteins, a sufficient number of reports are now available to enable some general observations to be made on the function and role of the carbohydrate moiety.

The role of the prosthetic group in conjugated proteins and of those amino acid residues modified following polypeptide synthesis is to enhance or add a function above that available from the coded amino acids. An obvious and excellent example is the presence of coordinately bonded electrophilic agents such as Mg^{2+} and Fe^{2+}. Our knowledge of carbohydrate chemistry suggests that the contribution of polysaccharide prosthetic groups will be derived from their ability to interact in a cooperative manner with other carbohydrates and proteins. In cooperative systems, the formation of each noncovalent bond enhances the formation of the next bond and so on, making possible the self-assembly of stable macromolecular structures. Taking this as a guide, together with evolutionary implications and the demonstrated functions of certain glycoproteins, it is reasonable to predict certain general functions for the carbohydrate moiety.

In structural proteins, the carbohydrate will provide additional covalent and dipole bonding groups to other polysaccharides and proteins. This will enhance the integrity of interfacial boundaries and, in the case of the weaker electrostatic interactions, provide the potential for a nonenzyme-requiring rearrangement of structure. Illustrations of these functions may be found in glycoproteins of the plant cell wall (53) and the intercellular matrix material of some algal colonies (20, 54). In catalytic proteins, a direct involvement in catalysis seems unlikely, and there are a number of observations on glycoenzymes to support this view (4). It is more probable that the carbohydrate functions in a structure-protective role by increasing the number of surface interactions with water and cellular structures. This bonding could enhance enzyme stability and also provide for attachment to membranes and formation of enzyme complexes. Bonded to receptor and carrier proteins, carbohydrates could provide the specificity for recognition of proteins and carbohydrates through hydrogen bonding. Perhaps this is the function of the carbohydrate in those glycoproteins that may be responsible for symbiont specificity (118), cell adhesion (119), sexual induction (44), fusion of mating types (120), and pollen-stigma incompatability (121).

To fully understand the significance of plant glycoproteins presupposes a knowledge of their structure, composition, and function which, of course, is not yet available. Therefore, the concepts suggested only serve to outline the potential significance of these constituents. At the same time, these authors believe they are more useful and cause less stress on our understanding of evolution than the suggestion by some authors that the carbohydrate moiety is a tolerated but nonfunctional acquisition.

Cell Wall Structure

The presence of a hydroxyproline-rich protein in cell walls was reported in 1960 by Dougall and Shimbayashi (122), working with tobacco callus cultures, and by Lamport and Northcote (123) with cell cultures of sycamore. Subsequently, a similar protein was demonstrated in the walls of normal tissue (124). Lamport speculated that this protein could be a structural component of cell walls with a role in extension and named it "extensin" (53). Further work on the composition showed that the hydroxyproline residues were glycosylated with arabinose (see under "Hydroxyproline") and that extensin was therefore a glycoprotein (16, 100).

Hydroxyproline-rich glycoproteins have now been demonstrated in the cell walls of a wide variety of normal and cultured tissue. For example, they were found in all green plants surveyed (13) (determined as hydroxyproline arabinosides) and in the Chlorophyta (125) (determined as hydroxyproline content of wall protein) with the exception of the Characeae (125, 126). It is important to realize that despite their wide distribution throughout the Planta, there is evidence that they are absent from certain cells within an organism. For instance, the walls of *Nitella* internodal cells (125, 126), barley (*Hordeum sativum* Pers.) aleurone cells (127), and wheat endosperm cells (128).

In addition to extensin, cell walls contain other hydroxyproline-containing proteins (86, 98), a hydroxyproline-rich peroxidase has been located in pea (*Pisum sativum* L.) epicotyl cell walls (98, 99), and hydroxyproline-poor proteins were detected in the hemicellulose fraction of leguminous pod walls (22), elongating tissue (129), and monocotyledons (52).

Lamport (53, 102, 130), expresses the view that "the primary cell wall of higher plants exists as a protein-glycan network" in which a "network of protein, pectic polymers, and xyloglucan serves to cross link the cellulose fibers of the wall" (39). However, it should be recognized that the bonds which stabilize the wall glycoprotein to other wall constituents have not been identified. A consideration of the available groups shows that bonding could involve covalent linkages, charge, and dipole interactions with potentially both the carbohydrate and protein moieties participating. Because of a lack of suitable groups, it is unlikely that either fluctuating dipoles or hydrophobic interactions are of primary importance. Despite the large number of interactions which are possible between glycoproteins and other wall constituents, most workers seem inclined to the view that one form of bonding predominates and that this determines the function of the glycoprotein.

Lamport suggests that the serine residues in extensin are involved in cross-linking other cell wall macromolecules through a glycosidic attachment (102) to an arabinogalactan polyuronide (24). This association between pectic polysaccharides and extensin has also been postulated by others (39, 46), and a model of the cell wall proposed by Keegstra et al. (39) shows a sequence of glycoprotein (Ser-*O*-Gal)-arabinogalactan-rhamnogalacturonan-xyloglucan hydrogen-bonded to cellulose. In contrast, the evidence from a number of other studies is not consistent with pectic polysaccharides being covalently bonded to extensin (7, 12, 17, 23, 26). For example, polyuronides and glycoprotein can be extracted separately with either 10% KOH, 6 M guanidine thiocyanate (26), or, after treatment, with pectin esterase (7). Furthermore, Knee (7) has suggested that glycoprotein-polyuronide associations may be artifacts of preparation. Results such as these are consistent with a link through the arabinose and galactose side chains of the glycoprotein directly to cellulose (11, 23, 131). Because there are no free hemiacetal groups, these bonds will be dipole:dipole (including hydrogen bonds) and not covalent.

In conclusion, while the occurrence of a hydroxyproline-rich glycoprotein in the cell wall is generally accepted, the structural role of extensin has still to be established. Perhaps the clearest indication of this has been stated by Albersheim (132, 133): "The protein component of the wall may also have a structural function. There are sugars attached to it and there is some evidence that rhamnogalacturonan molecules bind to these sugars. Since the structural role of the protein has not been established, however, and since a coherent model of the wall can be formulated without it, we have ignored it in our proposed structure. . . ." However, with our present knowledge about the role of cell wall polysaccharides one could apply the same reasoning with equal force to any of the wall constituents!

Cell Growth

As originally proposed in the extensin hypothesis (53), the cross-links between extensin and other wall polysaccharides are cleaved in a process mediated by auxin, causing an increase in extensibility of the wall and thereby enhancing elongation. Extensin does appear to influence the mechanical properties of the wall, because its removal or modification in oat coleoptiles (134), *Cladophora,* and *Chaetomorpha* (135) causes an increase in extensibility. There is also some evidence that the binding of extensin to the cell wall is weakened during elongation, as hydroxyproline-containing material is more easily removed from elongating walls of lupin (11).

There is disagreement on the question of whether extensin content increases during elongation or only at its cessation. Using elongating pea stems, Klis and Lamport (136) found an increase in extensin content and the number of glycosylated serine residues and they suggest, "during elongation the degree of cross-linking of the material being deposited in the cell wall increases." Similarly, Winter et al. (137) found that the period of maximum elongation coincided with the highest hydroxyproline content. However, there are many more reports which find increases in cell wall hydroxyproline only when cell elongation has ceased (26, 98, 138–141). In addition, when elongation is inhibited by ethylene (98, 99), benzimidazole, and 2-chloroethylphosphonic acid (138), there is an increase in wall hydroxyproline. When this increase is blocked by the addition of α,α'-dipyridyl, growth is no longer inhibited by either benzimidazole or 2-chloroethylphosphonic acid (138). These results are not conclusive, but they do support the idea of a causal relationship between extensin deposition and resistance to elongation.

Interpretation of these results for extensin is further complicated by the recent evidence that wall extension is accompanied by a loosening of xyloglucan which is initiated by an increase in hydrogen ions (142). Obviously, further investigation is needed before the role of extensin in the control of elongation growth can be assessed. It is perhaps ironic that persuasive evidence for the role of extensin in wall structure and elongation may come from two tissues in which extensin is absent! The cell walls of wheat endosperm (128, 143) and barley aleurone (127) both lack a hydroxyproline-rich glycoprotein equivalent to extensin. This observation is interesting when one considers that neither of these cells undergoes elongation and both have walls which become disorganized during germination.

Polysaccharide Synthesis

The synthesis of polysaccharides requires a primer whose nature and mechanism of synthesis has been controversial for 30 years. Recently a major advance in understanding the synthesis of these macromolecules has been accomplished by the discovery that starch and glycogen may be polymerized on a glycoprotein which becomes incorporated into the product.

Starch synthetases (EC 2.4.1.11) and phosphorylases (EC 2.4.1.1) are widely distributed in plants and animals and catalyze the synthesis of α-1,4-glucosyl bonds to form linear glucosidic chains. Some of the phosphorylases appear to be glycoproteins capable of synthesizing α-1,4-glucan chains of the amylose type without the addition of a primer (67, 144–146). It is postulated that this reaction is involved in the formation of starch de novo (67, 114–147) and occurs by the addition of glycosyl residues from glucose 1-phosphate directly to the glucan moiety of the enzyme (67, 144). This mechanism has previously been demonstrated in the mammalian system (148).

Starch synthetases occur as particulate forms bound to the starch granule and also in a soluble state. An interesting paper by Lavintman et al. (66) deals with a membranous starch synthetase from potato (*Solanum tuberosum* L.). They propose that a transglucosylase catalyzes the transfer of glucose from UDP-glucose to an endogenous protein, forming a glucoprotein, which then accepts further glucose with the use of ADP-glucose, UDP-glucose, or glucose 1-phosphate as substrate to produce α-1,4-glucan chains. A similar mechanism has also been demonstrated, again in potato, for a soluble synthetase (68). In this system, glucose is transferred from glucose 1-phosphate (rather than UDP-glucose) to a protein acceptor which serves as a primer for α-1,4-glucan synthesis, with the use of ADP-glucose, UDP-glucose, or GDP-glucose as glucosyl donors.

The possibility that involvement of a glycoprotein acceptor may be a general mechanism for synthesis of polysaccharides is suggested by the reports of Villemez (149) and Franz (70). A membrane-bound glucan synthetase prepared from mung bean seedlings (70) used UDP-glucose as substrate to catalyze the formation of a glycoprotein which, Franz suggests, possesses acceptor functions for the initiation and lengthening of glucan chains.

This recent work has initiated a totally new area of investigation for polysaccharide synthesis, with many questions remaining to be answered. Can the protein-bound β- and α-1,4-glucans from the in vitro systems be equated with starch and cellulose-bound protein in vivo? How many enzymes are involved in forming links to the protein and addition of glycosyl residues? Which amino acids are glycosylated? Is the primer a protein or a glycoprotein, perhaps synthesized in the Golgi complex and transported to the site where further glycosylation will occur?

Pathogenesis

The mechanism whereby plants resist or tolerate infection remains uncertain. In some infections, the pathogen will become locally established, but spread to other regions of the plant is prevented through the development of a hypersensitive response by the host. The nature of the barrier which successfully localizes the pathogen within an area around the point of entry is not agreed upon.

However, a number of studies show that cells around the hypersensitive area undergo structural modification, and it has recently been suggested (150, 151) that wound-induced redifferentiation of the cell wall may be responsible for localization. In support of this hypothesis, Kimmins and Brown (150) found that the wounding caused by transmitting tobacco necrosis virus or tobacco mosaic virus to French bean leads to the formation of a glycoprotein (9) which may be related to the events of cell wall thickening. It is also significant that despite identical wounding from the inoculation, less glycoprotein is obtained from leaves infected with a virus which is not localized (152). This suggests that intercellular spread of this virus is associated with supression of some aspects of cell wall modification. In what could be a related observation, Esquerré-Tugayé et al. (51, 153) have shown that a cell wall hydroxyproline-rich glycoprotein increases during fungal infection and that susceptibility and hydroxyproline level are related. In addition to their application to disease resistance, these observations support the view that increased levels of cell wall glycoproteins are a part of the general wound response to injury (154).

Certain plant viruses are transmitted by the seed, and an interesting proposal has been made by Partridge et al. (155) that specificity is determined by a glycoprotein of the virus coat protein. They examined two closely related viruses, cowpea mosaic, which is seed-transmitted, and bean pod mottle virus, which is not, and found that cowpea mosaic has a glycoprotein coat while the capsid of bean pod mottle virus is devoid of carbohydrate.

This association between glycoprotein in the capsid and seed transmission was also found within a group of unrelated viruses. Partridge et al. (155) suggest that the carbohydrate functions as a recognition site in attachment to the reproductive tissue. Although speculative, this hypothesis deserves further examination particularly in view of Heslop-Harrison's work on pollen incompatability (see under "Reproductive Physiology").

Cell Surface Interactions

Symbiosis A wide-ranging investigation is being pursued in an effort to increase the availability of fixed nitrogen to crops. One of the greatest obstacles to this search is the specificity existing between the *Rhizobium* bacterium and its legume host. A major advance toward understanding this problem has been made by Bohlool and Schmidt (118), who studied the reactions between soy bean lectins, which are glycoproteins, and soy bean *Rhizobia.* The host lectins combined specifically with 22 of 25 strains of the nodulating bacterium, but not with any of 23 non-nodulating strains. The lectins bound to the lipopolysaccharide *O*-antigen on the *Rhizobium* surface (156). This surface interaction could be the basis for the specificity that precedes development of the nitrogen fixation symbiosis.

Reproductive Physiology Breeding patterns in plants can be determined by compatibility relationships. Interspecific incompatibility prevents fertilization between types that are too remote, and intraspecific self-incompatibility prevents a union that is too related. In an incompatible response, germination of the pollen tube is inhibited on or near the stigma papilla surface, and a β-1,3-glucan is formed on the papilla adjacent to the rejected pollen. A study of the incompatibility response in the Cruciferae by Heslop-Harrison (121) and Heslop-Harrison and Knox (157) indicates that a protein fraction of the pollen sexine is responsible for triggering rejection by the stigma. Preliminary fractionation of the protein isolate shows nine electrophoretically distinct bands, two of which may be glycoproteins. In view of the recent investigations of cell surface interactions (reviewed by Albersheim and Anderson-Prouty (158)), it will be interesting to learn whether the triggering molecules on the pollen wall are lectins or glycosyl transferases.

Although there are now some indications that lectins are involved in cell surface recognition phenomena, a great deal of uncertainty remains about their role. This is largely due to the difficulties encountered in determining their cellular location. A significant advance in solving this problem has been accomplished by Clarke et al. (159). They have developed a high resolution technique with the use of fluorescein isothiocyanate-labeled immunoglobulin preparations which bind to the lectins in sections of plant tissue. Specificity is determined by the use of specific sugar inhibitors. Using this method with legume cotyledons, they conclude that concanavalin A and phytohemagglutinin occur in cytoplasmic sites and that the β-lectins isolated by Jermyn (160) and Jermyn and Yeow (161) are associated with the plasmalemma, cell walls, and intercellular spaces. This technique can be applied to most plant tissues and should be very useful in studies of plant lectins.

Frost Resistance

All arctic and temperate plants will at some time be exposed to temperatures low enough to cause frost damage. When this occurs, ice crystals will form in the intercellular spaces, producing a disruption of cell organization which, together with the difference in water potential, causes export of water from the cell. Some plants (hardy) are less damaged by low temperatures than others (non-hardy), being less susceptible to water loss and intracellular ice crystal formation. It has been suggested (162) that hardy plants may resist dehydration through the formation of hydrophilic proteins. In sufficient quantities, these would reduce the amount of water flux and inhibit crystal growth. Although increases in soluble protein levels have been found in frost-hardy plants (162–165), glycoproteins will be potentially more effective. For this reason, the reports by Williams (166–168) of glycoprotein changes in a hardy plant are worthy of attention. He has isolated from dogwood (*Cornus florida* L.) three glycoprotein fractions that constitute 5–7% of the cell sap, 37% of the osmolytes, and can render osmotically inactive 1–3 g of cell water/g of glycoprotein, equivalent to 30% of the total cell water. Formation of a new glycoprotein fraction during induction of freezing tolerance in locuse (*Robinia pseudoacacia* L.) has also recently been reported (169).

Table 2. Plant glycoenzymes

EC No.	Systematic name	Trivial name	Source	Carbohydrate (%)	Monosaccharides	Reference
1.11.1.7	Donor: H_2O_2 oxidoreductase	Peroxidase	*Raphanus raphanistroides*, Nakai	28	–	176
			Armoracia rusticans	18	Gal, Ara, Xyl, Fuc, Man, GalN, ManN	177
			Pisum sativum			98, 99
1.14.18.1	*p*-Diphenyl: O_2 oxidoreductase	Laccase	*Prunus persica*			45
3.2.1.22	α-D-Galactosidase					
	Galactohydrase I	α-D-Galactosidase I	*Vicia faba*	25		178
	Galactohydrase II	α-D-Galactosidase II	*Vicia faba*	3		178
3.2.3.1	Thioglucoside Glucohydrolase	Myrosinase, Sinigrinase	*Sinapis alba*	18	GlcN, hexose, pentose	42
3.4.22.3	–[a]	Ficin	*Ficus* sp.	6	GlcN, Man, Fuc, Xyl	61
3.4.22.4	–[a]	Stem bromelain II	*Ananas comosus*	2.4	D-Man, D-GlcNAc, L-Fuc, D-Xyl	63
		Stem bromelain III	*Ananas comosus*	2	D-Man, D-GlcNAc	63
3.4.22.5	–[a]	Fruit bromelain	*Ananas comosus*	3.3		179

[a]Systematic name not assigned.

The stress of freezing is partly one of dehydration, and it is an interesting parallel that tolerance to drought and dessication is also associated with the presence of water-binding macromolecules, for example, the colloidal polysaccharides of the intertidal seaweeds.

Germination

The "storage" proteins of leguminous seeds have been characterized as glycoproteins (47, 71, 170–172). It is suggested that the significance of the carbohydrate prosthetic group will be related to the stress tolerance and water-imbibing properties (170, 171).

Enzymes

An increasing number of enzymes in the Protista, Animalia, and Planta are being shown to contain carbohydrate prosthetic groups. Although examples have been found in all the enzyme classes, most of the known glycoenzymes are either hydrolases or oxidoreductases. A list of the glycoenzymes reported in plants is presented in Table 2.

The carbohydrate group increases the stability of the enzyme in vitro, but the function in vivo has not been established. It has been proposed (173) that extracellular proteins may be aided in their transport by the carbohydrate serving as a tag for membrane carrier proteins. While some role in transport cannot be excluded, it can certainly be discounted as a general explanation for the transport of extracellular proteins because some of these are not glycosylated (174). A direct role in catalysis also seems unlikely because modification (117, 175) or removal (63) of the carbohydrate moiety has been shown to have very little effect on activity.

REFERENCES

1. Lindberg, B. (1972). Methods Enzymol. 28B:178.
2. Aspinall, G. O., and Stephen, A. M. (1973). *In* G. O. Aspinall (ed.), MTP International Review of Science, Organic Chemistry, Series 1, Vol. 7, p. 285. University Park Press, Baltimore.
3. Spiro, R. G. (1972). Methods Enzymol. 28B:3. Academic Press, New York.
4. Spiro, R. G. (1973). *In* C. B. Afinsen, J. T. Edsall, and F. M. Richards, Advances in Protein Chemistry, Vol. 27, pp. 349–455. Academic Press, New York.
5. Lindberg, B., Lönngren, and Svensson, S. (1975). *In* R. S. Tipson and D. Horton (eds.), Advances in Carbohydrate Chemistry and Biochemistry, pp. 185–240. Academic Press, New York.
6. Knee, M. (1973). Phytochemistry 12:637.
7. Knee, M. (1975). Phytochemistry 14:2181.
8. Brown, R. G., and Kimmins, W. C. (1973). Can. J. Botany 51:1917.
9. Brown, R. G., Kimmins, W. C., and Lindberg, B. (1975). Acta Chem. Scand. 29:843.
10. Brown, R. M., Frank, W. W., Kleinig, H., Falk, H., and Sitte, P. (1969). Science 166:894.
11. Monro, J. A., Bailey, R. W., and Penny, D. (1974). Phytochemistry 13:375.
12. Monro, J. A., Bailey, R. W., and Penny, D. (1975). Carbohydr. Res. 41:153.
13. Lamport, D. T. A., and Miller, D. (1971). Plant Physiol. 48:454.
14. Heath, M. F., and Northcote, D. H. (1971). Biochem. J. 125:953.

15. Israel, H. W., Saltpeter, M. M., and Steward, F. C. (1968). J. Cell Biol. 39:698.
16. Lamport, D. T. A. (1969). Biochemistry 8:1155.
17. Cho, Y.-P., and Chrispeels, M. J. (1976). Phytochemistry 15:165.
18. Lamport, D. T. A., Katona, L., and Roerig, S. (1973). Biochem. J. 133:125.
19. Heath, M. F., and Northcote, D. H. (1973). Biochem. J. 135:327.
20. Miller, D. H., Lamport, D. T. A., and Miller, M. (1972). Science 176:918.
21. Miller, D. H., Mellman, I. S., Lamport, D. T., and Miller, M. (1974). J. Cell Biol. 63:420.
22. Selvendran, R. R., Davies, A. M. C., and Tidder, E. (1975). Phytochemistry 14:2169.
23. Selvendran, R. R. (1975). Phytochemistry 14:2175.
24. Mort, A., and Lamport, D. T. A. (1975). Plant Physiol. (Suppl.), 56:16.
25. Monro, J. A., Bailey, R. W., and Penny, D. (1972). Phytochemistry 11:1597.
26. Bailey, R. W., and Kauss, H. (1974). Planta 119:233.
27. Cleland, R. (1968). Plant Physiol. 43:865.
28. Yosizawa, Z., Sato, T., and Schmid, K. (1966). Biochim. Biophys. Acta 121:417.
29. Bradbury, J. H. (1956). Nature 178:912.
30. Bradbury, J. H. (1958). Biochem. J. 68:475.
31. Fincher, G. B., Sawyer, W. H., and Stone, B. A. (1974). Biochem. J. 139:535.
32. Yamagishi, T., Matsuda, K., and Watanabe, T. (1975). Carbohydr. Res. 43:321.
33. Fincher, G. B., and Stone, B. A. (1974). Aust. J. Biol. Sci. 27:117.
34. Selvendran, R. R. (1975). Phytochemistry 14:1011.
35. Boundy, J. A., Wall, J. S., Turner, J. E., Woychik, J. H., and Dimler, R. J. (1967). J. Biol. Chem. 242:2410.
36. Puztai, A., and Wall, W. B. (1969). Eur. J. Biochem. 10:523.
37. Brysk, M. M., and Chrispeels, M. J. (1972). Biochim. Biophys. Acta 257:421.
38. Mani, U. V., and Radhakri, A. N. (1974). Biochem. J. 141:147.
39. Keegstra, K., Talmadge, K. W., Bauer, W. D., and Albersheim, P. (1973). Plant Physiol. 51:188.
40. Ericson, M. C., and Chrispeels, M. J. (1973). Plant Physiol. 52:98.
41. Allen, A. K., and Newberger, A. (1973). Biochem. J. 135:307.
42. Bjorkman, R., and Jansen, J.-C. (1972). Biochim. Biophys. Acta 276:508.
43. Herth, W., Franke, W. W., and Stadler, J. (1972). Planta 105:79.
44. Kochert, M., and Yates, I. (1974). Proc. Natl. Acad. Sci. U.S.A. 71:1211.
45. Lehman, E., Harel, E., and Mayer, A. M. (1974). Phytochemistry 13:1713.
46. Puztai, A., Begbie, R., and Duncan, I. (1971). J. Sci. Food Agric. 22:514.
47. Puztai, A., and Watt, W. B. (1970). Biochim. Biophys. Acta 207:413.
48. Starr, R. C., and Jaenicke, L. (1974). Proc. Natl. Acad. Sci. U.S.A. 71:1050.
49. Moore, T. S. (1973). Plant Physiol. 51:529.
50. Scurfield, G., and Nicholls, P. W. (1970). J. Exp. Botany 21:857.
51. Esquerré-Tugayé, M.-T., and Mazau, D. (1974). J. Exp. Botany 25:509.
52. Burke, D., Kaufman, P., McNeil, M., and Albersheim, P. (1974). Plant Physiol. 54:109.
53. Lamport, D. T. A. (1965). *In* R. D. Preston (ed.), Advances in Botanical Research, Vol. 2, pp. 151–218. Academic Press, New York.
54. Lamport, D. T. A. (1974). *In* Macromolecules Regulating Growth and Development, 30th Symposium of the Society of Developmental Biology, p. 113. Academic Press, New York.
55. Goto, K., Murachi, T., and Takahashi, N. (1976). FEBS Lett. 62:93.
56. Talmadge, K. W., Keegstra, K., Bauer, W. D., and Albersheim, P. (1973). Plant Physiol. 51:158.
57. Feingold, D. S., and Fan, D. F. (1975). *In* F. Loewus (ed.), Biogenesis of Plant Cell Wall Polysaccharides, p. 69. Academic Press, New York.
58. Laborada, F., Archer, S. A., Fielding, A. H., and Byrde, R. J. W. (1974). J. Gen. Microbiol. 81:151.
59. Knee, M., Fielding, A. H., Archer, S. A., and Laborda, F. (1975). Phytochemistry 14:2213.
60. Karr, A. L. (1972). Plant Physiol. 50:275.
61. Friedenson, B., and Liener, I. E. (1974). Biochim. Biophys. Acta 342:209.
62. Murachi, T., Suzuki, A., and Takashi, N. (1967). Biochemistry 6:000.
63. Scocca, J., and Lee, Y. C. (1969). J. Biol. Chem. 244:4852.

64. Howe, M. L., and Barrett, J. T. (1970). Biochim. Biophys. Acta 215:97.
65. Lavintman, N., and Cardini, C. E. (1973). FEBS Lett. 29:43.
66. Lavintman, N., Tandecarz, J., Carceller, M., Mendiara, S., and Cardini, C. E. (1974). Eur. J. Biochem. 50:145.
67. Slabnick, E., and Frydman, R. B. (1970). Biochem. Biophys. Res. Commun. 38:709.
68. Tandecarz, J., Lavintman, N., and Cardini, C. E. (1975). Biochim. Biophys. Acta 399:345.
69. Pall, M. L. (1974). Biochem. Biophys. Res. Commun. 57:683.
70. Franz, G. (1975). Abstract of the 8th Cellulose Conference, Cellulose Research Institute, Suny, Syracuse, New York.
71. Ericson, M. C., and Chrispeels, M. J. Biochemistry, in press.
72. Koshiyama, I. (1969). Arch. Biochem Biophys. 130:370.
73. Lee, Y. C., and Scocca, J. R. (1972). J. Biol. Chem. 247:2753.
74. Lis, H., Sharon, N., and Katchalski, E. (1969). Biochim. Biophys. Acta 192:364.
75. Racusen, D., and Foote, M. (1971). Can. J. Botany 49:2107.
76. Sawai, H., and Morita, Y. (1970). Agric. Biol. Chem. 34:61.
77. Roberts, R. M., Cetorelli, J. J., Kirby, E. G., and Ericson, M. (1972). Plant Physiol. 50:531.
78. Roberts, R. M., Connor, A. B., and Cetorelli, J. J. (1971). Biochem. J. 125:999.
79. Goding, L. A., Bhatty, R. S., and Finlayson, A. J. (1970). Can. J. Biochem. 48:1096.
80. Zalik, S., and Jones, B. L. (1973). Annu. Rev. Plant Physiol. 24:47.
81. Boulter, D., Ellis, R. J., and Yarwood, A. (1972). Biol. Rev. 47:113.
82. Hoober, J. K. (1972). J. Cell Biol. 52:84.
83. Hoober, J. K. (1970). J. Biol. Chem. 245:4327.
84. Smith-Hohannsen, H., and Gibbs, S. P. (1972). J. Cell Biol. 52:598.
85. Pollard, J. K., and Steward, F. C. (1959). J. Exp. Botany 10:17.
86. Olson, A. C. (1964). Plant Physiol. 39:543.
87. Dashek, W. V., and Rosen, W. G. (1966). Protoplasma 61:192.
88. Sadava, D., and Chrispeels, M. J. (1970). Biochemistry 10:4290.
89. Steward, F. C., and Pollard, J. K. (1958). Nature 182:828.
90. Holleman, J. (1967). Proc. Natl. Acad. Sci. U.S.A. 57:50.
91. Chrispeels, M. J. (1970). Plant Physiol. 45:223.
92. Lamport, D. T. A. (1963). Fed. Proc. 22:647.
93. Lamport, D. T. A. (1963). J. Biol. Chem. 238:1438.
94. Stout, E. R., and Fritz, G. J. (1964). Plant Physiol. (Suppl.), 39:XX.
95. Stout, E. R., and Fritz, G. J. (1966). Plant Physiol. 41:197.
96. Sadava, D., and Chrispeels, M. J. (1971). Biochim. Biophys. Acta 227:278.
97. Chrispeels, M. J. (1970). Biochem. Biophys. Res. Commun. 39:732.
98. Ridge, I., and Osborne, D. J. (1970). J. Exp. Botany 21:843.
99. Ridge, I., and Osborne, D. J. (1971). Nature (New Biol.) 229:205.
100. Lamport, D. T. A. (1967). Nature 216:1322.
101. Gardiner, M., and Chrispeels, M. J. (1975). Plant Physiol. 55:536.
102. Lamport, D. T. A. (1974). *In* F. Loewus (ed.), Biogenesis of Plant Cell Polysaccharides, p. 149. Academic Press, New York.
103. Lucas, J. J., Waechter, C. J., and Lennerz, W. J. (1975). J. Biol. Chem. 250:1992.
104. Hsu, A.-F., Baynes, J. W., and Heath, C. E. (1974). Proc. Natl. Acad. Sci. U.S.A. 71:2391.
105. Forsee, W. T., and Elbein, A. D. (1973). J. Biol. Chem. 248:2858.
106. Alam, S. S., and Hemming, F. W. (1971). FEBS Lett. 19:60.
107. Sharma, C. B., Babczinski, P., Lehle, L., and Tanner, W. (1974). Eur. J. Biochem. 46:35.
108. Forsee, W. T., and Elbein, A. D. (1975). J. Biol. Chem. 250:9283.
109. Roberts, K., and Northcote, D. H. (1972). Planta 107:43.
110. Dashek, W. V. (1970). Plant Physiol. 46:831.
111. Gardiner, M. G., and Chrispeels, M. J. (1973). Plant Physiol. (Suppl.), 51:60.
112. Steward, F. C., Israel, H. W., and Salpeter, M. M. (1974). J. Cell Biol. 60:695.
113. Steward, F. C., and Chang, L. O. (1963). J. Exp. Botany 14:379.
114. Hashimoto, Y., Tsuiki, S., Nisizawa, K., and Pigman, W. (1963). Annu. N. Y. Acad. Sci. 106:233.
115. Arnold, W. N. (1969). Biochim. Biophys. Acta 178:347.

116. Gottschalk, A., and de St. Growth, S. F. (1960). Biochim. Biophys. Acta 43:513.
117. Pazur, J. H., Knull, H. R., and Simpson, D. L. (1970). Biochem. Biophys. Res. Commun. 40:110.
118. Bohlool, B. B., and Schmidt, E. L. (1974). Science 185:269.
119. Robertson, A., Cohen, M. H., Drage, D. J., Durston, A. J., Rubin, J., and Wonio, D. (1972). *In* L. S. Silvestri (ed.), Cell Interactions. North-Holland, Amsterdam.
120. Crandall, M., Lawrence, L. M., and Saunders, R. M. (1974). Proc. Natl. Acad. Sci. U.S.A. 71:26.
121. Heslop-Harrison, J. (1975). Proc. R. Soc. Lond. 190:275.
122. Dougall, D. K., and Shimbayashi, K. (1960). Plant Physiol. 35:396.
123. Lamport, D. T. A., and Northcote, D. H. (1960). Nature 188:665.
124. King, N. J., and Bayley, S. T. (1965). J. Exp. Botany 16:294.
125. Gotelli, I. B., and Cleland, R. (1968). Am. J. Botany 55:907.
126. Thompson, E. W., and Preston, R. D. (1967). Nature 213:684.
127. McNeil, M., Albersheim, P., Taiz, L., and Jones, R. L. (1975). Plant Physiol. 55:64.
128. Mares, D. J., and Stone, B. A. (1973). Aust. J. Biol. Sci. 26:793.
129. Cleland, R. (1971). Annu. Rev. Plant Physiol. 22:197.
130. Lamport, D. T. A. (1970). Annu. Rev. Plant Physiol. 21:235.
131. Northcote, D. H. (1972). Annu. Rev. Plant Physiol. 23:113.
132. Albersheim, P. (1975). Sci. Am. 232:80.
133. Albersheim, P. (1975). Presented at the XIIth International Botanical Congress, July 3–10, Leningrad, USSR.
134. Cleland, R. (1964). Planta 74:197.
135. Thompson, E. W., and Preston, R. D. (1968). J. Exp. Botany 19:690.
136. Klis, F. M., and Lamport, D. T. (1974). Plant Physiol. (Suppl), 55:15.
137. Winter, H., Meyer, L., Hengeveld, E., and Wiersma, P. K. (1971). Acta Botany Neerl. 20:489.
138. Sadava, D., and Chrispeels, M. J. (1973). *In* F. Loewus (ed.), Biogenesis of Plant Cell Wall Polysaccharides, p. 165. Academic Press, New York.
139. Sadava, D., Walker, F., and Chrispeels, M. J. (1973). Dev. Biol. 30:42.
140. Barnett, N. M. (1970). Plant Physiol. 45:188.
141. Cleland, R., and Karlsnes, A. (1967). Plant Physiol. 42:669.
142. Jacobs, M., and Ray, P. M. (1975). Plant Physiol. 56:373.
143. Mares, D. J., and Stone, B. A. (1973). Aust. J. Biol. Sci. 26:813.
144. Fredrick, J. F. (1971). Physiol. Plant 25:32.
145. Fredrick, J. F. (1974). Plant Sci. L. 3:183.
146. Fredrick, J. F. (1975). Plant Sci. L. 5:131.
147. Tsai, C. Y., and Nelson, O. E. (1968). Plant Physiol. 43:103.
148. Feigin, I., Fredrick, J. F., and Wolf, A. (1951). Fed. Proc. 10:181.
149. Villemez, C. L. (1970). Biochem. Biophys. Res. Commun. 40:636.
150. Kimmins, W. C., and Brown, R. G. (1973). Can. J. Botany 51:1923.
151. Wu, J. H. (1973). Virology 51:474.
152. Kimmins, W. C., and Brown, R. G. (1975). Phytopathology 65:1350.
153. Esquerré-Tugayé, M.-T., and Touzé, A. (1975). Abstracts of the XII International Botanical Congress, p. 491. July 3–10, Leningrad, U.S.S.R.
154. Chrispeels, M. J., Sadava, D., and Cho, Y.-P. (1974). J. Exp. Botany 25:1157.
155. Partridge, J. E., Shannon, L. M., Sumpf, D. J., and Colbaugh, P. (1974). Nature 247:391.
156. Albersheim, P., Ayers, A., Ebel, J., Valent, B., and Wolpert, J. (1975). Plant Physiol. (Suppl.) 56:23.
157. Heslop-Harrison, J., and Knox, R. B. (1974). Theoret. Appl. Genet. 44:133.
158. Albersheim, P., and Anderson-Prouty, A. J. (1975). Annu. Rev. Plant Physiol. 26:31.
159. Clarke, A. G., Knox, R. B., and Jermyn, M. A. (1975). J. Cell Sci. 19:157.
160. Jermyn, M. A. (1974). Proc. Aust. Biochem. Soc. 7:32.
161. Jermyn, M. A., and Yeow, Y. M. Aust. J. Plant Physiol., in press.
162. Siminovitch, D., Rheaume, B., Pomeroy, K., and Lepage, M. (1968). Cyrobiology 5:202.
163. Gerloff, E. D., Stahmann, M. A., and Smith, D. (1967). Plant Physiol. 42:895.
164. Jung, G. A., Shih, S. C., and Shelton, D. C. (1967). Plant Physiol. 42:1653.

165. Levitt, J. (1972). Responses of Plant to Environmental Stresses. Academic Press, New York.
166. Williams, R. J. (1972). Cyrobiology 9:313.
167. Williams, R. J. (1973). Plant Physiol. (Suppl.) 51:25.
168. Williams, R. J. (1974). Cyrobiology 11:555.
169. Brown, G. N., and Bixby, J. A. (1975). Physiol. Plant 34:187.
170. Puztai, A., and Duncan, I. (1971). Biochim. Biophys. Acta 229:785.
171. Racusen, D., and Foote, M. (1973). Can. J. Botany 51:495.
172. Ericson, M. C., and Chrispeels, M. J. (1975). Plant Physiol. (Suppl.) 56:59.
173. Eylar, E. H. (1965). J. Theoret. Biol. 10:89.
174. Winterbum, P. J., and Phelps, C. F. (1972). Nature (New Biol.) 236:147.
175. Yasuda, Y., Takahashi, N., and Murachi, T. (1971). Biochemistry 10:2624.
176. Morita, Y., and Kameda, K. (1959). Bull. Agric. Chem. Soc. Jap. 23:28.
177. Shannon, L. M., Kay, E., and Lew, J. Y. (1966). J. Biol. Chem. 241:2166.
178. Dey, P. M., and Pridham, J. B. (1969). Biochem. J. 113:49.
179. Ota, S., Moore, S., and Stein, W. H. (1964). Biochemistry 3:180.

International Review of Biochemistry
Plant Biochemistry II, Volume 13
Edited by D. H. Northcote

7
Functions of Ion Transport in Plant Cells and Tissues

E. A. C. MACROBBIE
Botany School, University of Cambridge, Cambridge, England

EXISTENCE OF ACTIVE ION TRANSPORT IN PLANT CELLS 212
Active Transport of Sodium and Potassium 213
Active Proton Extrusion 214
Cellular Functions of Proton Pump 216
Net Salt Accumulation 218
Plasmalemma Processes 219
Transfer from Cytoplasm to Vacuole 220
Cellular Functions of Salt Accumulation 223
Osmotic Regulation 223
Action Potentials 224

FUNCTIONS OF ION TRANSPORT PROCESSES IN TISSUES: PLANT MOVEMENTS 226
Stomatal Guard Cells 226
Leaf Movements: Endogenous Rhythms and Sleep Movements 234
Leaf Movements: Seismonasty 237
Insectivorous Plants 239
Dionaea 239
Drosera 240
Utricularia 240

ROLE OF IONS IN DEVELOPMENTAL PROCESSES 241
Fucoid Eggs 241
Regenerating *Acetabularia* Fragments 242

This chapter is concerned with functions of active ion transport in plants, both within plant cells and within the whole plant. After a discussion of the nature of the processes of active ion transport shown to exist in plant cells, a variety of plant processes are considered, for which ion movements are responsible. Before considering more specialized processes involving ion movements, this review discusses the functions of ion transport at the cellular level, most of which are common to all plant cells, rather than specific to particular types of cell. Certain basic roles are general, being concerned with the maintenance of suitable cytoplasmic conditions or with growth. A subsequent section is concerned with processes which are specific to particular types of cell, with a function at the plant or organ level rather than at the cellular level. Most of this discussion is concerned with two kinds of process. The first of these involves plant movements in which ion-mediated turgor changes are involved; examples include stomatal movements, sleep movements of leaves, motor action in the sensitive plant *Mimosa,* and movement of traps in various insectivorous plants. A second type of process in plants for which ion fluxes are responsible is that found in signaling; certain types of communication within the plant body are achieved by the propagation of electrical signals, i.e., by the occurrence of action potentials in excitable tissues. Roles for ion movements in various hormone responses, in a number of developmental processes, and in endogenous rhythms have also been proposed but are less well understood.

EXISTENCE OF ACTIVE ION TRANSPORT IN PLANT CELLS

Ion transport processes are essential in plant cells in two general ways–in the maintenance of suitable cytoplasmic conditions for metabolism and in the generation of the high internal osmotic pressure to allow turgor-driven cell expansion. Two aspects of cytoplasmic composition are essentially regulated, the K:Na ratio and the cytoplasmic pH, and although the primary transports responsible seem to be at the plasmalemma, the importance of transport from cytoplasm to vacuole is also indicated. Expansion growth in walled plant cells is primarily a process of vacuolation, of the creation and expansion of solute-filled compartments within, but separate from, the cytoplasm. This is essentially a process of vacuolar salt accumulation, because the major fraction of osmotic pressure is contributed by either inorganic or organic salts; whether the accumulation is of KCl or of K-carboxylate depends on the nature of the plant, the nature of the solutes available to the cells concerned, and their stage of development, but salts rather than sugars are concerned with the ordinary processes of vacuolation and expansion growth. Salts are also concerned with most of the secondary processes in which turgor changes play a role.

Most work has been concerned with measurements and interpretation of ion fluxes at the plasmalemma, and three transport systems have been identified at that membrane: 1) a cation transport system responsible for maintaining high internal K:Na; 2) a proton extrusion process with a role in pH regulation but also responsible for a number of processes of secondary active transport of other solutes; and 3) an active accumulation of anions (for example, chloride). The relations between these transport processes are still somewhat unclear. The existence of specific uptake processes for metabolically important anions, such as phosphate, nitrate, etc., will not be considered in this chapter, which will be restricted to the anion uptake associated with maintenance of turgor by uptake and vacuolar accumulation of salt; under some conditions, with high nitrate available, nitrate may be accumulated in the vacuole and contribute to the cell osmotic pressure, instead of being largely metabolized, but there is much less information available on the properties and characteristics of the nitrate transport system than on the chloride transport system.

Thus, what can be discussed represents only a part of the whole ion transport activity in any plant cell in that it is restricted to those ions and those membranes on which measurements are easily made. Most of our information refers to plasmalemma fluxes; information on vacuolar fluxes is scanty and ill-defined. Two areas of complete ignorance stand out. The first is that the question of intracellular fluxes has not yet been tackled. There is information on ion movements in chloroplasts in vitro, particularly in isolated thylakoids, but there are no in vivo measurements on chloroplasts in the intact cell. But what may be even more important in the overall ion balance in the cytoplasm may be the question of movements of ions into and out of the endomembrane system, particularly the endoplasmic reticulum. The second area in which vital information is simply not available is that of calcium fluxes; the fluxes considered in ion transport work—those of K^+, Na^+, Cl^-, and H^+—may be quantitatively most important in the cell, but it would seem likely that regulation of Ca^{2+} levels in various compartments of the cytoplasm may be qualitatively much more important. The role of intracellular membrane systems in regulating Ca^{2+} levels in the bulk cytoplasm of a range of animal cells (and in slime molds) is now recognized (1,2), and it is unfortunate that no information is yet available in plant cells.

Active Transport of Sodium and Potassium

In a very wide range of cells, the level of cytoplasmic sodium is maintained low, and active extrusion from the cytoplasm, both to the outside and to the vacuole, is indicated. In giant algal cells, the cytoplasmic concentrations and potentials can be measured, and the site of pumping can be allocated with more certainty, but what evidence is available from higher plants suggests that the same pattern holds also there (3, 4). Under many conditions, the electrogenic proton extrusion to be discussed in the next section maintains a membrane potential more negative than the potassium equilibrium potential, and a net potassium influx can be maintained passively from a low external concentration to a cytoplasmic concentration of 100 mM or higher. In these conditions, the level of potassium in the cell is not above its equilibrium concentration, and active transport of potassium, defined as transport against its electrochemical potential gradient, is not observed. However, the question of the mechanism of potassium influx, by passive, independent diffusion or by a carrier-mediated mechanism, is not settled by the thermodynamic definition, and the kinetics of potassium transport in many cells suggests carrier-mediated transport (5–8). At very low external concentrations of potassium, when only a low cell concentration could be achieved passively, there is evidence of active potassium influx, and in *Lemna* the activity of this transport system is regulated in response to the external level of potassium (9).

It is suggested that a high K/Na level in the cytoplasm is essential for the activity of many enzymes, in particular for protein synthesis (10). There is evidence that, even in halophytes, this sensitivity to high sodium remains (11), and that cytoplasmic sodium levels are maintained low in halophytes, even when very high vacuolar concentrations are observed. But beyond maintaining ionic conditions in the cytoplasm suitable for metabolism, no other functions for the cation transport system have been proposed, and it has not been implicated in secondary processes, either at a cellular or multicellular level. Thus, cotransport of solutes with sodium, although important in animal cells, does not seem to occur in plant cells or fungal cells (12).

Active Proton Extrusion

An active proton extrusion at the plasmalemma is now recognized as a major transport activity in plant cells, with a role in pH regulation and a number of secondary functions. There is both direct and indirect evidence for such a transport in many plant cells, and it is now considered to be a basic property of all plant cells, whose activity is essential for cell function. The direct evidence comes from observation of acidification of the medium, in spite of an adverse electrochemical potential gradient for proton extrusion on any credible value assigned to cytoplasmic pH. This is characteristic of conditions of excess cation uptake and has been recognized for many years, in both roots and storage tissues (13–17). The indirect evidence comes from the electrical behavior of the membrane and the evidence for a metabolically-powered electrogenic transport process transferring positive charge out of the cell. In the absence of any other suitable candidate for the charge transfer, the pH dependence of the process reinforces its identification with an active proton extrusion pump. The existence of such electrogenic transport in a range of plant cells was demonstrated by Higinbotham et al. (18) and is reviewed by Higinbotham and Anderson (19), Slayman (12), and MacRobbie (3).

The consequence of this process is the establishment of a more negative membrane potential and the maintenance of a pH gradient across the membrane, with the cytoplasm much more alkaline than would otherwise be the case. The diffusion potential across the plasmalemma is commonly of the order of 100 mV (inside negative), and the membrane potentials of −200 mV or so which are often observed are the consequence of proton extrusion.

The most extensive work on the characteristics of the proton pump has been done in *Neurospora crassa* by Slayman (20) and Slayman et al. (21, 22), who have shown that the decline in membrane potential on metabolic inhibition by a range of respiratory inhibitors parallels the decline in cell ATP levels. It appears, therefore, that a transport ATPase is involved, with the energy for proton transport derived from ATP hydrolysis. A similar conclusion is reached from work on the giant algal cells. Kitasato (23) was first to recognize the importance of H^+ influxes for the electrical properties of *Nitella,* but his hypothesis was later modified and extended by Spanswick (24–26). Spanswick suggested that proton movement controlled the membrane potential in Characean cells as a result of the operation of a voltage-dependent active proton extrusion pump, which provides the major charge-carrying species in the plasmalemma. The inhibitor sensitivity of the associated hyperpolarization of the membrane potential was consistent with ATP as the energy source; thus, in *Nitella* the transport could be driven by far-red light, in conditions in which cyclic photophosphorylation, but not noncyclic phosphorylation, could function.

Spanswick argues that the pump will extrude H^+ until the electrochemical gradient for proton excretion becomes too large for the free energy decrease in the driving reaction, when the pump will stall. Thus, the limit on the combination of pH gradient, ΔpH, and membrane potential, ΔE, is set by the energy available from the driving reaction of ATP hydrolysis. In the absence of processes tending to discharge these gradients, the pump will then stall. However, there are processes in the membrane by which the continued build-up of these gradients can be moderated, and the proton extrusion in the steady state will then be controlled by such dissipative processes. The charge transfer will be balanced by the net passive flux of ions, of which the passive influx of K^+ may be the major part. The pH balance can only be achieved by processes tending to produce H^+ inside, either in biochemical reactions such as organic acid synthesis, by membrane transport processes such as passive H^+ influx, by inward cotransport of H^+ with some other solute, or by exchange of internal OH^- for an external anion. Two ways in which this might be achieved have been suggested. In one, the charge balance is achieved by K^+ influx, whereas Cl^--OH^- exchange (or H^+Cl^- in a cotransport inward) is responsible for limiting the pH gradient; the net result is uptake of K^+ and Cl^- in equivalent amounts. In the second, by the activation of phosphoenolpyruvate (PEP) carboxylase the OH^- is removed internally by organic acid synthesis to generate oxalacetate, and thence malate or an alternative organic acid. In this form, acidification of the medium will be observed, and internally K^+ and malate will be accumulated. It should be noted, however, that neither of these sequences is possible as a continuing process unless salt is removed to the vacuole; in the absence of vacuolar accumulation, the driving forces for continued entry of K^+ and Cl^- will become inadequate, and the build-up of cytoplasmic malate will inhibit PEP carboxylase (27).

Evidence that the proton pump in *Chara corallina* is close to the stalled condition has been provided by recent measurements of the proton electrochemical potential gradient ($\Delta\bar{\mu}_{H^+}$) under various conditions by Smith and Walker (28, 29). They found that the cytoplasmic pH (pH_c) rose little with increasing pH outside (pH_o), fitting the relation $pH_c = 6.28 + 0.22\ pH_o$. At external pH values in the range 5–8, the measured $\Delta\bar{\mu}_{H^+}$ agreed well with half the value estimated for the free energy of hydrolysis at the cytoplasmic pH. $\Delta\bar{\mu}_{H^+}$ remained constant at 27 kJ/mol over the external range pH 5–7 but fell at higher external pH. They argue that these results indicate the operation of a proton-pumping ATPase in the membrane, with a stoichiometry of $2H^+$:1 ATP, held close to the stalled equilibrium condition.

Figures are also available for cytoplasmic and vacuolar pH in *Phaeoceros laevis* (30). The cytoplasm appears to maintain a more or less constant pH of 6.6–6.8 over the external range 4.7–6.7 in the light, and at pH 5.7 in light or dark. At an external pH of 5.7, this provides a proton gradient $\Delta\bar{\mu}_{H^+}$ of 24–27 kJ/mol, similar to the figure already quoted for *Chara.*

The existence of proton transport from cytoplasm to vacuole is also clearly indicated, because the cell vacuole is considerably more acid than the cytoplasm, and the tonoplast potential is small (10–20 mV, with the vacuole positive with respect to the cytoplasm). In *Phaeoceros,* Davis (30) found that the vacuole pH changed markedly from light to dark; at an external pH of 5.7, the vacuolar pH was 5.8 in the light but 6.6 in the dark, giving values of $\Delta\mu_{H^+}$ across the tonoplast of 5.7 kJ/mol in light but only 2.0 kJ/mol in the dark. Thus, considerable light activation of the proton transport from cytoplasm to vacuole is indicated. In *Chara* the proton gradient at the tonoplast is significantly higher, with a vacuolar pH of 5.5 (31), and $\Delta\bar{\mu}_{H^+}$ of some 10–15 kJ/mol (28). Hence, proton transport from cytoplasm to vacuole may also be important in regulating cytoplasmic pH, but there is no information on the nature or characteristics of the process.

Cellular Functions of Proton Pump

The functions of proton pumps in cells have been considered by Raven and Smith (32–34), who suggest that the mechanism arose as a means of regulating cytoplasmic pH, but evolved to serve other functions also (35). Their suggestion is that the proton pump is part of a biophysical pH-stat by which excess H^+ generated in common conditions of growth may be excreted from the cell; such conditions include growth with carbohydrate or CO_2 as the C source, or with NH_4^+ as N source. With nitrate as N source, nitrate reduction in the cytoplasm will offset this by producing OH^-, and the net effect will depend on the growth conditions. Such a biophysical pH-stat may complement the operation of the biochemical pH-stat proposed by Davies (36, 37), in which an increase in the cytoplasmic pH leads to activation of PEP-carboxylase and generation of H^+, whereas a decrease leads to activation of malic enzyme and generation of OH^-. Raven and Smith (38) have recently considered the operation of these mechanisms in the higher plant as a whole, where metabolic processes may be divided between roots and shoots and where large exchanges of H^+ between symplasm and apoplasm in the leaves and shoots would produce secondary problems. They argue that the forms in which nitrogen is delivered to the leaf in the xylem have adjusted to minimize the rate of generation of H^+ in leaf cells and the consequent proton excretion to the free space. But in roots, or in free-living cells in direct contact with the external environment, proton extrusion at the plasmalemma does offer a mechanism for avoiding the consequences of a steady production of H^+ in metabolism.

The second major role proposed for the proton pump is in generating the driving force for secondary active transport of other solutes, such as sugars and amino acids. Thus, evidence is available that glucose uptake in *Chlorella* involves the cotransport of glucose and H^+, with the energy for asymmetric movement provided by the proton gradient created by the proton pump (39, 40). Uptake of amino acids and sugars by yeast (41, 42) also shows characteristics of this type of transport. The role of the proton gradient in a glucose transport system is best established in *Neurospora,* which allows both electrical and flux measurements and direct measurements of ATP levels (43). This work makes it clear that the electrical effects of initiating glucose transport are consistent with secondary active transport, a cotransport of glucose and H^+. Recent work on sucrose uptake by *Ricinus* cotyledons (44) also suggests the operation of a proton cotransport system and may provide an explanation of the very high pH typical of sieve tube exudates. Thus, there is evidence available, from a number of solutes and tissues, for a role of the proton gradient in other processes of solute uptake, and it seems likely that the mechanism is more widespread than is yet recognized.

A role of the proton gradient in driving chloride uptake at the plasmalemma has also been suggested, but here the nature of the link is less clear. Smith (45) revived, with modification, older concepts of salt uptake as a double exchange process with his suggestion that chloride uptake in giant algal cells, which is strongly light-dependent, was achieved by Cl^--OH^- exchange. There is evidence suggesting that chloride uptake is sensitive to both internal and external pH, but neither the nature of the light activation nor the energy supply for chloride uptake is clearly established. The evidence for Cl^--OH^- exchange includes the inhibition of Cl^- influx at high pH and its stimulation by ammonium ions (45, 46); such effects are also seen in pretreatment experiments at high pH or with ammonium, both of which stimulate Cl^- influx. But the more recent work in which cytoplasmic pH was measured makes it clear that the proton gradient is too small to provide the energy source and not very much affected by external pH or by light (28, 29). The implication is that, although Cl^- transport may involve cotransport with H^+ or exchange for OH^-, the proton gradient is unlikely to provide the sole driving force, and some other biochemical energy source is required. It has been suggested that a stoichiometry of $2H^+:1Cl^-$ might avoid these difficulties, but the amount of extra Cl^- influx produced by pretreatment with ammonium is too high to be explained in these terms (47). The measured stoichiometry of extra Cl^- transport to potential OH^- storage in the pretreatment period was found to be close to 1:1, which is surprisingly high given competing dissipative processes, even if the true value is 1:1; this may instead suggest an activation of the transport by internal pH changes rather than a substrate effect.

Net Salt Accumulation

The ability of vacuolate plant cells to accumulate solutes in a central vacuole, occupying some 90% of the total volume in the mature cell, is perhaps their most striking characteristic. It is a property of plant cells which sets them apart from animal cells. A major fraction of the cell's osmotic pressure is always contributed by salts, but there are two basic patterns of accumulation, depending on the conditions and the phase of growth of the cell.

Cram (48) has compared the contributions of inorganic salts, organic salts, and sugars in algae, glycophytic higher plants, and halophytic higher plants. Although chloride (balanced largely by potassium) makes up the major fraction of the osmotic pressure in algae, this is not true in higher plants. In glycophytes, the osmotic pressure is typically 13–17 atm., but, of this, chloride (with balancing cation) makes up only 3–23%; the major contribution is of organic salts, commonly as malate. In plants growing exposed to very high salt levels (mangroves, halophytes, or salt marsh plants), chloride makes up 50–66% of the higher osmotic pressure of 30–33 atm., and sodium is also high. But, in general, accumulation of either KCl (with smaller amounts of NaCl) or of K-carboxylate is responsible for the generation of cell turgor, with minor contributions from sugars or amino acids.

Steward and Mott (49, 50) have stressed this versatility in the pattern of vacuolar accumulation and have compared the behavior of carrot cells in culture in a range of conditions. They drew attention to the changing pattern during growth; in young cells with incipient vacuoles the osmotic pressure is largely contributed by potassium plus organic acid anions, and it is only in mature cells growing in the presence of chloride that KCl is accumulated to high levels. By contrast, in algae, in which a great deal of experimental work has been done, the process of vacuolar accumulation is basically one of inorganic salt accumulation, largely in the form of KCl.

Steward and Mott stress a further point—that the emphasis should be placed on solute accumulation in cell vacuoles, on the secretion of solutes in internal vacuoles, with the creation of vacuolar volume, and that the particular solute concerned is of secondary importance. This is an important concept, but one which has not yet been translated into a well-defined molecular mechanism.

In each pattern of vacuolar salt accumulation, two processes are involved—uptake of ions at the plasmalemma and transfer from cytoplasm to vacuole. In the accumulation of KCl, the uptake of both K^+ and Cl^- and their subsequent transfer to the vacuole is involved. In the accumulation of K-carboxylate, which occurs in the presence of an anion which cannot easily be absorbed, such as sulfate, or in the presence of one which is metabolized, such as bicarbonate, we observe instead the exchange of K^+ for H^+ at the plasmalemma and the transfer to the vacuole of K^+ with internally synthesized organic acid anion, such as malate. As has already been argued, neither of the uptake processes at the plasmalemma can continue unless the removal of salt from cytoplasm to vacuole also occurs. Hence, although two parts of the overall process are involved—the events at the plasmalemma and the transfers from cytoplasm to vacuole—the two may not operate independently. It is important to consider the two separately, but also to consider their interrelations. In practice, there is very little information on tonoplast fluxes or on the nature of tonoplast transport processes, and most work has been concerned with the mechanisms involved at the plasmalemma. It is only in giant algal cells, in which separation of cytoplasm and vacuole is possible, that attempts to make direct measurements of tonoplast fluxes are possible. It is perhaps unfortunate, therefore, that such cells show only one of the two patterns of vacuolar salt accumulation, that involving chloride accumulation.

Plasmalemma Processes It has been suggested that the proton extrusion pump is closely involved with the plasmalemma processes in both types of accumulation. In the accumulation of K-carboxylate, it is suggested (3, 14, 15, 51, 52) that the primary proton extrusion at the plasmalemma both provides the electrical driving force for K^+ entry and increases the cytoplasmic pH; it is argued that the pH change activates PEP-carboxylase (and also increases the level of HCO_3^-, the correct form of its substrate), thereby generating oxalacetate, and thence malate. In the alternative form of KCl accumulation, it has also been argued that proton extrusion is a primary driving force and that Cl^--OH^- exchange is a secondary consequence. Some of the difficulties in identifying the proton gradient as the energy source for chloride transport have been discussed in the previous section. Smith and Walker's (28, 29) measurements of $\Delta\bar{\mu}_{H^+}$ in *Chara* make it clear that chloride uptake can occur against its electrochemical potential gradient, at high rates, in conditions in which $\Delta\bar{\mu}_{Cl}$ is greater than $\Delta\bar{\mu}_{H^+}$. Evidence against a stoichiometry of $2H^+$:$1Cl^-$ have already been cited (47), and the conclusion seems to be clear: even if the mechanism of Cl^- uptake at the plasmalemma involves cotransport with H^+(or exchange with OH^-), some further energy input is required.

The question of the nature of this energy input is still unclear. In two of the giant algal cells it seems that it is not ATP. In both *Nitella translucens* and *Hydrodictyon africanum,* chloride influx is strongly light-dependent but will not work in conditions of cyclic electron flow, in far-red light, or in the presence of low concentrations of dichlorophenyldimethylurea (DCMU) (53–56). Cyclic phosphorylation is capable of driving other plasmalemma transport processes in these conditions, including the proton pump (26), and hence cyclically produced ATP is capable of reaching the plasmalemma. It would seem, therefore, that chloride transport in these cells is not simply ATP-dependent; this is consistent with the observations that chloride transport is relatively insensitive to uncouplers. The simplest hypothesis is that some energy input other than ATP is involved, and Raven and Glidewell (57) have suggested a scheme involving an oxidation-reduction-powered Cl^- pump in the plasmalemma. The only alternative would be to suggest that conditions for cyclic electron flow are inhibitory to chloride influx for some other reason, connected with the conditions for export of ATP from the chloroplast. Two exchange systems are involved in the export of ATP and reducing power out of the chloroplast—a phosphate translocator (carrying 3-phosphoglycerate, glyceraldehyde 3-phosphate, dihydroxyacetone phosphate, and inorganic phosphate) and a dicarboxylate translocator (carrying oxalacetate and malate) (58, 59). Under cyclic conditions, it is necessary to return both the carbon of phosphoglycerate and the reducing equivalents to the chloroplast, relying on light energy to supply only the ATP. This differs from the needs for export of ATP in noncyclic conditions, in which the phosphate translocator may simply exchange cytoplasmic inorganic phosphate for chloroplast triose phosphate, and there is no requirement to return phosphoglycerate also. It may be that the operation of the shuttle system in cyclic conditions, with return of the doubly charged anion, phosphoglycerate^{2-}, is only possible with conditions of low cytoplasmic pH, which in turn inhibits chloride influx. In *Chara* the situation is even less clearcut, because there is no evidence that cyclic phosphorylation can even power transport in the plasmalemma (60). This may reflect simply an inability to export ATP from the chloroplasts in cyclic conditions in *Chara,* but it means that there is no possibility of distinguishing between ATP and some other form of energy coupling in *Chara.*

Thus, the nature of the light effect on chloride influx is not established, but neither the effect of light on cytoplasmic pH nor on ATP levels seems likely to explain the light activation. In this connection, the observation by Takeda and Senda (61) that the light activation of chloride influx in *Nitella,* which is dependent on noncyclic conditions of electron flow, is not transmitted from a lighted region of the cell to a darkened region is interesting.

The evidence suggests, therefore, that the activity of the proton pump will affect chloride uptake through its influence on cytoplasmic pH and potential, but that the proton gradient is not adequate to power the chloride transport and that coupling to some other biochemical reaction is also required; the nature of this further process is still unclear. That chloride influx in turn affects cytoplasmic pH is the simplest explanation of Cram's observation (62) that external chloride inhibits malate synthesis in the cytoplasm—a process of Cl^--OH^- exchange would be expected to prevent the rise of cytoplasmic alkalinity to which the activation of PEP-carboxylase is attributed.

Transfer from Cytoplasm to Vacuole Very little information is available on the nature and characteristics of ion transfer processes from cytoplasm to vacuole. Measurements of cytoplasmic ion concentrations are limited to a very few systems, and measurements of clearly defined tonoplast fluxes pose even more difficulty. In various higher plant tissues, estimates for cytoplasmic concentrations and tonoplast fluxes have been made by kinetic analysis of the time course of tracer efflux from labeled tissue (3, 63). Such estimates depend on assumptions about the arrangement of kinetic compartments within the cell and on steady fluxes over an extended period of hours, and hence the figures obtained are somewhat uncertain. The use of chloride-sensitive microelectrodes in the cytoplasm of young, nonvacuolate cells in mung bean root tips is likely to give a more reliable estimate of cytoplasmic concentrations (64), but it has not yet proved feasible in mature higher plant cells. Even in giant algal cells, from which cytoplasmic samples can be obtained, there is still argument about the values for ion concentrations in particular phases of the cytoplasm and for the ionic gradients between cytoplasm and vacuole. Nevertheless, it is quite clear that both sodium and chloride are transferred from cytoplasm to vacuole against their electrochemical potential gradients and that active transport is therefore demanded. Whether these are independent transport systems or are related has not yet been established, and it is indeed only with respect to chloride transport from cytoplasm to vacuole that a study of the characteristics of the process has been attempted.

Two species have been used for flux measurements; the published figures for chloride concentration in the flowing cytoplasm of *N. translucens* are 65–87 mM (65, 66), whereas for *C. corallina* the figures are 10–20 mM (67, 68). Active transport of chloride is indicated both into the cytoplasm at the plasmalemma and from cytoplasm to vacuole. In both species, the chloride concentrations in the chloroplast layer are considerably higher (130–240 mM), thus ion compartmentation within the cytoplasm must be considered in the interpretation of tracer movements.

The measurement of a flux from cytoplasm to vacuole demands the measurement of two quantities–the rate of tracer transfer to the vacuole and the specific activity of the tracer in the cytoplasmic compartment from which it moves to the vacuole. In practice, what the experiments provide is the time course of the arrival of tracer chloride in the vacuole over the early stages of labeling, and from these kinetics workers must deduce the number and nature of the kinetic compartments involved in vacuolar transfer. Such measurements have been made in *N. translucens, Tolypella intricata* (69–71), and *C. corallina* (72–74). In all three, it is clear that two components of vacuolar transfer contribute to the vacuolar kinetics. The interpretation of the fast phase is still open to doubt, and there are two possibilities. If the fast phase represents a genuine cytoplasmic compartment, it can only be a very small one. Although Walker and Pitman (63) have argued that the fast phase should be identified with the bulk cytoplasm, outside the plastids, this explanation is not tenable in *Nitella.* Measurements of the time course of rise of specific activity in the flowing cytoplasm, deduced from the rate of transfer of tracer to the other end of a cell only half immersed in radioactive solution, make it clear that the bulk cytoplasm must be identified with the slow component of vacuolar transfer (75). Hence, the fast component, if it is to be attributed to a real cytoplasmic compartment, can be only a very small membrane-bound phase within the bulk cytoplasm, and the endoplasmic reticulum would seem to be the obvious candidate. The alternative explanation is that the fast component of vacuolar transfer reflects the transfer of tracer to the vacuole which is associated with the occurrence of action potentials on cutting open the cell for analysis. This explanation explains the peculiarities of the fast phase as measured–its extreme variability, its clustering in groups, its independence of the slow component–and is in many ways a more satisfactory explanation. It is, however, not settled by clear experimental evidence which view should be favored.

What is important for understanding the process of chloride transport to the vacuole are the properties of the slow phase of vacuolar transfer. In *Nitella,* for which measurements have been made in a range of conditions, there are two unexpected features of the kinetics which tell us something of the nature of the process involved (69, 71, 75). The two features established are the complete lack of discrimination between chloride and bromide in the transfer from cytoplasm to vacuole and an apparent link between transfer to the vacuole and influx at the plasmalemma. Inability to distinguish between Br^- and Cl^- would not be expected of any active transport mechanism transporting halide ions as such, and indeed a marked preference for chloride over bromide is shown by the plasmalemma mechanism. It might, however, be more consistent with a mechanism of salt transfer rather than single anion transfer. The apparent link between influx and vacuolar transfer is also difficult to explain. What is established is in fact a relation between the rate constant for exchange of the cytoplasm and the influx. This rate constant is determined by the ratio of the fluxes out of the phase concerned and its content; the flux to the vacuole, which is much larger than the efflux to the outside, is the determining flux. If the concentration of chloride in the phase concerned changed in response to an increase in influx at the plasmalemma, then the rate constant for vacuolar transfer would be expected to fall, if the tonoplast flux remained the same, or to remain the same, if both tonoplast flux and concentration rose proportionally. An increase in the rate constant for vacuolar transfer implies rather that the transfer to the vacuole rises as the plasmalemma influx rises, in the absence of a concentration change. The behavior is more consistent with a regulatory system, in which the cytoplasmic concentration is maintained constant, and the flux out of the cytoplasm is increased in response to any potential increase in content arising from increased inflow at the plasmalemma. The process of vacuolar transfer seems to be more directly controlled by the rate of change of cytoplasmic content, rather than by the content itself. This could be a direct reflection of changing salt levels in the cytoplasm, or it might possibly be an indirect effect mediated through consequent pH changes in the cytoplasm and regulation of vacuolar transfer by pH. If it is an indirect effect, it may be that the plasmalemma processes are in fact controlled by the processes of vacuolar transfer. It was argued earlier that in the absence of other processes dissipating the proton gradient (ΔpH and ΔE) the plasmalemma proton pump will increase $\Delta\bar{\mu}_{H^+}$ to a limit set by the free energy of the driving reaction, when it will stall; if K^+ entry and Cl^--OH^- exchange can occur and arrest the steady rise in $\Delta\bar{\mu}_{H^+}$, then the proton pump will turn over slowly, well below its maximum rate, at a rate set by such dissipative processes. As this ensues, $\Delta\bar{\mu}_{K^+}$ $(\bar{\mu}_{out} - \bar{\mu}_{cyt})$ will fall, and $\Delta\bar{\mu}_{Cl^-}(\bar{\mu}_{cyt} - \bar{\mu}_{out})$ will rise, until a point is reached at which further entry by K^+ and Cl^- is limited by their removal to the vacuole. In this state, the processes of vacuolar transfer may exert control on the plasmalemma fluxes.

Hence, the question of the control of plasmalemma and tonoplast fluxes is an interesting one, but unsolved, and the interpretation of the apparent link between chloride influx and transfer of chloride to the vacuole is important for our understanding of the overall process.

The kinetics of chloride transfer to the vacuole does not define the nature of the process involved, but it is not easily explained in terms of a model of single ion pumps in a static tonoplast membrane. It is perhaps easier to envisage a process of salt transfer to the vacuole in the creation of new vacuoles within the cytoplasm and their discharge to the central vacuole. In the original process of vacuolation, the vacuole is generally considered to arise from endoplasmic reticulum by the formation of small vesicles from it and their subsequent fusion. Some of these may be lysosomes, and it has been argued that the vacuole should be considered as part of the lysosomal compartment of the cell (76, 77). But formation and discharge of nonlysosomal vesicles from endoplasmic reticulum is also argued, and the occurrence of the enzymes NADH-cytochrome *c* oxido-reductase and NADPH-cytochrome *c* oxidoreductase in the tonoplast membrane (78–80) is consistent with an origin from endoplasmic reticulum. It is likely that the maintenance of the vacuole in the mature cell involves the same processes as were involved in its original formation. On these grounds, it can be argued that the processes of salt transport concerned in vacuolar accumulation are likely to be associated with the endoplasmic reticulum. In this case, the membrane flow processes may well be controlled by the consequences of a rise in solute content of the endoplasmic reticulum, with subsequent swelling of the organelle. This type of mechamism might provide a degree of common control on all vacuolar transfer processes. The nature of the vacuolar transfer processes remains to be established, but it seems clear that the ion fluxes should not be considered in isolation, divorced from the processes of vacuole formation and of creation of new vacuolar volume.

Cellular Functions of Salt Accumulation

Osmotic Regulation The primary function of salt accumulation is to generate turgor, to create the driving force for expansion growth. The question of the control of turgor generation and the regulatory characteristics of the ion transport processes have recently been reviewed by Cram (81) and is clearly important for cell function. In some instances, turgor as such is controlled–for example, in a number of marine algae which can grow in a range of salinities. Thus, in *Valonia ventricosa,* a marine species having giant cells, turgor is kept constant over a considerable range of external osmotic pressure by the sensitivity of active K^+ influx to hydrostatic pressure (82, 83). Effects of turgor on both active and passive K^+ fluxes in *V. utricularis* are also suggested by Zimmermann and Steudle (84). In another marine alga, *Codium decorticatum,* in which turgor regulation is also observed, the effect of pressure is exerted on the active influx of chloride (85). In *Nitella flexilis,* on the other hand, the control seems to be not of turgor directly but of internal osmotic pressure or internal chloride; because *Nitella* lives strictly in fresh water, the effect in nature is equivalent. In *Nitella* half-cells with abnormally high or low osmotic pressure produced by experimental manipulation, the original osmotic pressure can be restored by net uptake or net loss of salts (86–89). Nakagawa et al. (90) showed that this regulation was achieved by a dependence of the net KCl flux on the difference between the cell's osmotic pressure and its "correct" value. Hence, the processes of salt accumulation are subject to controls appropriate to their role in turgor generation.

In higher plant cells also, there appear to be controls on the ion fluxes, and in salt-loaded tissues the further influx is reduced. Thus, Cram (91) showed that in carrot and barley roots the vacuolar content of chloride + nitrate appeared to regulate the influx of Cl^- at the plasmalemma, and in barley the transfer from cytoplasm to vacuole was also reduced in ($Cl + NO_3$)-loaded tissue. In this respect, chloride and nitrate were equivalent, and Cram suggested an effect on a common nonselective transport site at the tonoplast. There are also effects on other ion fluxes in salt-loaded tissue; Pitman et al. (92) showed that in salt-loaded barley roots both Cl^- influx and H^+ efflux were much reduced, and an increased selectivity for potassium over sodium was also shown. This does suggest that processes at the tonoplast do have important feedback effects on the plasmalemma ion transport processes.

In the control of cytoplasmic volume, ion transfer from cytoplasm to vacuole must also have a role in the osmotic regulation of the cytoplasm. The excretion of solute from the cytoplasm is essential in animal cells, having no rigid cell wall, to prevent Donnan swelling of the cytoplasm (93); in a walled plant cell, this is not a problem with respect to the external environment, but entry of water from the vacuole to the cytoplasm would remain a problem unless solute of some kind were transported from cytoplasm to vacuole. Hence, vacuolar salt transfer is essential to prevent Donnan swelling in the cytoplasm.

Action Potentials The ion gradients set up by the salt accumulation system are responsible for conferring excitability on many cells, thereby creating the possibility of electrical signaling. The functions of such activities in plants are considered in a later section, but the basic phenomenon may be considered at this stage.

The action potential in Characean cells has been the subject of study for many years, and some of the processes involved are now well recognized, although the nature of the structural changes in the membrane responsible for the drastic changes in its permeability properties remain obscure. In response to electrical or mechanical stimulation, there are transient changes in membrane properties which are seen as electrical perturbations which can propagate along the cell (94, 95). If the cell is depolarized below a threshold level, an action potential ensues as an all-or-none response. The cell depolarizes in about 0.1 s, from its normal resting potential of −120 to −180 mV to a value near zero (but usually still negative). The resistance also falls markedly. Both changes are transient, and the cell returns to its normal state within seconds. Findlay and Hope (96) showed that the response involved electrical changes at both plasmalemma and tonoplast. The tonoplast action potential occurs later than the plasmalemma spike, is much slower, and involves an increase in the potential of the vacuole with respect to the cytoplasm from its resting value of +10 mV to about +50 mV. At the plasmalemma, the changes arise from a great increase in chloride permeability; chloride, therefore, moves out, depolarizing the membrane, the driving force for potassium movement is then disturbed, and a large efflux of potassium follows (97, 98). The net Cl^- and K^+ are many times larger than are required to carry the currents during the rising and falling phases of the spike, having values of 500–3,000 pmol cm^{-2} $impulse^{-1}$, and hence, during the action potential, there is a net outflow of KCl very much larger than the net ion current.

It is presumed that Cl^- movement is also responsible for the tonoplast action potential, which involves a net positive current from cytoplasm to vacuole; net chloride movement from vacuole to cytoplasm, in response to an increase in chloride permeability, seems to be the only possible charge carrier.

Therefore, it appears that both plasmalemma and tonoplast are capable of phase transitions resulting in very large permeability changes in response to external stimulation of various kinds. Stimuli include pressure changes (99) and temperature shocks as well as electrical stimulation, and trains of repetitive action potentials can be produced spontaneously by the removal of Ca^{2+} from the outer surface of the cell (199, 101). The effects of Ca^{2+} are complex, because cells may also be unexcitable in the absence of Ca^{2+} (102), and the membrane permeability to chloride is strongly calcium-dependent (98). In *Nitella* the action potential can be propagated through the node to the neighboring internodal cells (103), indicating that the plasmodesmatal connections through the node provide a suitably low resistance path. While this is of no obvious benefit to the Characean cells, such propagation has clearly identified functions in other plant systems and will be considered later.

One secondary consequence of the action potential in Characean cells may be noted—the stoppage of cytoplasmic streaming at the peak of the action potential (104). This occurs on excitation in the presence of Ca^{2+} in the medium, but not in solutions of Mg^{2+} (105), which suggests that although Ca^{2+} influx may not carry the current during the action potential, there may nevertheless be associated Ca^{2+} influx and this may have marked cytoplasmic effects. Tazawa and Kishimoto (106) showed that this stoppage of streaming arises from a disappearance of motive force, rather than from a change in viscosity. Again this has no obvious significance in Characean cells, but it does suggest that marked changes in cytoplasmic Ca^{2+} might, in appropriate conditions, induce intracellular processes (for example, involving membrane fusion) as a result of the initiation of an action potential propagating through a tissue.

FUNCTIONS OF ION TRANSPORT PROCESSES IN TISSUES: PLANT MOVEMENTS

A number of plant movements involve turgor changes in specific groups of cells, which are mediated through ion movements. In some of these, the processes involved can now be described, but the control of such processes and the means by which the appropriate environmental signals are translated into ion fluxes are not understood. Because some of these processes are phytochrome-controlled or involve endogenous rhythms, they are of considerable interest for the understanding of much more general plant processes.

Stomatal Guard Cells

One of the processes in which ion movements are now known to play a major role is the opening and closing of stomata, controlled by turgor changes in the stomatal guard cells. Stomatal movements have been reviewed recently (107–109) and neither the detailed evidence for the role of ion movements nor the complex environmental factors whose interplay results in the feedback loops by which stomatal aperture is controlled will be reconsidered in this chapter. The aim of this section is simply to consider the events shown to take place in guard cells, in the light of present knowledge of such ion transport processes, and to consider to what extent the processes taking place in cells of the stomatal complex, both guard cells and subsidiary cells, resemble the typical activities of plant cells in general, and to what extent they are special.

The sequence of changes leading to increased turgor in stomatal guard cells and consequent stomatal opening is now recognized, but as Raschke (107) points out, the nature of the primary process responsible for the initiation of this sequence is still unclear. Histochemical detection of potassium by precipitation with sodium cobaltinitrite gave evidence for high potassium in the guard cells of open stomata, but not in closed, as early as 1905 (110). However, recognition of the importance of potassium in the mechanism came much later—from Imamura (111) in 1943, from Yamashita (112), more conclusively from Fujino (113) (first published in English in 1967), and independently from Fischer (114, 115). The marked accumulation of potassium in guard cells during stomatal opening is now established in some 50 species, including examples from both pteridophytes and gymnosperms, as well as a wide range of angiosperms (116, 117). Quantitative estimates of potassium movements are available from only a few species, and most of these figures relate to the behavior of guard cells in the isolated epidermal strips of three species, *Vicia faba, Commelina communis,* and *Zea mays.*

The most detailed information is available for epidermal strips of *V. faba,* selected or treated to provide living guard cells surrounded by only dead epidermal cells. *V. faba* does not have clearly differentiated subsidiary cells in a stomatal complex, although in the intact leaf the neighboring epidermal cells may nevertheless play a role in stomatal movements. But isolated guard cells do function and can be made to open and close in response to the usual environmental stimuli. Opening of such stomata depends on the supply of potassium from the medium, and potassium uptake associated with the opening has been demonstrated by staining, by electron microprobe analysis (118–121), by flame photometry (122), and by measurement of tracer fluxes (115, 123). Opening is associated with an uptake of about 2 pmol of potassium/guard cell or about 0.2–0.4 pmol of potassium/μm of aperture, and concentration changes of 400–800 mM potassium are estimated (119, 122, 124). In *Vicia* very little of the potassium uptake is balanced by chloride uptake (whether in isolated epidermal strips or in the intact leaf); the ratio of Cl:K in the uptake was measured as 0.05 by Humble and Raschke (119), or 0.1–0.5 (mean 0.2) by Pallaghy and Fischer (125). Instead of chloride uptake, an equivalent amount of organic acid anion is synthesized within the guard cell, and protons are excreted to the medium. Raschke and Humble (126) showed an extrusion of 0.2–1.1 pEq of H^+/μm of aperture, consistent with a 1:1 K^+-H^+ exchange at the guard cell plasmalemma. The organic acid content of guard cells has been shown to increase with opening; Allaway (124) found an increase of 1.2 pEq of malate/guard cell, Pearson (127) found a higher figure of 4 pEq of malate/guard cell, and Pallas and Wright (128) measured increases per guard cell of 0.5–1.2 pEq of malate, 2–3.5 pEq of citrate, and 2–2.5 pEq of glycerate, giving a total increase of 6 pEq of organic acid anion/ guard cell, more than adequate to balance the potassium uptake. Therefore, the pattern that emerges for the *Vicia* stomate is that under opening conditions the guard cell is able to achieve high rates of salt accumulation of the second pattern discussed earlier and to maintain high salt levels within the vacuole; that is, it is capable of K^+-H^+ exchange at the plasmalemma and of vacuolar accumulation of K^+ and organic acid anion synthesized within the cytoplasm. The changes in pH of the guard cells, which is discussed later, are also in the same direction as those in barley root cells showing this pattern of salt accumulation (14, 15), with the vacuole going more alkaline as KA is accumulated. The guard cell fluxes in such isolated epidermal strips are also of the order seen in other tissues; Fischer (123) found K influxes from 10 mM KCl of 11–16.5 pmol cm^{-2} s^{-1} at the start of opening process, reducing to 2–3 pmol cm^{-2} s^{-1} after 5 hr, when the net flux was small; thus he suggests that the potassium influx to the guard cell may be regulated by turgor and that the control is exerted, as in other tissues, by a reduction in influx at high turgor. Thus, the behavior of the guard cell in opening conditions is similar to that of other plant cells, but in closing conditions the guard cell, for some reason, loses its ability to accumulate salt. Interest must center on the question of which component process is in primary control of the overall process and on the nature of the inhibition leading to impaired salt accumulation.

The work on *V. faba* establishes the ion transport capabilities of the guard cells themselves and shows that they can function in isolation. The work done on *Commelina* and on *Zea* shows that, in species in which there are well-marked subsidiary cells associated with the guard cells, events in the subsidiary cells may also play a role in stomatal movements. The behavior of stomata in isolated epidermal strips of *Commelina* is more variable than in *Vicia,* depending on the degree of integrity of the subsidiary cells and surrounding epidermal cells. The potassium concentration required in the medium to support the opening of stomata in epidermal strips of *Commelina* is much higher (about 67 mM or even higher) than that required in *Vicia.* However, Squire and Mansfield (129) found that, if the subsidiary cells were killed by treatment at low pH, then the stomata would open in much lower concentrations of external potassium (10 mM); they suggest, therefore, that the guard cell normally receives its salt via the subsidiary cell, with very restricted access from the external solution. This is also suggested by the observations of Willmer and Mansfield (130) on the uptake of neutral red by cells in the epidermis. Whereas in *Vicia* neutral red was taken up immediately by both guard cells and epidermal cells, the behavior in *Commelina* was markedly different. In *Commelina,* epidermal cells took up neutral red immediately, but guard cells became stained only after a delay of some 10 min, and movement from epidermal cells to guard cells during this period took place even if neutral red was removed from the outside; such movement took place through subsidiary cells which themselves remained unstained. Similar observations were made by Guyot and Humbert (131) in *Anemia.*

The role of subsidiary cells in stomatal opening and closing in *Commelina* is most clearly seen in the work of Penny and Bowling (132, 133, 134), who measured pH, potassium, and chloride activity in the various cells of the stomatal complex by the direct insertion of ion-sensitive electrodes in the intact leaf. Their figures are shown in Table 1, and they allow calculation of the driving forces for potassium and chloride movement during both opening and closing. Penny and Bowling (132) suggest active uptake of potassium by the guard cells and inner lateral subsidiary cells during opening, but by the outer lateral subsidiary cells and epidermal cells during closing. Clearly, all cells of the complex are concerned and not simply the guard cells. They calculate the flux across the wall between the guard cell and inner lateral subsidiary cell during opening to be 150–190 pmol cm^{-2} s^{-1}, and suggest that this is too high to be a transmembrane flux and more likely represents a symplastic movement. They also suggest that the pH changes associated with the potassium movements are small enough to indicate movement also of mobile anions across the stomatal complex. Their results also show active movement of chloride out of the guard cells and inner lateral subsidiary cells in closed stomata, but into these cells in open stomata. Their results, therefore, suggest a shuttle of potassium salt between guard cells plus inner subsidiary cells and between outer subsidiary cells plus neighboring epidermal cells, the shuttle being responsible for the changes in guard cell turgor associated with opening/closing.

Table 1. Changes in pH, potassium activity, and chloride activity in the guard cells of *Commelina communis* associated with stomatal opening

Cell	Open stomata			Closed stomata		
	K (mM)	Cl (mM)	pH	K (mM)	Cl (mM)	pH
Guard cell	448	121	5.60	95	33	5.19
Inner lateral subsidiary	293	61.5		159	35.5	
Outer lateral subsidiary	98	47	5.56	199	55	5.78
Terminal subsidiary	169			289		
Epidermal	73	86	5.11	448	117	5.74

From Penny and Bowling (132, 133) and Penny et al. (134).

A shuttle of salt between guard cells and subsidiary cells is also clearly indicated in the stomata of *Z. mays,* with its graminaceous type stoma having two clearly defined subsidiary cells associated with two dumbbell-shaped guard cells. Histochemical staining suggests shuttling of both potassium and chloride within the stomatal complex (135), but the clearest indications come from microprobe analysis. Raschke and Fellows (136) found that the total amount of K^+ and Cl^- in the stomatal complex did not change with opening/closing, but that there was a marked shift in their distribution within the complex; in light (open stomata), 91% of the potassium and 77% of the chloride was in the guard cells, whereas in closed stomata only 30% of the potassium and 20% of the chloride was in the guard cells. The rates of transfer into the guard cell during the early stages of opening were 9.5 fmol of K^+ min^{-1} and 5.1 fmol of Cl^- min^{-1}. Hence, a large fraction of the potassium movement was balanced by chloride, in contrast to the results with *Vicia* where very little chloride movement was involved. Some indication of the path was provided by the time course of concentration changes in different regions of the cells. In the dark, potassium and chloride were higher in the cytoplasm of the subsidiary cells than in the vacuole, particularly in or near the nucleus. After 5 min in light, potassium and chloride had moved from this region into the tubular parts of the guard cells, with a depletion region in the subsidiary cells next to the guard cells. After 30 min of light, potassium and chloride had accumulated in the bulbous ends of the guard cells. Somewhat similar indications of intracellular compartmentation were found by Pallaghy (120), also using microprobe analysis, with both potassium and chloride high in the bulbous ends of open guard cells, whereas the tubular regions were high in potassium but not in chloride. The implication would be that opening is associated with the vacuolar accumulation of KCl in the guard cell, but the stomatal complex is self-contained and only internal redistribution of ions is in fact involved.

Thus, open guard cells in different species are capable of one or the other of the normal patterns of ion accumulation seen in plant cells, either of both K^+

and Cl^- from outside the cell or of K^+ from the outside solution, taken up in exchange for H^+ extruded and accumulated in the vacuole together with an organic acid anion synthesized in the cell. Considered as an ion transport process, there seem to be two possibilities for the primary process in the sequence, but our interpretation rests heavily on the view taken of the path of salt movement. If the movement is apoplastic and takes place across the plasmalemmas of the cells concerned (i.e., between cell wall and cytoplasm of guard cells and neighboring epidermal cells), then the overall process could be controlled at two places. The H^+ extrusion might be primary, and the changes in cytoplasmic pH and electric potential consequent on the initiation or stimulation of proton extrusion would then lead to potassium uptake and either Cl^-/OH^- exchange at the plasmalemma or organic acid synthesis in the cytoplasm. But the continued operation of either form of accumulation does depend on removal of either KCl or KA to the vacuole, and it is possible that control is exerted at the stage of vacuolar transfer. If, however, in the intact system the movement is symplastic between guard cells and subsidiary cells (or neighboring epidermal cells in the absence of clearly identifiable subsidiary cells), then it would seem that the control must rest with the process of salt transfer from cytoplasm to vacuole. Raschke (107, 137) suggests that proton extrusion may be the primary process, but this does imply that in the intact system, as well as in the isolated guard cells of *Vicia,* apoplastic transport is involved.

The question of plasmodesmata between mature guard cells and subsidiary or neighboring epidermal cells is still the subject of argument. Plasmodesmatal connections between immature guard cells and subsidiary cells have been frequently reported, but there are few reports of such connections in mature guard cells. Claims of connections persisting in mature guard cells have been made by Fujino and Jinno (138) (*Commelina*), Pallas and Mollenhauer (139) (*V. faba*), and by Litz and Kimmins (140) (*Nicotiana tabacum, Phaseolus vulgaris, Datura stramonium*), but other authors suggest that protoplasmic connections are lost before maturity, and the discussion by Carr (141) in a recent symposium on plasmodesmata suggests the weight of evidence is against symplastic connection between mature guard cells and neighboring cells. However, two sets of physiological observations already discussed—movement of neutral red through subsidiary cells into guard cells (129, 131) and the amount of potassium shuttling in *Commelina* (132)—would suggest that protoplasmic connections are likely. If they exist, such pores must be plugged in a way that allows them to support a pressure gradient, because the osmotic pressure of guard cells is higher than that in subsidiary cells in a number of species (142); in closed stomata the difference was 2–5 atm in the species measured, but increased to 4–26 atm after opening. Therefore, it is uncertain whether the transport path between guard cell and subsidiary cell involves plugged pores in plasmodesmata or transmembrane transport in the absence of a symplastic path, and the relative probabilities of control by proton pumping on the one hand or by tonoplast transport processes on the other hand cannot be assessed by considerations of this kind.

If, however, the other changes in the state of the guard cells which are found to be associated with opening/closing are considered, in comparison with "normal" behavior of plant cells, then the importance of control of tonoplast fluxes may be suggested. Before discussing these changes, it should first be considered what is different about guard cell behavior; the conclusion must be, not that they allow stomata to open, but that they show the responses involved in stomatal closure. The behavior of guard cells in open stomata is not unlike that in any other plant cell in showing net salt accumulation of one of the two typical patterns discussed earlier—the vacuolar accumulation of either K^+ and Cl^-(both from outside the cell), or of K^+ from outside (taken up in exchange for H^+ at the plasmalemma), with an organic acid anion (synthesized in the cytoplasm). Guard cells seem to compete effectively with subsidiary (or neighboring epidermal) cells for ions, but relatively minor differences in the transport properties of the two types of cell could achieve this. If, for example, the uptake processes in the subsidiary cell were more sensitive to turgor than those in the guard cell, then flux equilibrium between the two cells would be achieved with higher salt content (and turgor) in the guard cell. What is odd about guard cells is that, in response to various external factors, of which high CO_2 seems to the most important in normal function, the cells lose the ability to accumulate salts in the vacuole, and they lose turgor. Thus, the critical question is not why stomata open, which requires only normal behavior of the guard cells, but rather what goes wrong with the ability of the guard cells to accumulate salts and why the sequence of events involved in stomatal closure then ensues. It is in producing closure that the guard cells show abnormal behavior of a type not observed in other cells, and attention should, therefore, be concentrated on the changes in guard cells during the process of stomatal closure. Recent concentration of attention on the role of potassium fluxes in opening/closing of guard cells may have diverted attention from other associated changes which were earlier recognized and which may be of equal or greater importance.

Marked changes in the appearance of cellular contents of guard cells have been recognized for many years and have been described by a number of authors (143–145); the early observations are summarized by Stålfelt (146). The changes include differences between open and closed guard cells in streaming, in form of plasmolysis (concave in the guard cells of open stomata, but convex plasmolysis in those of closed stomata), in the shape of the nucleus and form of its contents, and in the relative volumes of nucleus, cytoplasm, and vacuole. Very large pH changes in guard cells on opening/closing have also been recognized for many years as a result of observations with indicator dyes; in closed stomata the guard cells are much more acid (pH 4–5), whereas the pH in the guard cell rises markedly on opening (pH 6–7.4) (147–149). The direct measurements of guard cell pH in *Commelina* by Penny and Bowling (133) (see Table 1) confirm the early observations. Scarth (147) observed that the pH change in light preceded stomatal opening and also that stomata opened in ammonia, provided the internal pH rose to about 7.4, whereas they closed in acetic acid. It was, therefore, suggested that development of cytoplasmic acidity was responsible for the changes involved in stomatal closure and that cytoplasmic alkalinity resulted in the opening process. Most of the pH measurements relate to species in which potassium is largely balanced by organic anion. Dayanandan and Kaufman (117) found significant amounts of chloride movements in *Ophioglossum engelmanni, Ginkgo, Plantago rugelii, Avena sativa,* and *Z. mays,* but of these, pH measurements are only available for *Ophioglossum;* in this species the figures are similar to those in *Commelina,* with the guard cell pH going from 5.6 to 6.4 on opening, whereas the subsidiary cell pH falls from 6.9 to 5.8. Until measurements of Cl/K stoichiometry and pH changes in one such species are available, it is not possible to decide whether processes other than proton extrusion associated with organic acid synthesis are responsible for the internal pH changes. Inhibition of proton extrusion may be the primary process leading to the decrease in cell pH, as Raschke (107) suggests, but consideration of the other internal changes suggests that the role of vacuolar transport may also be important in the primary processes.

The changes in guard cell structure have recently been examined at the electron microscope level, and the results emphasize the peculiarity of the closed guard cell. Perhaps the most striking between open and closed guard cells is in the relative proportions of vacuole and cytoplasm (131, 150, 151). The guard cells of open stomata have the normal appearance of the mature plant cell, with a large vacuole with dispersed contents and only a thin cytoplasmic layer. In contrast, the closed guard cell has numerous small vacuoles, 0.5–2 μm in diameter, with dense granular contents also containing small vesicles. This is characteristic of very young cells or of secretory cells, but is quite abnormal for a mature plant cell, and the impression is of a failure of the process of vacuole formation under closing conditions. It may be that inhibition of proton extrusion leads to acidification of the cytoplasm and thence to an inhibition of the vacuolar transport, but, equally, the failure of vacuolar transport may be primary. As was discussed earlier, failure to remove salt from cytoplasm to vacuole will lead to inhibition of the plasmalemma transport in either pattern of salt accumulation.

Other changes in the state of the guard cells have also been described, but their significance and relationship to the ion movements are not yet clear. Fujino (113) described very marked changes in guard cell ATPase activity associated with opening/closing of stomata in epidermal strips of *Commelina.* However, the specificity of the enzyme for ATP was not investigated, and it is not clear how this activity is related to the acid phosphatase activity reported in guard cells by Sorokin and Sorokin (152). Fujino found very high ATPase activity in closed guard cells, but not in open, and the effects of a variety of agents on aperture paralleled their effects on ATPase activity. Thus, the enzyme was inhibited by 10^{-5} M *p*-chloromercuribenzoate (PCMB), which also inhibited stomatal closure; the effect was reversed by the addition also of 10^{-3} M cysteine. ATPase activity was strongly activated by 1 mM Ca^{2+}, but not by Mg^{2+}, and this led to stomatal closure. (The effect of Ca^{2+} seems to depend on species; Thomas (153) found closure induced by 1 mM Ca^{2+} in tobacco, whereas *Vicia* does not seem to be as sensitive.) NaF inhibited stomatal closure, as did 10^{-3} M ethylenediaminetetraacetic acid (EDTA), whereas ATP promoted stomatal opening. Thomas (154) also found stomatal opening promoted by added ATP in epidermal strips of tobacco. However, recent work by Lüttge et al. (155) makes it clear that such effects of added ATP may reflect its ability to chelate Ca^{2+} rather than its provision of an energy source, with a consequent release of a Ca^{2+}-inhibition of K^+ uptake.

In spite of some uncertainties of interpretation, the conclusion from Fujino's work seems clear—that the activation of "ATPase" activity in guard cells is closely associated with stomatal closure. At the moment, neither the mechanism of such activation nor its causal relations to the ion transport processes have been established. Fujino (113) also reports a number of other cytochemical changes in guard cells on opening/closing. He claims that inorganic phosphate is high in closed guard cells, but not in open cells, and that free Ca^{2+} is high in open guard cells, but not in closed cells. He suggests hydrolysis of organic phosphate to produce the increased level of inorganic phosphate in closing and binding of a very large fraction of cell Ca^{2+} under the same conditions. The specificity of one of his methods for the cytochemical detection of Ca^{2+} is doubtful, because he uses silver nitrate (used otherwise as an indicator of chloride—or other silver salts under some conditions), but the alternative method using pyrogallol may be more reliable. Electron microprobe analysis (120, 136) shows no change in total calcium within the guard cell, therefore the effect must reflect a marked change in calcium binding. The results suggest that further study of the three major changes claimed by Fujino, using more precise cytochemical methods, might be rewarding. A release of Ca^{2+} from the vacuole in open guard cells to the enlarged cytoplasm in closed guard cells might result in much increased binding, but the nature of the reactions involved in the ATPase activity and its consequences for vacuolar transport remain obscure; it may be that the failure of vacuolar transport is a consequence of activation of this cytoplasmic enzyme. The activity detected by Fjuino may in fact reflect the same enzyme activity as seen by Sorokin and Sorokin (152); they report very high acid phosphatase activity associated with spherosomes (lysosomes) in closed guard cells, but not in open. Such activation may be triggered by a decrease in cytoplasmic pH consequent on inhibition of proton pumping in closing conditions and leading to permeability changes in lysosomal membranes and unmasking of latent enzyme activity; Changes in the vacuolar transport processes may then lead to loss of the central vacuole, but changes in the free Ca^{2+} in the cytoplasm may well be critical both for the activation of enzyme activity and for the ability of the cell to maintain its vacuolar volume. It should, however, be noted that Willmer and Mansfield (156) say they could not detect ATPase activity in the closed guard cells of *Commelina,* although they give no details of their methods.

One other change in guard cell conditions may be noted—the changes in the content of unbound water (157). By measurements of the absorption of microwave radiation, a very clear correlation was found between stomatal aperture and "unbound" water. Stomatal opening is associated with a very large increase in the amount of unbound water in guard cells of *Tradescantia virginiana;* on closing, the guard cell volume decreased from 47 pliter to 23 pliter, but the "unbound" water fell by 63 pliter/guard cell. The authors argue that this reflects movement of water between vacuole and cytoplasm and changes in the state of bound water in the wall.

It may be argued, therefore, that the most striking change in guard cells as they close is in their ability to maintain a vacuole, i.e., to remove salts from the cytoplasm and sequester them in vacuoles, and that such vacuolar transport processes may be of critical importance. The precise nature of the various changes in the cytoplasm on initiation of stomatal closing and their sequential relations needs to be established.

Leaf Movements: Endogenous Rhythms and Sleep Movements

Many leaves show so-called sleep movements (nyctinasty), in which the leaf closes up at night and opens during the day. In the species in which this has been investigated, it is clear that such movements reflect the redistribution of potassium within pulvinar tissue at the base of the leaf or leaflet.

The phenomenon has been most thoroughly studied in two legumes, *Albizzia julibrissin* and *Samanea saman,* by Satter, Applewhite, Galston, and others (158–167). The control of such movements is complex, but swelling or shrinking of motor cells on each side of a central vascular strand in the pulvinus can always be implicated. Such volume changes are always the result of uptake or loss of potassium in the cells concerned (with movement of an undisclosed ion for charge balance). In *Albizzia,* the leaflets open when the ventral (adaxial) motor cells (extensor cells) swell as a result of potassium uptake and the dorsal (abaxial) motor cells (flexor cells) shrink as a result of potassium loss. The reverse movements occur on closing. In *Samanea,* the direction of leaflet movement differs, and hence the disposition of the motor cells with respect to the axis is the reverse. Therefore, the terms "extensor" and "flexor" are now to be preferred. Measurements of potassium by electron microprobe analysis (159, 165–167) make it clear that stores of potassium outside the motor cells must also be involved, rather than a simple shuttle from extensor to flexor motor cells and vice versa.

Two aspects of behavior have been investigated—1) the response to light-on or light-off signals involved in the nyctinastic movement and 2) the existence of an endogenous rhythm of opening/closing which persists for several cycles in constant darkness.

The properties of the endogenous rhythm have been established in a number of studies and are summarized by Satter et al. (165). They distinguish six phases of the cycle, starting from the light-off signal, the transfer to continuous dark.

The early opening phase (8.5–11.5 hr) is achieved by the uptake of potassium by the ventral motor cells. It is only in the second half of the opening phase (11.5–14 hr) that loss of potassium from the dorsal motor cells takes place. In both these stages, opening is temperature-sensitive and sensitive to respiratory inhibition, suggesting that both active uptake of potassium by the ventral cells and active extrusion of potassium from the dorsal cells are involved. Rhythmic opening is inhibited by cycloheximide in a preopening phase (6–8.5 hr) and in the first opening stage; the authors suggest that protein synthesis of particular proteins involved in the active potassium uptake by the ventral cells is required and that these are made just before opening. Cycloheximide has no effect on rhythmic closure (taking place in the 19–21-hr period of the cycle), which is also insensitive to temperature and respiratory inhibition; it is suggested that rhythmic closure involves passive leakage of ions rather than active transport. In *Samanea* (166, 167), which has very large pulvini, the timing of potassium changes has been established in detail, and it is clear that both extensor and flexor cells have an endogenous rhythm of potassium content, the two being out of phase. But in the early stages of opening in this species also, there is an increase of potassium in the extensor cells before any change in the flexor cells; it is only subsequently that the content of the flexor cells falls. The rhythmic changes are most marked in the extensor cells, a finding which is consistent with previous observations of Palmer and Asprey (168, 169) that pulvini with extensor cells intact, but flexor cells removed, still showed a pronounced rhythm. By contrast, pulvini in which only the flexor cells remained showed very little rhythm, but still responded to light/dark changes.

Rhythms of leaf movement have also been studied in other species, for example, in clover (170, 171). In this species also, large changes in potassium distribution across the pulvinule are responsible for the turgor changes and hence the movement. Potassium influxes are high in cells in the expanded state relative to values in the contracted state, and the phase of the influx cycle was found (171) to be about 3 hr in advance of the leaf oscillation and the light/dark cycle. Oscillations of the membrane potential were also associated with the potassium oscillations.

These observations on rhythmic movements in leaves and associated rhythms of ion transport activity in clearly defined cells are of interest in view of the recent suggestion (172) that biological clocks may be very directly related to the existence of feedback oscillators in membranes. The suggestion is that the distribution of proteins in the membrane may be sensitive to specific ion gradients across the membrane, but may in turn form transport channels which influence such gradients. Scott and Gulline (171) suggest that the electric field in a mosaic membrane will have lateral components and that the field may, therefore, provide the link between the forces leading to formation or dispersal of ion transport channels and the consequences of their operation.

Nyctinastic leaf closure, in response to leaf darkening, has also been much studied. Again, the movement results from potassium redistribution and is of more general interest in that it represents a phytochrome-controlled process in which membrane processes are primary. The role of phytochrome in nyctinastic closure was first established by Fondeville et al. (173) in *Mimosa;* closure starts within 5 min of the light-off signal and is promoted by prior exposure to red light, an effect reversed by prior exposure to far-red. The involvement of phytochrome control in *Albizzia* was established by Jaffe and Galston (174) and has since been studied in detail by Galston's group (159–162, 167) and by McEvoy and Koukkari (175). Their studies show that the nyctinastic closure promoted by Pfr (the far-red absorbing form of phytochrome) is a temperature-sensitive process, sensitive to respiratory inhibitors and to aminophylline and also affected by divalent cations ($CaCl_2$ promotes closure, whereas EDTA inhibits closure). This suggests potassium redistribution by processes of active ion transport, whereas rhythmic closure seemed to involve passive leakage of ions rather than active uptake or secretion. In *Samanea* (167), it was shown that Pfr has effects on potassium levels throughout the pulvinus, although the flexor cells respond most to light/dark changes. Phytochrome also affects the rhythmic movement, and the overall behavior is the result of complex interactions between the endogenous rhythms of ion pumps and membrane permeabilities in two sets of motor cells, and of their responses to environmental light signals. The light-promoted opening, in contrast to the dark closure, has an action spectrum with a peak in the blue (176, 177) and seems to be under the control of the so-called high energy reaction. Potassium secretion from the dorsal motor cells in *Albizzia* is suggested as the first response, and Mg^{2+} treatment can substitute for the light effect.

The nature of the molecular changes in the membrane in all these processes remains unclear, but they are of general interest as accessible examples of control by phytochrome or by endogenous rhythms and of their interaction.

The fine structure of the motor cells in *Albizzia* has been investigated by Satter et al. (158), who comment on three odd features of these cells. The first is that they are multivacuolate—a highly unusual condition for mature plant cells. They contain a very large number of small vacuoles (0.2 μm to several μm in size), and this is particularly marked in those cells showing the largest volume changes. Large numbers of lysosomes were also present in the multivacuolate cells, which were said to give the appearance of coalescence and growth to form vacuoles; the authors suggest that the control of the sequestration of lysosomal enzymes may be important in the turgor changes. A high concentration of fibrils in the cytoplasm was also noted. Because the fibrils involved in protoplasmic streaming in other cells were subsequently identified as actin (178, 179), it would be interesting to establish whether the same identification can be made in motor cells. The peculiar multivacuolate condition of the motor cells might suggest that ion transport processes between cytoplasm and vacuole are critical in the primary events leading to the turgor changes and that plasmalemma processes may be secondary—as was suggested by some of the processes in stomatal guard cells discussed in the previous section.

Leaf Movements: Seismonasty

A number of plants, of which *Mimosa pudica* L., the sensitive plant, has been most studied, show rapid leaf movements in response to certain external stimuli. In the doubly compound leaf of *Mimosa,* the leaflets fold up, and the pinnae and the leaf collapse, in response to mechanical stimulation, heat or cold shocks, wounding, or electrical stimulation.

There are three parts to this process, namely the nature of the receptor process, the nature of the conduction of the stimulus from the point of origin, and the nature of the motor process resulting from the stimulus. Work in *Mimosa* has been concerned with the last two, and there is no firm information on the nature of the receptor. It seems likely that pressure changes in touch-sensitive cells are responsible for initiation of the response, similar to the action potential initiated in Characean cells in response to mechanical stimulation, but the cells responsible have not been studied. The lower surface of the main pulvinus is most sensitive to touch, but sensitivity to other sorts of stimulus is found in cells in every part of the leaf and stem.

The conduction of excitation in *Mimosa* has been reviewed by Sibaoka (180, 181) and by Pickard (182), who give references to an extensive early literature. Two sorts of conduction are distinguished (183). The first is a rapid conduction by means of propagated action potentials in elongate parenchyma cells in the phloem and protoxylem, identified by Sibaoka (184). Such cells have large negative resting potentials (−150 mV), good protoplasmic connections to other cells, and produce large propagating spikes on stimulation. Toriyama (185) (by the potassium staining technique) showed that potassium was lost from such cells in phloem parenchyma in the petiole as a result of passage of the stimulus, appearing in extracellular spaces. The nature of the anion involved in such propagated action potentials in higher plants has not been established.

The second form of conduction is much slower and is seen as the slow movement of an electrical response, a variation potential. This is produced by wounding and involves the movement of an unidentified wound substance in the xylem, producing changes in membrane potential of the surrounding cells. The nature of the wound substance is not known, although myoinositol has been considered as a candidate. Although the action potential involved in rapid conduction does not move through pulvini, the variation potential can, and it may also trigger an action potential in cells as it passes, which then moves ahead of the slow response.

The arrival of either type of propagating response in a pulvinus may result in its collapse, the necessary condition being the triggering of an action potential in the pulvinus by the petiolar signal. The pulvinar action potential and its relation to the turgor changes in the motor cells thereby initiated were studied by Oda and Abe (186). The action potential in the pulvinus has a fast-rising phase of 40–70 mV with an extended plateau, in contrast to the usual rapidly falling spike. The bending force is generated, and movement starts within 10–20 ms of the peak of the pulvinar action potential, i.e., within 70–120 ms of its initiation, and the force reaches a maximum within 1–2 s. Oda and Abe found that excitable cells in the main pulvinus were confined to the lower half. This contrasts with Aimi (187, 188), who produced bending by electrical stimulation even when only the upper or only the lower half of the motor tissue was left.

Associated with the combined processes in the pulvinus of excitation and mechanical response is a movement of potassium out of pulvinar parenchyma cells into the very marked intercellular spaces characteristic of the tissue. This was shown by potassium staining by Toriyama (189–191), who observed fluid containing potassium, nitrate, and tannins in spaces which were air-filled before stimulation. Allen (192) also showed that ^{42}K efflux from the excised pulvinus was greatly increased on stimulation. What is not clear is the extent to which potassium loss should be attributed to each of the two processes, the excitation and electrical response, and to the turgor changes in the motor cells themselves. Greatly increased efflux of potassium is likely to be a feature of the action potential in higher plant cells, as in Characean cells, although firm measurements on this point are not available. The collapse of the pulvinus is brought about by the loss of turgor of the motor cells in the lower half of the pulvinus. The extent to which this loss of turgor is brought about by a large increase in potassium permeability is not easy to assess. The loss seems to be both too rapid and too extreme to be simply attributed to osmotic water loss in response to an outflow of solutes induced by permeability changes of the kind seen in Characean cells during the action potential. Various suggestions of contractile processes have been made, but these are ill-defined.

A number of very marked changes in the structural state of the motor cells on stimulation have been observed, which are presumably related to the mechanism of turgor changes. In mature plants, the motor cells of the main pulvinus have very large tannin vacuoles within the central vacuole, which show marked changes on stimulation. However, the existence of such a tannin vacuole does not seem to be essential to the mechanism, because young plants show seismonastic movement before the development of the tannin vacuoles (193). On stimulation, the shape and size of the tannin vacuole change, there are changes in the appearance of the fibrillar contents of the main vacuole, and there is overall shrinkage of the cell. Before stimulation, the tannin vacuole is large and spherical, whereas it appears shrunken, crushed, and of irregular outline after stimulation. The vacuolar contents include fibrils 20–30 nm thick and 200–400 nm long before stimulation, which aggregate after stimulation (189–191, 193–199). Observed in live sections of the pulvinus, the recovery process takes about 30 min and is accelerated by EDTA; the spherical shape is restored in 10 mM $CaCl_2$ without any swelling, which can take place in the presence of KCl (197); this suggests ion pumping to restore the vacuolar contents lost in the response to stimulation. Similar changes have been described in the motor cells of the tertiary pulvinule, with a time course corresponding to that of the nyctinastic movement (198). Toriyama and Jaffe (199) give evidence for large intracellular movements of calcium associated with these vacuolar changes, for the release of Ca^{2+} from the tannin vacuole to the main vacuole on stimulation. They suggest that this may be the cause of the changed state of the vacuolar fibrillar contents, and they speculate on the possibility of its being a contractile protein. Suggestions that contractile vacuoles exist in the motor cells and are important in the turgor loss have been made by a number of authors (194, 195, 200). The most detailed observations are those of Weintraub (200), who observed living motor cells in sections of the tertiary pulvini as they closed slowly (in prechilled plants with slow responses or by the gradual increase in external NaCl). In the prechilled plants, closure was in two stages. In the initial rapid phase of a few seconds, the whole cell contracted, drawing in the wall, and fewer small vacuoles were visible in the cytoplasm after this stage. In the subsequent slow phase (about 15 min), he observed the disappearance of small vacuoles already present and the appearance and subsequent disappearance of new ones. The collapse of the motor cells induced by NaCl was extended over 10–15 min, and similar changes were observed. He suggested that the small vacuoles observed represented contractile vacuoles involved in the excretion of vacuolar contents from the cell.

The rate and extent of the loss of turgor suggest that some active expulsion is involved, rather than a simple leakiness. It remains to be established whether either of the mechanisms discussed is in fact involved—either the contraction (Ca-triggered?) of a contractile protein network, perhaps producing an increased internal pressure to extrude water and solutes through a leaky membrane, or the operation of an excretion system involving contractile vacuoles.

Insectivorous Plants

Ion-mediated processes have also been shown to be involved in the activities of several insectivorous plants in a number of ways—in the setting of the trap system, in its firing, in the receptor process, and in the conduction of the stimulus. An extensive early literature on such plants is summarized by Lloyd (201).

Dionaea *Dionaea,* the Venus-flytrap, is perhaps the best studied of these systems, for which some information is available on the nature of the receptor process, the conduction of the signal, and the motor response. In response to mechanical stimulation of trigger hairs on the leaf lobes, the leaf snaps shut, trapping the stimulating insect between the two lobes for subsequent digestion. Closure may also be induced by electrical stimulation. Jacobson (202) showed that bending of the trigger hairs produced an electrical response, i.e., a receptor potential, in the hair; if the receptor potential goes above a threshold, then an action potential is initiated and travels across the surface of the leaf lobe. At normal temperatures, at least two propagated action potentials are required for shutting of the trap, but the number required depends on their spacing in time, and some sort of memory of previous excitations persists in the lamina for considerable times. A great deal of information on the conduction of the action potential in *Dionaea* is collected in two early papers by Burdon-Sanderson (203, 204). The closure of the trap seems to be the result of loss of turgor in the inner epidermal cells in the hinge region in response to the arrival of the propagated action potentials, but there is no detailed information on the changes.

The receptor potential which triggers this activity arises as a depolarization of certain cells in hair, i.e., on their deformation (205), and the structural characteristics of the hair cells are therefore of interest. In the sensitive region of the hair, various types of cell were distinguished in an electron microscopical study by Williams and Mozingo (206). There appeared to be good protoplasmic connections in peripheral cells in the basal podium of the hair, as would be required for transmission of an action potential. In the sensitive cells, at the indentation region of the hair, two features were marked in the fine structure– 1) the existence of numerous vacuole-like structures, with protein bodies or an anastomosing system of cisternae, and 2) the existence of concentric whorls of endoplasmic reticulum, often around the vacuolar structures. It is presumed that these represent the structures by which a pressure change on deformation is transduced to an electrical response, and then to a propagated action potential. It is suggested that the whorled endoplasmic reticulum is responsible for transmission within the cell, and it would seem to be implied that the membrane is pressure-sensitive, producing an electrical response on mechanical deformation in a way reminiscent of the action potentials produced in Characean cells by mechanical deformation.

Drosera In *Drosera,* the leaf has a number of tentacles which secrete mucilage and are capable of folding around an insect, trapping it. The responses of *Drosera* tentacles to mechanical stimulation by an insect caught in the mucilage also involve electrical changes. Again, as in *Dionaea,* the immediate response to mechanical stimulation is the production of a receptor potential in the head. When the receptor potential reaches a threshold, then a train of action potentials ensues which travels from the head to the base of the tentacles (207–209). The size and duration of the receptor potential determines the number and spacing of the action potentials, which in turn determines the extent of the bending response of the tentacle. Tentacles may also be stimulated electrically or chemically. The response differs from that in *Dionaea* in that a train of action potentials, rather than a single action potential, is initiated by a single receptor potential, and in the extreme slowness of the action potentials.

Utricularia *Utricularia* is a water plant with animal-catching bladders which act as traps which can be triggered by mechanical stimulation of sensitive hairs near a trapdoor with a valve-like action. The trap is set by pumping out water and solutes from the interior of the bladder, leaving it under a reduced pressure with the trapdoor shut. On triggering the trap, the door opens and water sweeps in carrying the animal. The means by which contents are pumped out in resetting the trap after firing have recently been investigated by Sydenham and Findlay (210, 211). They suggest that there is active transport of chloride from the lumen to an intermediate space, creating an osmotic gradient for following water flow. They suggest hydrostatic pressure as the final driving force between this intermediate space and the outside. The exudate emerges from a region close to the mouth of the trap, and sets of specialized cells in that region are thought to be involved. The cells include a number of two-armed hairs on the inner surface of the bladder near the mouth, stalk cells, or specialized epidermal cells on the outer bladder wall in the same region. The authors suggest that the chloride uptake into the cells is the primary active process, driving subsequent fluxes.

ROLE OF IONS IN DEVELOPMENTAL PROCESSES

There is now evidence in a few systems that ion fluxes and their consequences may be important in the control of development, particularly in the origin of morphogenetic polarity within cells. This is likely to be a more general mechanism, but there are rather few systems in which the processes concerned are accessible to experiment. Two systems have given evidence of ionic involvement—developing fucoid eggs and regenerating fragments of the giant marine alga, *Acetabularia.*

Fucoid Eggs

The processes concerned in the development of the fucoid egg after fertilization and in the transition to a highly polarized two-cell stage have been studied in some detail. Marked changes of the membrane properties of the zygote are a feature of its early development. The unfertilized egg is wall-less and is in osmotic equilibrium with its sea water environment. Its membrane is nonselective, and the membrane potential is low (−20 mV). In the first 2 hr after fertilization, the cell wall is formed, and in the next 6–12 hr membrane properties change drastically. The membrane potential increases to −80 mV, the membrane becomes highly selective (its permeability to potassium increasing while that to sodium decreases), the cell will not leak chloride, and an inward chloride pump is developed (212–214). Over the 2–8-hr. period between cell wall formation and germination, the osmotic pressure rises by 180–300 mOSM, an increase which can be accounted for by the KCl accumulation observed (215). The osmotic pressure increases before germination; during expansion growth potassium and chloride continue to increase, but without further increase in osmotic pressure. In the course of this development, the apolar zygote differentiates into two very different regions and then into cells, with a tip-growing rhizoid cell and a thallus cell. The polarity, i.e., the differentiation into a cell with a rhizoidal pole, can be set by a wide range of external agents (216), but the first sign of such polarization is seen as the establishment of an electrical polarity within the cell (the existence of a current through the cell) several hours before any asymmetric growth is observed (217). Jaffe, therefore, suggested that such current might act to amplify polarization within the cell by inducing asymmetric movement of intracellular organelles, and this idea has been developed and expanded in subsequent work. The extracellular field around developing *Fucus* eggs has been mapped in detail, and the nature of the charge-carrying ions is now established. There is both a steady current through the eggs (positive inward at the tip of the rhizoid cell and outward through the basal part of the rhizoid cell and the thallus cell) and a pulse component; at the two cell stage, about 25% of the total current flow through the cell is in the form of large current pulses, about 100-s duration and 3–10 $\mu A\ cm^{-2}$ amplitude (218, 219). The existence of a calcium current through the cells is established by flux measurements (220), and the possible role of intracellular calcium gradients in control of polarized cell development is discussed (221, 222). Although the intracellular gradient may be most influenced by calcium transport, it is shown in recent work that the pulses are initiated by transient increases in membrane permeability to Cl^-, i.e., they reflect permeability changes of the kind seen in Characean action potentials. Thus, Nuccitelli and Jaffe (223) have shown that although entry of calcium may be important in triggering the pulse, the effect is to open transient chloride channels, allowing net efflux of chloride. In the consequent depolarization of the membrane, a net K^+ efflux occurs, so that, as in the Characean action potential, a net efflux of KCl is the characteristic event. It is suggested that such pulses act as osmotic regulators in the adjustment of cell turgor. The effects of environmental changes on the pulses are consistent with these suggestions. Small changes in salt concentration outside have marked effects on the frequency of pulsing, and in depolarized cells in high external K, in which reversed pulsing is observed, the embryos may burst on entry of considerable amounts of KCl. The importance of calcium in control of the cell processes is evident–in its influence on the membrane permeability to other ions and the associated electrical events, in its presumed effects within the cell, in establishing electrical gradients for movement of cell organelles or charged molecules, and perhaps also in regulating processes of membrane fusion.

Regenerating *Acetabularia* Fragments

Another example in which establishment of a morphogenetic polarity is correlated with patterns of electrical activity is in the regeneration of anucleate stalk segments of *Acetabularia.* These resume growth at one end, and the determination of the site of localized growth has been studied by Novak and Bentrup (224). They found that the establishment of a membrane potential difference between the two ends of the cell preceded any visible sign of regeneration, that regeneration took place at the end at which the membrane potential was most negative, and that abolishing such an electrical polarity by voltage clamping inhibited regeneration. They also observed spontaneous spikes at the presumptive regenerative pole arising from transient changes in membrane conductance during action potentials, so that in this system also impulse propagation is associated with the setting-up of polarity. The action potential in *Acetabularia* also involves a large efflux of chloride (225) as in other systems considered. Novak and Bentrup (224) suggest that both the potential gradient within the cell and the propagating impulses, as a means of conveying positional information, are important in the control of morphogenesis.

REFERENCES

1. Baker, P. F. (1972). Prog. Biophys. Mol. Biol. 24:177.
2. Braetz, R., and Komnick, H. (1970). Cytobiology 2:457.
3. MacRobbie, E. A. C. (1975) Curr. Top. Membranes Transport 7:1.
4. Raven, J. A. (1975). *In* D. A. Baker and J. L. Hall (eds.), Ion Transport in Plant Cells and Tissues, pp. 125–160. North Holland Publishing Company, Amsterdam.
5. Barber, J. (1968). Biochim. Biophys. Acta 163:141.
6. Slayman, C. L., and Slayman, C. W. (1968). J. Gen. Physiol. 52:424.
7. Poole, R. J. (1966). J. Gen. Physiol. 49:551.
8. Epstein, E. (1972). Mineral Nutrition of Plants: Principles and Perspectives. John Wiley and Sons, New York.
9. Young, M., and Sims, A. P. (1972). J. Exp. Botany 23:77.
10. Lubin, M. (1964). *In* J. F. Hoffman (ed.), The Cellular Functions of Membrane Transport, pp. 193–211. Prentice-Hall, Englewood Cliffs, New Jersey.
11. Flowers, T. J. (1975). *In* D. A. Baker and J. L. Hall (eds.), Ion Transport in Plant Cells and Tissues, pp. 309–334. North Holland Publishing Company, Amsterdam.
12. Slayman, C. L. (1974). *In* U. Zimmermann and J. Dainty (eds.), Membrane Transport in Plants, pp. 107–119. Springer-Verlag, Berlin.
13. Jackson, P. C., and Adams, H. R. (1963). J. Gen. Physiol. 46:369.
14. Hiatt, A. J. (1967). Plant Physiol. 42:294.
15. Hiatt, A. J., and Hendricks, S. B. (1967). Z. Pflanzenphysiol. 56:220.
16. Kirkby, E. A. (1969). *In* I. H. Rorison (ed.), Ecological Aspects of the Mineral Nutrition of Plants, pp. 215–235. Blackwell, Oxford.
17. Pitman, M. G. (1970). Plant Physiol. 45:787.
18. Higinbotham, N., Graves, J. S., and Davis, R. F. (1970). J. Membrane Biol. 3:210.
19. Higinbotham, N., and Anderson, W. P. (1974). Can. J. Botany 52:1011.
20. Slayman, C. L. (1965). J. Gen. Physiol. 49:93.
21. Slayman, C. L., Lu, C. Y.-H., and Shane, L. (1970). Nature 226:274.
22. Slayman, C. L., Long, W. S., and Lu, C. Y.-H. (1973). J. Membrane Biol. 14:305.
23. Kitasato, H. (1968). J. Gen. Physiol. 52:60.
24. Spanswick, R. M. (1972). Biochim. Biophys. Acta 288:73.
25. Spanswick, R. M. (1973). *In* W. P. Anderson (ed.), Ion Transport in Plants, pp. 113–128. Academic Press, London.
26. Spanswick, R. M. (1974). Biochim. Biophys. Acta 332:387.
27. Ting, I. P. (1968). Plant Physiol. 43:1919.
28. Walker, N. A., and Smith, F. A. (1975). Plant Sci. Lett. 4:125.
29. Smith, F. A., and Walker, N. A. J. Exp. Botany 27:451.
30. Davis, R. F. (1974). *In* U. Zimmermann and J. Dainty (eds.), Membrane Transport in Plants, pp. 197–201. Springer-Verlag, Berlin.
31. Hirakawa, S., and Yoshimura, H. (1964). Jap. J. Physiol. 14:45.
32. Raven, J. A., and Smith, F. A. (1973). *In* W. P. Anderson (ed.), Ion Transport in Plants, pp. 271–278. Academic Press, London.
33. Raven, J. A., and Smith, F. A. (1974). Can. J. Botany 52:1035.
34. Raven, J. A., and Smith, F. A. (1976). Curr. Adv. Plant Sci. 8:649.
35. Raven, J. A., and Smith, F. A. (1976). J. Theor. Biol. 57:301.
36. Davies, D. D. (1973). *In* B. V. Milborrow (ed.), Biosynthesis and its Control in Plants, pp. 1–20. Academic Press, London.
37. Davies, D. D. (1973). Symp. Soc. Exp. Biol. 27:513.
38. Raven, J. A., and Smith, F. A. (1976). New Phytol. 76:415.
39. Komor, E., and Tanner, W. (1974). *In* U. Zimmermann and J. Dainty (eds.), Membrane Transport in Plants, pp. 209–215. Springer-Verlag, Berlin.
40. Komor, E., and Tanner, W. (1974). J. Gen. Physiol. 64:568.
41. Eddy, A. A., and Nowacki, J. A. (1971). Biochem. J. 122:701.

42. Seaston, A., Inkson, C., and Eddy, A. A. (1973). Biochem. J. 134:1031.
43. Slayman, C. L., and Slayman, C. W. (1974). Proc. Natl. Acad. Sci. U.S.A. 71:1935.
44. Hutchings, V. M. Manuscript in preparation.
45. Smith, F. A. (1970). New Phytol. 69:903.
46. Smith, F. A. (1972). New Phytol. 71:595.
47. Jayasuriya, H. D. (1975). Ph.D. thesis, University of Cambridge.
48. Cram, W. J. (1976). *In* U. Lüttge and M. G. Pitman (eds.), Transport in Plants II, Part A, Cells, pp. 284–316. Springer-Verlag, Berlin.
49. Steward, F. C., and Mott, R. L. (1970). Intl. Rev. Cytol. 28:275.
50. Mott, R. L., and Steward, F. C. (1972). Ann. Botany 36:915.
51. Jacoby, B., and Laties, G. G. (1971). Plant Physiol. 47:525.
52. Smith, F. A., and Raven, J. A. (1974). *In* U. Zimmermann and J. Dainty (eds.), Membrane Transport in Plants, pp. 380–385. Springer-Verlag, Berlin.
53. MacRobbie, E. A. C. (1965). Biochim. Biophys. Acta 94:64.
54. MacRobbie, E. A. C. (1966). Aust. J. Biol. Sci. 19:363.
55. Raven, J. A. (1967). J. Gen. Physiol. 50:1627.
56. Raven, J. A. (1968). J. Exp. Botany 19:233.
57. Raven, J. A., and Glidewell, S. M. (1975). New Phytol. 75:205.
58. Heber, U. (1974). Annu. Rev. Plant Physiol. 25:393.
59. Walker, D. A. (1974). *In* D. H. Northcote (ed.), M.T.P. Int. Rev. Sci. Plant Biochemistry, Vol. 11, p. 1. University Park Press, Baltimore.
60. Smith, F. A., and Raven, J. A. (1974). New Phytol. 73:1.
61. Takeda, J., and Senda, M. (1974). Plant Cell Physiol. 15:957.
62. Cram, W. J. (1974). J. Exp. Botany 25:253.
63. Walker, N. A., and Pitman, M. G. (1976). *In* U. Lüttge and M. G. Pitman (eds.), Transport in Plants II, Part A, Cells, pp. 36–52. Springer-Verlag, Berlin.
64. Gerson, D. F., and Poole, R. J. (1972). Plant Physiol. 50:603.
65. Spanswick, R. M., and Williams, E. J. (1964). J. Exp. Bot. 15:193.
66. Hope, A. B., Simpson, A., and Walker, N. A. (1966). Aust. J. Biol. Sci. 19:355.
67. Coster, H. G. L. (1966). Aust. J. Biol. Sci. 19:545.
68. Tazawa, M., Kishimoto, U., and Kikuyama, M. (1974). Plant Cell Physiol. 15:103.
69. MacRobbic, E. A. C. (1969). J. Exp. Botany 20:236.
70. MacRobbie, E. A. C. (1971). J. Exp. Botany 22:487.
71. MacRobbie, E. A. C. (1973). *In* W. P. Anderson (ed.), Ion Transport in Plants, pp. 431–446. Academic Press, London.
72. Findlay, G. P., Hope, A. B., and Walker, N. A. (1971). Biochim. Biophys. Acta 233:155.
73. Walker, N. A., and Bostrom, T. E. (1973). *In* W. P. Anderson, (ed.), Ion Transport in Plants, pp. 447–458. Academic Press, London.
74. Walker, N. A. (1974). U. Zimmermann and J. Dainty (eds.), Membrane Transport in Plants, pp. 173–179. Springer-Verlag, Berlin.
75. MacRobbie, E. A. C. (1975). J. Exp. Botany 26:489.
76. Matile, P., and Moor, H. (1968). Planta 80:159.
77. Berjak, P. (1972). Ann. Botany 36:73.
78. Matile, P. (1966). Z. Naturforsch. 21b:871.
79. van der Wilden, W., Matile, P., Schellenberg, M., Meyer, J., and Wiemken, A. (1973). Z. Naturforsch. 28c:416.
80. Leigh, R. A., and Branton, D. Manuscript in preparation.
81. Cram, W. J. (1976). *In* U. Lüttge and M. G. Pitman, (eds.), Transport in Cells II, Part A. Cells, pp. 284–316. Springer-Verlag, Berlin.
82. Gutknecht, J. (1968). Science 160:68.
83. Hastings, D. F., and Gutknecht, J. (1974). *In* U. Zimmermann and J. Dainty (eds.), Membrane Transport in Plants, pp. 79–83. Springer-Verlag, Berlin.
84. Zimmermann, U., and Steudle, E. (1974). J. Membrane Biol. 16:331.
85. Bisson, M. A., and Gutknecht, J. (1975). J. Membrane Biol. 24:183.
86. Kamiya, N., and Kuroda, K. (1956). Protoplasma 46:423.
87. Tazawa, M. (1961). Protoplasma 53:227.
88. Tazawa, M., and Nagai, R. (1960). Plant Cell Physiol. 1:255.
89. Tazawa, M., and Nagai, R. (1966). Z. Pflanzenphysiol. 54:333.

90. Nakagawa, S., Kataoka, H., and Tazawa, M. (1974). Plant Cell Physiol. 15:457.
91. Cram, W. J. (1973). J. Exp. Botany 24:328.
92. Pitman, M. G., Courtice, A. C., and Lee, B. (1968). Aust. J. Biol. Sci. 21:871.
93. Dainty, J. (1968). *In* J. B. Pridham (ed.), Plant Cell Organelles, pp. 40–46. Academic Press, London.
94. Hope, A. B., and Walker, N. A. (1975). The Physiology of Giant Algal Cells. University Press, Cambridge.
95. Findlay, G. P., and Hope, A. B. (1976). *In* U. Lüttge and M. G. Pitman (eds), Transport in Plants II, Part A, Cells, pp. 53–92. Springer-Verlag, Berlin.
96. Findlay, G. P., and Hope, A. B. (1964). Aust. J. Biol. Sci. 17:62.
97. Gaffey, C. T., and Mullins, L. J. (1958). J. Physiol. 144:505.
98. Mullins, L. J. (1962). Nature 196:986.
99. Kishimoto, U. (1968). Ann. Rep. Biol. Works Fac. Sci. Osaka Univ. 16:61.
100. Kishimoto, U. (1966). Plant Cell Physiol. 7:547.
101. Kishimoto, U. (1968). Plant Cell Physiol. 9:539.
102. Hope, A. B. (1961). Aust. J. Biol. Sci. 14:312.
103. Spanswick, R. M., and Costerton, J. W. F. (1967). J. Cell Sci. 2:451.
104. Sibaoka, T., and Oda, K. (1956). Sci. Rep. Tohoku Univ. 22:157.
105. Barry, W. H. (1968). J. Cell Physiol. 72:153.
106. Tazawa, M., and Kishimoto, U. (1968). Plant Cell Physiol. 9:361.
107. Raschke, K. (1975). Annu. Rev. Plant Physiol. 26:309.
108. Thomas, D. A. (1975). *In* D. A. Baker and J. L. Hall, (eds.), Ion Transport in Plant Cells and Tissues, pp. 377–412. North-Holland Publishing Company, Amsterdam.
109. Hsiao, T. C. (1976). *In* U. Lüttge and M. G. Pitman (eds.), Transport in Plants II, Part B, Tissues and Organs, pp. 195–221. Springer-Verlag, Berlin.
110. Macallum, A. B. (1905). J. Physiol. 32:95.
111. Imamura, S. (1943). Jap. J. Botany 12:251.
112. Yamashita, T. (1952). Sieboldia 1:51.
113. Fujino, M. (1967). Sci. Bull. Fac. Educ. Nagasaki Univ. 18:1.
114. Fischer, R. A. (1968). Science 160:784.
115. Fischer, R. A., and Hsiao, T. C. (1968). Plant Physiol. 43:1953.
116. Willmer, C. M., and Pallas, J. E. (1973). Can. J. Botany 51:37.
117. Dayanandan, P., and Kaufman, P. B. (1975). Am. J. Botany 62:221.
118. Sawney, B. L., and Zelitch, I. (1969). Plant Physiol. 44:1350.
119. Humble, G. D., and Raschke, K. (1971). Plant Physiol. 48:447.
120. Pallaghy, C. K. (1973). Aust. J. Biol. Sci. 26:1015.
121. Willmer, C. M., and Pallas, J. E. (1974). Nature 252:126.
122. Allaway, W. G., and Hsiao, T. C. (1973). Aust. J. Biol. Sci. 26:309.
123. Fischer, R. A. (1972). Aust. J. Biol. Sci. 25:1107.
124. Allaway, W. G. (1973). Planta 110:63.
125. Pallaghy, C. K., and Fischer, R. A. (1974). Z. Pflanzenphysiol. 71:332.
126. Raschke, K., and Humble, G. D. (1973). Planta 115:47.
127. Pearson, C. J. (1973). Aust. J. Biol. Sci. 26:1035.
128. Pallas, J. E., and Wright, B. G. (1973). Plant Physiol. 51:588.
129. Squire, G. R., and Mansfield, T. A. (1972). New Phytol. 71:1033.
130. Willmer, C. M., and Mansfield, T. A. (1969). New Phytol. 68:363.
131. Guyot, M., and Humbert, C. (1970). C. R. Acad. Sci. (D.) (Paris) 270:2787.
132. Penny, M. G., and Bowling, D. J. F. (1974). Planta 119:17.
133. Penny, M. G., and Bowling, D. J. F. (1975). Planta 122:209.
134. Penny, M. G., Kelday, L. S., and Bowling, D. J. F. (1976). Planta 130:291.
135. Pallaghy, C. K. (1971). Planta 101:287.
136. Raschke, K., and Fellows, M. P. (1971). Planta 101:296.
137. Raschke, K. (1975). Planta 125:243.
138. Fujino, M., and Jinno, N. (1972). Sci. Bull. Fac. Educ. Nagasaki Univ. 23:101.
139. Pallas, J. E., and Mollenhauer, H. H. (1972). Am. J. Botany 59:504.
140. Litz, R. E., and Kimmins, W. C. (1968). Can. J. Botany 46:1603.
141. Carr, D. J. (1976). *In* B. E. S. Gunning and A. W. Robards (eds.), Intercellular Communication in Plants: Studies on Plasmodesmata, pp. 243–289. Springer-Verlag, Berlin.

142. Meidner, H., and Edwards, M. (1975). J. Exp. Botany 26:319.
143. Weber, F. (1930). Protoplasma 9:128.
144. Heller, F. O., and Resch, A. (1967). Planta 75:243.
145. Heller, F. O., Kausch, W., and Trapp, L. (1971). Naturwissenschaften 58:419.
146. Stålfelt, M. G. (1956). *In* W. Ruhland (ed.), Encyclopedia of Plant Physiology, III, Water Relations of Plants, pp. 351–426. Springer-Verlag, Berlin.
147. Scarth, G. W. (1932). Plant Physiol. 7:481.
148. Small, J., and Maxwell, K. M. (1939). Protoplasma 32:272.
149. Pekarek, J. (1934). Planta 21:419.
150. Humbert, C., and Guyot, M. (1972). C. R. Acad. Sci. (D.) (Paris) 274:380.
151. Humbert, C., Lougnet, P., and Guyot, M. (1975). C. R. Acad. Sci. (D.) (Paris) 280:1373.
152. Sorokin, H. O., and Sorokin, S. (1968). J. Histochem. Cytochem. 16:791.
153. Thomas, D. A. (1970). Aust. J. Biol. Sci. 23:961.
154. Thomas, D. A. (1971). Aust. J. Biol. Sci. 24:689.
155. Lüttge, U., Schöch, E. V., and Ball, E. (1974). Aust. J. Plant Physiol. 1:211.
156. Willmer, C. M., and Mansfield, T. A. (1970). New Phytol. 69:983.
157. Sheriff, D. W., and Meidner, H. (1975). J. Exp. Botany 26:315.
158. Satter, R. L., Sabnis, D. D., and Galston, A. W. (1970). Am. J. Botany 57:374.
159. Satter, R. L., Marinoff, P., and Galston, A. W. (1970). Am. J. Botany 57:916.
160. Satter, R. L., and Galston, A. W. (1971). Plant Physiol. 48:740.
161. Satter, R. L., and Galston, A. W. (1971). Science 174:518.
162. Satter, R. L., Marinoff, P., and Galston, A. W. (1972). Plant Physiol. 50:235.
163. Satter, R. L., Applewhite, P. B., Kreis, D. J., and Galston, A. W. (1973). Plant Physiol. 52:202.
164. Applewhite, P. B., Satter, R. L., and Galston, A. W. (1973). J. Gen. Physiol. 62:707.
165. Satter, R. L., Applewhite, P. B., and Galston, A. W. (1974). Plant Physiol. 54:280.
166. Satter, R. L., Geballe, G. T., Applewhite, P. B., and Galston, A. W. (1974). J. Gen. Physiol. 64:413.
167. Satter, R. L., Geballe, G. T., and Galston, A. W. (1974). J. Gen. Physiol. 64:431.
168. Palmer, J. H., and Asprey, G. F. (1958). Planta 51:757.
169. Palmer, J. H., and Asprey, G. F. (1958). Planta 51:770.
170. Scott, B. I. H., and Gulline, H. F. (1972). Aust. J. Biol. Sci. 25:61.
171. Scott, B. I. H., and Gulline, H. F. (1975). Nature 254:69.
172. Njus, D., Sulzman, F. M., and Hastings, J. W. (1974). Nature 248:116.
173. Fondeville, J. C., Borthwick, H. A., and Hendricks, S. B. (1966). Planta 69:357.
174. Jaffe, M. J., and Galston, A. W. (1967). Planta 77:135.
175. McEvoy, R. C., and Koukkari, W. L. (1972). Physiol. Plant. 26:143.
176. Fondeville, J. C., Schneider, M. J., Borthwick, H. A., and Hendricks, S. B. (1967). Planta 75:228.
177. Evans, L. T., and Allaway, W. G. (1972). Aust. J. Biol. Sci. 25:885.
178. Palevitz, B. A., Ash, J. F., and Hepler, P. K. (1974). Proc. Natl. Acad. Sci. U.S.A. 71:363.
179. Williamson, R. E. (1974). Nature 248:801.
180. Sibaoka, T. (1966). Symp. Soc. Exp. Biol. 20:49.
181. Sibaoka, T. (1969). Annu. Rev. Plant Physiol. 20:165.
182. Pickard, B. G. (1973). Bot. Rev. 39:172.
183. Houwink, A. L. (1935). Rec. Trav. Bot. Neerl. 32:51.
184. Sibaoka, T. (1962). Science 137:226.
185. Toriyama, H. (1962). Cytologia 27:431.
186. Oda, K., and Abe, T. (1972). Bot. Mag. (Tokyo) 85:135.
187. Aimi, R. (1960). Bot. Mag. (Tokyo) 73:412.
188. Aimi, R. (1963). Bot. Mag. (Tokyo) 76:374.
189. Toriyama, H. (1954). Cytologia 19:29.
190. Toriyama, H. (1955). Cytologia 20:367.
191. Toriyama, H. (1958). Bot. Mag. (Tokyo) 71:11.
192. Allen, R. D. (1969). Plant Physiol. 44:1101.
193. Toriyama, H. (1955). Bot. Mag. (Tokyo) 68:203.
194. Datta, M. (1957). Nature 179:253.

195. Dutt, A. K. (1957). Nature 179:254.
196. Toriyama, H., and Sato, S. (1971). Cytologia 36:359.
197. Toriyama, H., and Komada, Y. (1971). Cytologia 36:690.
198. Setty, S., and Jaffe, M. J. (1972). Planta 108:121.
199. Toriyama, H., and Jaffe, M. J. (1972). Plant Physiol. 49:72.
200. Weintraub, M. (1951). New Phytol. 50:357.
201. Lloyd, F. E. (1942). The Carnivorous Plants p. 352. Ronald Press Company, New York.
202. Jacobson, S. L. (1965). J. Gen. Physiol. 49:117.
203. Burdon-Sanderson, J. (1892). Philos. Trans. R. Soc. Lond. 173:1.
204. Burdon-Sanderson, J. (1888). Philos. Trans. R. Soc. Lond. 179:417.
205. Benolken, R. M., and Jacobson, S. L. (1970). J. Gen. Physiol. 56:64.
206. Williams, M. E., and Mozingo, H. N. (1971). Am. J. Botany 58:532.
207. Williams, S. E., and Pickard, B. G. (1972). Planta 103:193.
208. Williams, S. E., and Pickard, B. G. (1972). Planta 103:222.
209. Williams, S. E., and Spanswick, R. M. (1972). Plant Physiol. (Suppl.) 50:64.
210. Sydenham, P. H., and Findlay, G. P. (1973). Aust. J. Biol. Sci. 26:1115.
211. Sydenham, P. H., and Findlay, G. P. (1975). Aust. J. Plant Physiol. 2:335.
212. Bentrup, F. W. (1970). Planta 94:319.
213. Weisenseel, M. A., and Jaffe, L. F. (1972). Dev. Biol. 27:555.
214. Robinson, K. R., and Jaffe, L. F. (1973). Dev. Biol. 35:349.
215. Allen, R. D., Jacobsen, L., Joaquin, J., and Jaffe, L. F. (1972). Dev. Biol. 27:538.
216. Jaffe, L. J. (1968). Adv. Morphog. 7:295.
217. Jaffe, L. J. (1966). Proc. Natl. Acad. Sci. U.S.A. 56:1102.
218. Nuccitelli, R. C., and Jaffe, L. F. (1974). Proc. Natl. Acad. Sci. U.S.A. 71:4855.
219. Nuccitelli, R. C., and Jaffe, L. F. (1975). J. Cell Biol. 64:636.
220. Robinson, K. R., and Jaffe, L. F. (1975). Science 187:70.
221. Jaffe, L. F., Robinson, K. R., and Nuccitelli, R. (1974). Ann. N. Y. Acad. Sci. 238: 372.
222. Jaffe, L. F., Robinson, K. R., and Nuccitelli, R. (1974). *In* U. Zimmermann and J. Dainty (eds.), Membrane Transport in Plants, pp. 226–233. Springer-Verlag, Berlin.
223. Nuccitelli, R., and Jaffe, L. F. (1976). Dev. Biol. 49:518.
224. Novak, B., and Bentrup, F. W. (1972). Planta 108:227.
225. Gradmann, D., Gottfried, W., and Gläsel, R. M. (1973). Biochim. Biophys. Acta 323:151.

Index

Abscisic acid, 112
Acer pseudoplatanus, see Sycamore
Acetabularia, regeneration, 242
Action potential, salt accumulation and, 224–226
Adenosine 5′-diphosphate
  effect on sucrose synthetase activity, 95
  respiratory control and, 68
Adenosine 5′-diphosphate-glucose, substrate, of sucrose synthetase, 88, 90
Adenosine 5′-monophosphate
  effect on sucrose phosphate synthetase activity, 104
    on sucrose synthetase activity, 95
Adenosine triphophatase
  guard cell closure and, 232–233
  proton transport and, 215
Adenosine 5′-triphosphate
  and chloride uptake, 219–220
  effect on sucrose phosphate synthetase activity, 104
    on sucrose synthetase activity, 95
  osmotic regulation and, 130
  respiratory control and, 68
Aegopodium podagraria, plastocyanin, 4
Alanine, osmotic agent, 128, 135
Albizzia julibrissin, leaf movements, 234, 236
Albizzia sp, osmotic regulation and, 124, 137
Albugo tragopogonis, effect on host metabolism, 157
Alditol, osmotic agents, 124
Alfalfa, phytoalexin in, 166, 169
Algae, *see also* specific species
  osmotic regulation and 123–126, 214–226
  plastocyanin, 3, 4
Allium cepa, see Onion
Alternaria brassicicola, spore germination, 143
1-Amino-4-(2-amino-3-hydroxy-propoxy)-*trans*-but-3-enoic acid, 153
γ-Aminobutyric acid, osmotic agent, 124, 135
Ammonium chloride, uncoupler, 69
Anabaena variabilis
  plastocyanin, 4, 12, 15, 20–21, 33
    molecular weight, 10
    spectral properties, 9
Ananas comosus, see Pineapple
Anemia sp., stomatal opening in, 228
Antimycin A, 133
Antirrhinum majus, plastocyanin, 5
Anthriscus sylvestris, plastocyanin, 4, *see also* Parsley
Aphanomyces euteiches, phytoalexin and, 168
Apoplastocyanin, preparation of, 16
Apple
  antimicrobial compounds in, 162–163, 172–173
  glycoprotein, 188–189
Arabanase, 147–148
α-L-Arabinofuranosidase, 147
Arabinose
  in galacturonans, 145, 147
  in plant glycoprotein, 185–186, 188, 192–193
  in spinach plastocyanin, 18
Arabitol, osmotic agent, 124, 134–135
Arginine, in plastocyanin, lack of, 19
Armaracia lapathifolia, umecyanin, spectral properties, 13
Ascorbate oxidase, 2, 14
Ascorbic acid, 8
Asparagine, protein-carbohydrate linkage and, 195
Auxin, 113
Avenacin, 159–160
Avenacinase, 159–160
Avena sativa, see Oat
Azurin, 12–13, 16, 25, 28

Bacillus sp., osmotic regulation and, 124
Back reaction, spontaneous
in photosystem I, 47–48
in photosystem II, 51–54
Bacteria, *see also* specific species
osmotic regulation and, 124, 135
plant inhibitors of, 159ff
purple nonsulfur, reaction centers in, 43–45
purple photosynthetic, proton translocation, 72
Bacteriochlorophyll, 44
Bacteriophaeophytin, 44
Bacterium aeroideae, protoplast death and, 150
Banana, Cavendish, antifungal compound in, 162
Barium, effect on sucrose synthetase, 90, 94
Barley
antifungal compounds in, 160
chloride accumulation, 224
fungal infection and, 155–156, 158
glycoprotein, lack of, 199–200
plastocyanin, 31
sucrose phosphate phosphatase in, 105
sucrose phosphate synthetase in, 99, 103
sucrose synthetase in, 84
Bean, *see* specific species
Bean pod mottle virus, 202
2(3)-Benzoxazolinone, 161
Benzoic acid, role in resistance, 171–173
Benzylviologen, electron acceptor, 56–57
Beta vulgaris, plastocyanin, 4, 34
Biotroph
definition of, 142
effect on host metabolism, 155–159
Blechnum spicant, plastocyanin, 4
Blue protein, spectral properties, 13
Bordetella bronchiseptica, azurin, spectral properties, 13
Botrytis cinerea
macerating enzymes of, 150
plant inhibitors of, 161, 170
spore germination, 143
Botrytis fabae, plant inhibitor of, 170
Botrytis spp., spore germination, 143
Botrytis tulipae, plant inhibitors of, 161
Brassica oleracea, see Cauliflower
Bremia lactucae, 174
cell wall-degrading enzymes of, 151
Broad bean, *see Vicia faba*
Bromelian, 192–194, 204
Bumilleriopsis, cytochrome 553, 3

C550, 50–53
Cabbage, fungal infection of, 143
Calcium
effect on sucrose phosphate phosphatase activity, 107
on sucrose synthetase, 90–91, 94
osmotic agent, 125
role in fucoid egg development, 242
Capsella bursa-pastoris, plastocyanin, 5, 20, 33
Carbohydrate, *see also* specific compounds
metabolism, effects of infection on, 156–158
Carbonylcyanide-*p*-trifluoromethoxyphenyl hydrazone, 57, 133
β-Carotene, 50
chloride accumulation, 224
glycoprotein, 188–189, 193
salt accumulation, 218
sucrose phosphate phosphatase in, 105–106
Castor bean
invertase in, 110
sucrose phosphate phosphatase in, 105–106, 110
sucrose phosphate synthetase in, 99, 110
sucrose synthetase in, 84, 110
Catechol, structural formula, 162
d-Cathecin, 163
Cation, divalent, *see also* specific ion
effect on sucrose phosphate synthetase activity, 103–104
on sucrose synthetase, 90–91, 94–95
Cation transport, *see* specific ion
Cauliflower, plastocyanin, 3, 5
Cellulase, 147–149, 189, 194
Cellulose, chemical structure, 147
Cell wall
degradation, 144–151
enzymes causing, 145–147
role of glycoprotein in, 198–200
Centaurea niger, plastocyanin, 6
Ceratocystis fimbriata, 173

Ceruloplasmin, copper content, 14
CF., 50, 53
Chaetomorpha sp., osmotic regulation and, 124–125
Chara corallina
  chloride flux, 221–223
  proton pump, 216
Chara sp.
  action potential, 224–226
  chloride uptake, 219–220
Chemiosmotic hypothesis, 42
Chenopodium album, plastocyanin, 4, 10, 12–13, 16, 18, 23
Chitinase, 189
Chlamydomonas reinhardtii
  glycoprotein, 193
  plastocyanin, 8–10, 16–18, 32
Chlorella ellipsoidea, plastocyanin, 2, 8–9
Chlorella fusca, plastocyanin, 4, 8–10, 19–20
Chlorella sp.
  glucose transport, 217
  plastocyanin, 21, 33
Chloride
  osmotic agent, 124–125, 136, 218–223
  transport, 217
Chlorogenic acid, 162, 171–172
p-Chloromercuribenzoate, 12
  inhibitor of sucrose phosphate synthetase, 101
    of sucrose synthetase, 87
Chlorophyll
  photoreaction with plastocyanin, 15
  synthesis, sucrose level and, 113
Chlorophyll *a*, 46, 50
Chlorophyll *b*, 46, 50
Chloroplast
  plastocyanin, 7, 28–29
  ratio of plastocyanin to p700, 49
  sucrose phosphate synthetase in, 100
  sucrose synthetase in, 84
  Tris-washing of, 52
Chondroitin sulfate, 91
Chromous ion, reduction of plastocyanin by, 9
Chrysanthemum vulgare, plastocyanin, 6
Chymotrypsin, 88, 189
Cirsium vulgare, plastocyanin, 6
Cladogram, molecular, of plastocyanin sequences from Compositae, 34–35
Cladosporium cucumerinum, phytoalexin identification and, 165
Clostridium sp., osmotic regulation and, 124
Clover
  leaf movements, 235
  plastocyanin, 5
Cobalt, 16
Codium decorticatum, turgor regulation, 224
Codium sp., osmotic regulation and, 124–125
Collagenase, 189
Colletotrichum circinans, plant inhibitors of, 162
Colletotrichum lindemuthianium
  hemicellulase in, 147
  phytoalexin studies with, 167
  plant inhibitor of, 164
Coltsfoot, fungal infection of, 156–157
Commestrol, 167
Commelina communis, stomatal guard cell movement in, 226–232
Convolvulus sp., plastocyanin, 6
Copper
  ligands, in azurin, 24, 27–28
    in plastocyanin, 9, 12, 14, 23–28
    in stellacyanin, 25
    in superoxide dismutase, 25–27
  in plastocyanin, 2, 8–11, 16–18, 23–27
Corn, *see also* Maize; *Zea mays*
Corn
  glycoprotein, 188, 193
  plastocyanin, 17–18
Cornus florida, see Dogwood
p-Coumaric ester, role in resistance, 172–173
p-Coumarylagmatine, structural formula, 160
Cowpea mosaic virus, 202
Crataegus monogyna, plastocyanin, 5
Cronatium fusiforme, spore germination, 143
Cronartium ribicola, cell wall-degrading enzymes of, 151
Cucumber
  fungal infection and, 158
  plastocyanin, 5, 10, 13
Cucumis sativus, see Cucumber
Cucurbita pepo, see Marrow
Cusacyanin, 14, 28

Cycas revoluta, plastocyanin, 4
Cyclohexanetetrol, osmotic agent, 124
Cyclotella meneghiniana, osmotic regulation, 135
β-Cystathionase, 153
Cysteine
  in plastocyanin, 19–27
  sucrose synthetase purification and, 85, 87
Cytidine 5′-diphosphate-glucose, substrate, of sucrose synthetase, 88
Cytisis ballendieri, plastocyanin, 5
Cytochrome *b*-559, 50, 53–54, 63
Cytochrome *c*, evolutionary rate, 32, 35–36
Cytochrome c_2, 45
Cytochrome *c*-551, 58
Cytochrome *f*
  interaction with plastocyanin, 9, 29
  interchain position, 60–63
  localization, 57–58
  oxidation-reduction reaction, 59–60

2,4-D, *see* 2,4-Dichlorophenoxyacetic acid
Datura stramonium, plasmodesmata in, 230
Daucus carota, see Carrot
Daviesia latifolia, plastocyanin, 5
DBMIB, *see* 2,5-Dibromomethyliso-propyl-*p*-benzoquinone
DCMU, *see* 3-(3,4-Dichlorophenyl)-1,1-dimethylurea
Delphinium ajacis, plastocyanin, 6
Dendryphiella sp., osmotic regulation and, 124, 134
Development, role of ions in, 241–243
Diadzein, 167
Diaminodurene, 57, 58, 69
2,5-Dibromomethylisopropyl-*p*-benzoquinone, 58–59
2,6-Dichlorophenolindophenol, 52, 54, 57, 69
2,4-Dichlorophenoxyacetic acid, 112–113
3-(3,4-Dichlorophenyl)-1,1-dimethylurea, 29, 50, 52–54, 56, 67, 69, 133, 219
Diethylamine, 12, 15
Digitalis purpurea, plastocyanin, 5
Digitonin, 29, 57
Dihydroxyacetone reductase, 134
3,4-Dihydroxybenzaldehyde, 162
Dimethylformamide, 15
Dimethylsulfoxide, 15
Dinitrophenylhydrazine, 84
Dionaea, 239–240
Diphenolindophenol ascorbate, 29, 31
Diphenylcarbazide, 54
Diplodia viticola, hemicellulase in, 147
Disease, pathogenic
  biochemistry of, 141–180
  host metabolism changes during, 155–159
  host resistance, 159–175
Dithiothreitol, 17, 87
Dogwood, frost resistance, 203
DPIP, *see* Diphenolindophenol ascorbate
Drosera, 240
Dryopteris filix-mas, plastocyanin, 4
Dry rot, pathogenesis, 145
Dunaliella parva, osmotic regulation, 133–134
Dunaliella sp., osmotic regulation, 124, 133–134
Dunaliella tertiolecta, osmotic regulation, 133–134
Dunaliella viridis, osmotic regulation, 133

Echium plantadinium, plastocyanin, 5
Ethylenediaminetetraacetic acid
  inhibitor, of sucrose phosphate phosphatase, 105
  stabilizer, of sucrose synthetase, 87
Elastase, 189
Electron transport
  between photosystems, 55–67
  control of, 67–70
  photosynthetic, role of plastocyanin, 29–31
Endocellulase, 148
Endo-β-1,4-galactanase, 148
Endoglucanase, 189
Endopolygalacturonase, 189
  inhibitor of, 164
Endopolygalacturonate lyase, 148
Endoxylanase, 148
Enhancement effect, 42
Enteromorpha intestinalis, plastocyanin, 4
l-Epicatechin, 163

Epilobium angustifolium, plastocyanin, 6
Episclareol, 161
Equisetum arvense, plastocyanin, 4
Erwinia carotovora
  cell wall-degrading enzyme in, 150–151
  phytoalexin and, 171
Erwinia chrysanthemi, cell wall-degrading enzymes in, 151
Erysiphe cichoracearum, effect on host metabolism, 158
Erysiphe graminis, effect on host metabolism, 155, 156, 158
Escherichia coli, proton translocation, 72
Exciton coupling, 44
Exolyase, 146
Exopolygalacturonase, 146–149
Extensin, 192–193, 198–200
Euglena, plastocyanin in, 3

FII particles, 49–50
Fagopyrum esculentum, plastocyanin, 4
FCCP, *see* Carbonylcyanide-*p*-trifluoromethoxyphenylhydrazone
Fenugreek, sucrose synthetase in, 84
Ferrodoxin, photosystem I and, 46–47
Ferrodoxin-NADP reductase, 46
Ferulic ester, role in resistance,172–173
Ficin, 204
Field bean, sucrose phosphate synthetase in, 99
Flax, fungal infection and, 158
Formamide, 15
French bean, *see Phaseolus vulgaris*
Frost resistance, 203–205
Fructose
  inhibitor, of sucrose synthetase, 91, 96, 98
  substrate, of sucrose synthetase, 98
Fructose-6-phosphate, substrate, of sucrose synthetase, 89
Fungi
  osmotic regulation and, 124
  plant inhibitors of, 159–175
Fucoid egg, ion fluxes and, 241–242
Fucose, in plant glycoprotein, 194
Fusarium caeruleum, pectic enzyme of, 146, 150
Fusarium graminearum, spore germination, 144
Fusarium oxysporum f.sp. *cubense,* inhibitors of, 163
Fusarium oxysporum f.sp. *lycopersici,* cell wall-degrading enzymes in, 149–150, 164
Fusarium oxysporum f.sp. *vasinfectum,* pectic enzyme of, 146
Fusarium roseum "Avenaceum"
  cell wall-degrading enzymes in, 147–148
  plant inhibitors of, 163
Fusarium solani f.sp. *phaseoli,* phytoalexin and, 168, 170
Fusarium solani f.sp. *pisi,* phytoalexin and, 168

Gaeummanomyces graminis var *avenae,* avenacinase, 159
Galactomannanase, 147–148
Galactanase, 147–149
O-α-D-Galactopyranosyl-(1→1)-glycerol, osmotic agent, 127–131
Galactose
  in galacturonans, 145–146
  in plant glycoprotein, 185–186, 188–189, 193
D-Galactosamine, in plant glycoprotein, 194
α-D-Galactosidase, 204
α-Galactosyl-(1,1)-glycerol, osmotic agent, 124
Galactosyltransferase, 130–132
D-Galacturonic acid, in plant glycoprotein, 194
Galium aparine, plastocyanin, 4, 10
D-Galsucrose, 89
Germination
  fungal spore, *see* Spore
  seed, glycoproteins and, 205
Gibberellic acid, 112
Ginkgo, stomatal guard cell movement in, 232
Gladiolus sp., fungal infection, 144
Gloeosporium musarum, plant inhibitor of, 162
β-(1→3)-Glucan, 127
Gluconic acid, effect on sucrose synthetase activity, 96
δ-Gluconolactone, effect on sucrose synthetase activity, 96, 99

Glucosamine, effect on sucrose synthetase activity, 96
D-Glucosamine, in plant glycoprotein, 194
β-Glucosidase, 99
Glucose
  effect on sucrose synthetase activity, 96
  in plant glycoprotein, 193–194
  solution, physical parameters, 82
  in spinach plastocyanin, 18
  transport, 217
Glucose-6-phosphate dehydrogenase, 134
Glutamic acid, osmotic agent, 124, 135
Glyceollin, *see* Hydroxyphaseollin
Glycerol, osmotic agent, 124, 133–135
Glycerol dehydrogenase, 134
Glycerol-3-phosphate dehydrogenase, 134
Glycine
  osmotic agent, 135
  in plastocyanin, 19
Glycine max, see Soybean
Glycinin, *see* Hydroxyphaseollin
Glycoenzymes, 204–205
Glycogen phosphorylase, 99
Glycoprotein
  alternative names, 184
  amino acid constituents, 187, 190–192
  biological significance, 197–205
    in cell growth, 200
    in cell surface interactions, 202–203
    in cell wall structure, 198–200
    in frost resistance, 203–205
    in germination, 205
    in pathogenesis, 201–202
    in polysaccharide synthesis, 200–201
  biosynthesis, 195–197
  carbohydrate constituents, 192–194
  determination of purity, 189–190
  extraction, 185–190
    with acetic acid-sodium chlorite, 188
    with alkali, 185–186
    enzymic, 189
    with hot water, 188
    by hydrazinolysis, 186–188
  localization, 197
  structure, 190–195
Gossypium sp., osmotic regulation and, 124
Grana, photosystem in, 55
Gramicidin, pH equalization, 69
Grape, sucrose phosphate phosphatase in, 105
Growth
  glycoproteins and, 200
  ion transport and, 212, 223
Guanidinium chloride, 12, 14, 15
Guizotia abyssinica, plastocyanin, 6

Helianthus annuus, plastocyanin, 6
Helianthus sp., osmotic regulation and, 124
Helminthosporium carbonum
  host-specific toxin of, 152
  plant inhibitors of, 171
Helminthosporium maydis, host-specific toxin of, 152–153
Helminthosporium sacchari, host-specific toxin of, 152
Helminthosporium sativum
  plant inhibitors of, 160
  spore germination, 143
Helminthosporium victoriae, host-specific toxin of, 152
Helminthosporoside, 152
Hemagglutinin, soybean, 194
Hemicellulase, 147, 189
Hemicellulose, chemical structure, 147
Heracleum mantegazzianum, plastocyanin, 4
Heracleum sphondylium, plastocyanin, 4
Hieracium sp., plastocyanin, 5
Histidine, in plastocyanin, 19, 25–27
Hormone, interaction with sucrose, 111–113
Hordatine, structural formula, 160
Hordeum sativum, see Barley
Hordeum sp., osmotic regulation and, 124
Hydrazine, glycoprotein extraction, 186–188
Hydrodictyon africanum, chloride uptake, 219

p-Hydroxybenzoic acid, role in resistance, 172
Hydroxycinnamic acid, role in resistance, 171
Hydroxyphaseollin
  accumulation, 167–169
  structure, 166
1-α-Hydroxyphaseollone, accumulation, 168
Hydroxyproline
  in glycoprotein, 185ff
  protein-carbohydrate linkage and, 194–195
Hypersensitive cell death, 164–175

IF, *see* *O*-α-D-Galactopyranosyl-(1→1)-glycerol
Impatiens glandulifera, plastocyanin, 6
Incompatibility response, role of glycoprotein in, 202–203
Indoleacetic acid, 112, 136
Inhibitor, of pathogenic organisms, definition, 159
Insectivorous plants, 239–241
Invertase, 83, 105, 110
Ion transport, 211–243, *see also* Osmotic regulation; specific ions
  active, 212–226
  functions of, 226–243
Iron
  nonheme, in *R. Spheroides* reaction center, 44
  in photosystem I proteins, 47–48
5-Isobutyroxy-β-ionone, spore germination inhibitor, 143
Isocitrate dehydrogenase, 134
Isofluoridoside, *see* *O*-α-D-Galactopyranosyl-(1→1)-glycerol

Jerusalem artichoke, sucrose synthetase in, 84–96

Kievitone
  accumulation, 167–169
  structure, 166
Kinetin, 91, 112
5-Keto-D-fructose, substrate, of sucrose synthetase, 89

Laccase, 204
  spectral properties, 14
Lactuca sativa, see Lettuce
Ladino clover, sucrose phosphate synthetase in, 99, 103
Lauryldimethylamine oxide, 43, 46
Lavatera arborea, plastocyanin, 6
Leaf movement, 234–239
Lemna sp., active potassium transport in, 214
Leptosphaerulina briosiana, phytoalexin and, 169
Lettuce
  fungal infection of, 151
  hypersensitive cell death, 174
  plastocyanin, 5, 10, 20, 33
Leucocyanidin, 163
Leucoindophenol, 58
Levanbiose, substrate, of sucrose synthetase, 89
Lignification, role in resistance, 172
Lincomycin, 31
Liriodendron tulipifera, plastocyanin, 5
Locust
  frost resistance, 203
  plastocyanin, 5
Lonicera periclymenum, plastocyanin, 4
Lubimin, structure, 166
Lucerne, *see* Alfalfa
Lupinus sp.
  glycoprotein, 186, 188
  plastocyanin, 5
Lycopersicon esculentum, see Tomato

Maceration, definition, 144–145, 150
Maize
  antifungal compounds in, 161
  fungal toxin and, 152–153
  invertase in, 110
  sucrose phosphate phosphatase in, 105–106, 110
  sucrose phosphate synthetase in, 99, 110
  sucrose synthetase in, 84–86, 92, 94, 110
Magnolia X *soulangeana,* plastocyanin, 5
Magnesium
  effect on sucrose phosphate phosphatase activity, 105, 107

on sucrose phosphate synthetase activity, 103
on sucrose synthetase, 90–91, 94–95
osmotic agent, 124
Malate
accumulation, 218
osmotic agent, 125, 136–137
Malus domestica, see Apple
Manganese
effect on photosystem II, 53–54
on sucrose phosphate phosphatase activity, 107
on sucrose phosphate synthetase activity, 103
on sucrose synthetase, 90–91, 94
iron substitute, 44
Mannanase, 147–149
Mannitol
osmotic agent, 124, 134–135
in plant glycoprotein, 194
Mannose, in plant glycoprotein, 194
Mansonone, 162
Marrow, plastocyanin, 3, 4, 8–10, 12–13, 17–18, 20, 23
Medicago sativa, phytoalexin in, 166
Medicarpin
accumulation, 169
structure, 166
Melampsora lini
effect on host metabolism, 158
spore germination, 143
Membrane structure, fluid-mosaic model, 42
Mercaptoethanol, 87, 101
Mercurialis perenis, plastocyanin, 6, 10, 20, 33
Mercuric acetate, 12, 16
Mercuric chloride, 60
Mercury, inhibitor, of plastocyanin, 30
6-Methoxy-2(3)-benzoxazolinone, 161
2′-Methoxyphaseollinisoflavan
accumulation, 168
structure, 166
Methylamine, 52
Methyl-3,4-dimethoxycinnamate, spore germination inhibitor, 143
Methylferulate, spore germination inhibitor, 143
Methylviologen, 54, 61, 63
Miamensis avidus, osmotic regulation, 135
Mimosa pudica, seismonasty in, 237
Mimosa sp., leaf movements, 236
Mildew, *see* specific causative organism
Mitchell hypothesis, 68, 71–72
Monilicolin A, 169
Monilinia fructicola, 164, 169
Monoamine oxidase, spectral properties, 14
Monochrysis sp., osmotic regulation and, 124
Muhlenbergia montana
invertase in, 110
sucrose enzymes, 110
Mung bean, *see Phaseolus aureus*
Myrosinase, 204

Necrotroph, definition, 142
Nectria galligena, 172
Neurospora crassa, proton pump in, 215, 217
Nicotiana glutinosa, antifungal compounds in, 161
Nicotiana sp.
fungal infection of, 143
glycoprotein, 189, 198
sucrose phosphate phosphatase in, 105
sucrose phosphate synthetase in, 99
sucrose synthetase in, 84
Nicotiana tabacum
plasmodesmata in, 230
plastocyanin, 5
Nigericin, uncoupler, 70
Nitella flexilis, turgor regulation, 224
Nitella sp.
action potential, 225
glycoprotein, absence of, 199
proton transport, 215
Nitella translucens, chloride transport, 219–223
Nitrogen, metabolism, effect of fungal infection on, 158
Nucleic acid metabolism, effect of fungal infection on, 158–159
Nucleoside diphosphate sugar, synthesis, 108
Nyctinasty, 234–236

Oats
antifungal substance in, 159
fungal toxin and, 152

Oats, *continued*
glycoprotein, 186
osmotic regulation and, 124, 136
stomatal guard cell movement in, 232
sucrose phosphate phosphatase in, 105
Ochromonas malhamensis, osmotic regulation, 127–133
Ochromonas sp., osmotic regulation and, 124
Onion
antifungal compounds in, 162
fungal infection, 144, 150–151
Ophioglossum engelmanni, stomatal guard cell movement in, 232
Orange, fungal infection in, 150
Ornithine carbamoyl transferase, 154
Oryza sativa, see Rice
Osmotic regulation
agents for, 123–124
biochemistry of, 119–140, 223–224
in expanding cells, 136–137
feedback loop for, 121–123
in higher plants, 135–138
pressure-sensing mechanism, 125–126
salt accumulation and, 223–224
transport mechanisms in algae, 123–125

P430, 47
P680
in photosystem II, 53
properties, 50
P690, 30
P700, 30
interchain position, 60–61
oxidation-reduction potential, 62
in photosystem I, 46–49
structure, 46–47
P870, 44–45
Papain, 88, 192
Parsley, plastocyanin, 12–14, 17
Parsnip, sucrose phosphate phosphatase in, 105
Pastinaca sativa, plastocyanin, 4
Pea
glycoprotein, 199
invertase in, 110
phytoalexin of, 164–165, 168–169, 173
plastocyanin, 5, 12–13, 15
sucrose phosphate phosphatase in, 105, 110
sucrose phosphate synthetase in, 99, 110
sucrose synthetase in, 84–85, 88, 92, 97
Pectic acid, chemical structure, 145
Pectic fraction, enzymes degrading, 145–147
Pectin, chemical structure, 145
Pectinase, 189
Pectin methyl esterase, 146
Pectin methyl*trans*eliminase, 146, 150
Pectinic acid, chemical structure, 145
Penicillium digitatum, cell wall-degrading enzyme in, 150
Penicillium italicum, cell wall-degrading enzymes in, 150
Pennisetum purpureum
invertase in, 110
sucrose enzymes in, 110
Pennisetum typhoide, carbohydrate metabolism, 112
Pepsin, 189
Periconia circinata, host-specific toxin of, 152
Peronospora tabacina, spore germination, 143
Peroxidase, 204
role in resistance, 174
Petroselinum sp., plastocyanin, 10
pH
effect on plastocyanin, 9, 12, 15
and respiratory control, 68–71
Phaeoceros laevis, proton pump, 216
Phaeodactylum sp., osmotic regulation and, 124
Phaeodactylum tricornutum, osmotic regulation and, 135
Phaseollidin
accumulation, 167–169
structure, 166
Phaseollin
accumulation, 167–169
chemical structure, 165
structure, 59
Phaseollinisoflavin
accumulation, 167–169
structure, 166
Phaseolus coccineus, glycoprotein, 186
Phaseolus aureus
plastocyanin, 5, 35
glycoprotein, 186–187, 189, 193–194

sucrose phosphate synthetase in, 99
sucrose synthetase, 84–86, 88, 90, 92, 94–95, 98
Phaseolus sp., plastocyanin, 20, 33
Phaseolus vulgaris
antifungal compound in, 164
bacterial toxin and, 154
fungal infection, 148–149, 157
glycoprotein, 186, 193–194
phytoalexin of, 164–169, 173
plasmodesmata in, 230
Phaseotoxin, 154, 169, 176
o-Phenanthroline, 29, 45, 67
Phenol, storage of, 163
Phenolase, role in resistance, 174
Phenol compounds, biosynthesis of, 171–174
Phenylalanine, fine structure bands and, 9, 14
Phenylalanine ammonia lyase, role in resistance, 173
β-Phenylglycoside, effect on sucrose synthetase activity, 96–97
Phoma betae, spore germination, 143
Phoma herbarum var *medicagensis,* phytoalexin and, 169
Phosphate, inorganic, effect on sucrose phosphate phosphatase, 107
Phosphoenolpyruvate carboxylase, 215
Phosphorylation, substrates, effects on electron transport, 67–70
Photophosphorylation, uncouplers, effects on electron transport, 67–70
Photosynthesis, fungal infection and, 155–156
Photosystem I, 45–49
coupling to photosystem II, 55–67
localization in chloroplast, 55
plastocyanin and, 29
transmembrane model, 47–49
Photosystem II, 49–54
coupling photosystem I, 55–67
localization in chloroplast, 55
plastocyanin and, 29–31
transmembrane model, 51–54
Phyllosticta maydis, host-specific toxin of, 153
Phytoalexin
definition, 164
identification methods, 165
Phytochrome, leaf movements and, 235–236
Phytophthora erythroseptica, pectic enzyme of, 146, 150
Phytophthora infestans
cell wall-degrading enzymes in, 146, 148, 151
effect on host metabolism, 156–157
plant inhibitors of, 162, 170–175
Phytophthora megasperma var *sojae,* phytoalexin and, 169
Phytuberin
accumulation, 171
structure, 166
Pine, sucrose synthetase in, 84
Pineapple, glycoprotein, 193–194
Pinosylvin, 162
Pinus monticola, fungal infection in, 151
Pisatin
accumulation, 168
chemical structure, 165
Pisum sativa, see Pea
Plantacyanin, spectral properties, 13
Plantago rugelii, stomatal guard cell movement in, 232
Plasmalemma, ion transport at, 218–220
Plasmodesmata, 230
Plastocyanin, 1–40
amino acid composition, 8–9, 12, 19
biosynthesis, 31–32
carbohydrate content, 18–19
copper binding site, 23–27
criterion of purity, 7–8
interchain position, 60–61
effect of denaturing agents on, 14–15
function, 28–31, 60
inhibitors of, 30
interchain electron transfer and, 56, 58–61
isoelectric point, 15
localization, 2–3, 28–29, 48–49
molecular evolution, 32–35
molecular weight, 10–11, 16–18
oxidation-reduction properties, 8–9, 62
and photosystem I, 46, 48–49
primary structure, 19–23
purification, 3–8

Plastocyanin, *continued*
role in photosynthetic electron transport, 29–31
secondary structure, 23
sedimentation coefficients, 17
sources of, 4–6
spectral properties, 9–14
stability, 14–15
structural homology with type I copper proteins, 27–28
structure, 16–28
surface properties, 14–15
tertiary structure, 23
Plastoquinol-1, oxidation, mechanism of, 58–60
Plastoquinone, 50–51, 55–57
localization, 57
pool
electron transfer to, 63–67
function, 56
size, 56
Platymonas sp., osmotic regulation and, 124
Plocamium coccineum, cytochrome *f,* 61
Polygalacturonase, 146, 148–149
Polygalacturonate hydrolase, 146, 150
Polygalacturonate lyase, 146–149
Polylysine
inhibitor, of plastocyanin, 30
plastocyanin localization, 48–49
Polymorphism, in plastocyanin, 35
Polyphenol oxidase, 2
Polysaccharide synthesis, role of glycoprotein in, 200–201
Porphyridium aerugineum, plastocyanin in, 3
Potassium
accumulation, 218–223
active transport of, 213–214
flux, and guard cell movement, 226–229
osmotic agent, 124–128, 133, 135–137
Potassium cyanide, 60
Potassium fluoride, effect on sucrose phosphate phosphatase, 107
Potato
antimicrobial compounds in, 162–163, 170–175
fungal infection of, 146, 148, 150–151, 156–157
plastocyanin, 3, 5, 21–22
starch synthetase of, 201
sucrose phosphate phosphatase in, 105, 108
sucrose phosphate synthetase in, 99–104
sucrose synthetase in, 84–86, 89, 93–94
Proline
osmotic agent, 124, 135, 137–138
in plastocyanin, 19
Pronase, 189
Protein synthesis, 195
Protocatechuic acid, structural formula, 162
Proton
active transport, 214–218
pump, 52
artificial, 58
cellular functions of, 216–218
transfer, by plastoquinone, 56
Prunus serrulata splendens, plastocyanin, 5
Pseudomonas aeruginosa, azurin, 13
Pseudomonas fluorescens, azurin, 13
Pseudomonas glycinea, 167
Pseudomonas phaseolicola, toxin of, 154, 169
Pseudomonas saccharophila, sucrose phosphorylase, 83
Pteridium aquiliaceae, plastocyanin, 4
Puccinia antirrhini, uredospore, 143
Puccinia graminis f.sp. *tritici*
resistance to, 161, 173
uredospore, 143, 151
Puccinia helianthi, uredospore, 143
Puccinia poarum, effect on host metabolism, 156–157
Puccinia recondita f.sp. *tritici,* spore germination, 161
Puccinia sorghi, uredospore, 143
Puccinia striiformis, effect on host metabolism, 156–158
Pulvinus, movement of, 137, 226–241
Pyricularia oryzae, effect on host metabolism, 156, 174
Pyridine, 12, 15
Pyrophosphate
effect on sucrose phosphate phosphatase, 107
on sucrose synthetase, 91
Δ'-Pyrroline-5-carbonic acid, osmotic agent, 135
Pyruvate kinase, sucrose synthetase assay and, 84

Q
  fluorescence quencher, 51–54
  oxidation-reduction potentials, 65
  role in electron transfer, 62–67
Quiesone, spore germination inhibitor, 143

Raffinose, 89
Rape, sucrose phosphate synthetase in, 99, 102
Red kidney bean, *see Phaseolus vulgaris*
Respiration, fungal infection and, 155–156
L-Rhamnose
  in galacturonans, 145
  in plant glycoprotein, 194
Rhizobitoxine, 153–155
Rhizobium japanicum, nonspecific toxin of, 153
Rhizoctonia solani
  cell wall-degrading enzymes in, 148–149
  phytoalexin and, 167–168
  plant inhibitors of, 163
Rhodopseudomonas palustris, plastocyanin, lack of, 3
Rhodopseudomonas spheroides, R-26 mutant, reaction centers in, 43–45
Rhus vernicifera, stellamycin, 13
Ribes sanguinium, plastocyanin, 6
Ribonuclease, fungal infection and, 159
Ribonucleic acid, role in hypersensitivity, 175
Rice
  blue protein, spectral properties, 13
  fungal infection and, 156, 174
  glycoprotein, 188
  invertase in, 110
  starch synthesis in, 108
  sucrose phosphate phosphatase in, 105, 110
  sucrose phosphate synthetase in, 99–101, 103–104, 110
  sucrose synthetase in, 84–86, 92, 110
Ricinus, sucrose uptake, 217
Rishitin
  accumulation, 170–171
  structure, 166
Rishitinol, structure, 166
Robinia pseudoacacia, see Locust
Rumex acetosella, plastocyanin, 4, 20, 33
Runner bean, *see Phaseolus coccineus*
Rust, *see also* specific causative organism
  control of, 162
Ruta graveolans, plastocyanin, 6
Rye
  antifungal compounds in, 161
  fungal infection of, 143

[^{35}S]-diazoniumbenzene sulfonate, plastocyanin labeling with, 49, 54
Saccharese, *see* Sucrose synthetase
Saccharomyces sp., osmotic regulation and, 124
Salicine, inhibitor, of sucrose synthetase, 96
Salicyclic acid, role in resistance, 172
Samanea saman, leaf movements, 234–236
Samanea sp., osmotic regulation and, 137
Sambucus nigra, plastocyanin, 4, 10, 20, 33
Sandal, glycoprotein, 189
Santalum album, see Sandal
Sativan, structure, 166
Scenedesmus acutus, plastocyanin, 4, 17
Schlerotinia fructigena
  pectic enzyme of, 147, 150
  plant inhibitors of, 162–163, 172
Sclareol, 161
Sclerotium cepivorum
  macerating enzymes of, 150
  sclerotial germination, 144
Sclerotium rolfsii
  cell wall-degrading enzymes in, 147–148
  plant inhibitor of, 164
Sclerotium sp., germination, 144
Scopolin, 171
Semiquinone, 44, 51
Senecio squalidis, fungal infection and, 157
Senecio vulgaris, plastocyanin, 6
Serine, protein-carbohydrate linkage and, 195

Serology, with plastocyanins, 35
Silicomolybdate, 53, 54
Silicotungstate, 53
Sinigrinase, 204
Sodium
  active transport of, 213–214
  osmotic agent, 124–125, 135–136
Sodium cyanide, 16
Sodium diethyldithiocarbamate, 85
Sodium dithionate, 8
Sodium dodecylsulfate, 46
Soft rot, pathogenesis, 144
Solanum crispum, plastocyanin, 5
Solanum sp., plastocyanin, 20, 33
Solanum tuberosum, see Potato
L-Sorbose, substrate, of sucrose synthetase, 89
Sorghum, fungal toxin and, 152
Soybean
  bacterial toxin and, 153, 167
  glycoprotein, 194
  phytoalexin in, 166
Spinach
  plantacyanin, 13
  plastocyanin, 2–4, 8–20, 33
  sucrose phosphate phosphatase in, 105
  sucrose phosphate synthetase in, 99
  sucrose synthetase in, 84
Spinacia oleracea, see Spinach
Spore, fungal
  germination inhibitors, 143, 160–162
    stimulators, 143–144
Starch synthesis, 108–109
Starch phosphorylase, 201
Starch synthetase, 113, 201
Stellacyanin, 13, 16, 25, 28
Stellaria medica, plastocyanin, 6
Stemphyllium botryosum, phytoalexin and, 169
Stomata
  guard cells of, movement, 226–234
  osmotic regulation and, 137
Streptomyces scabies, plant inhibitor of, 162
Stroma lamellae, photosystem in, 55
Stromatinia gladioli, sclerotial germination, 144
Suberin, accumulation, 173
Subtilisin, 189
Sucrose
  chemical properties, 80–83
  free energy of hydrolysis, 81
  inhibitor, of sucrose phosphate phosphatase, 107
  metabolism, enzymes of, 83–107 *see also* specific enzymes
  role in plant metabolism regulation, 112–113
  solution, physical parameters, 82
  synthesis, 107–111
  transport, 217
Sucrose-6-phosphate, synthesis, 102, 107
Sucrose phosphate phosphatase
  historical background, 105
  isolation, 105–106
  methods of assay, 105
  occurrence, 105
  stability, 106
  substrate specificity, 106
Sucrose-6-phosphate phosphohydrolase, *see* Sucrose phosphate phosphatase
Sucrose phosphate synthetase
  catalytic reaction, 102
  chemical properties, 101
  historical background, 99
  isolation, 100
  kinetics, 102–105
  methods of assay, 100
  occurrence, 99–100
  physical properties, 100–101
  physiological role, 107–111
  purity, degree of, 100
  stability, 101
  substrate specificity, 101–102
Sucrose phosphorylase, 83, 98–99
Sucrose synthetase
  alternative names, 83
  catalytic reaction, equilibrium of, 89
  chemical properties, 87–88
  historical background, 83
  inhibitors, 87, 91, 95
  isolation, 85
  kinetics, 90–99
  methods of assay, 84–85
  metal ions and, 90
  molecular weight, 86
  occurrence, 84
  physical properties, 85–87
  physiological role, 107–111
  purity, degree of, 85
  stability, 87
  structure, 87

substrate affinities, 90
substrate specificity, 88–89
Sulfate ion, osmotic agent, 124
Superoxide dismutase
copper ligands in, 25–26
spectral properties, 14
Sugar beet
fungal infection of, 143
sucrose phosphate synthetase in, 99
sucrose synthetase in, 84–86, 89, 93, 95
Sugar cane
fungal toxin and, 152
invertase in, 110
sucrose phosphate phosphatase in, 105–106, 110
sucrose phosphate synthetase in, 99, 103, 110
sucrose synthetase in, 84, 110
Sweet potato
sucrose phosphate synthetase in, 99, 103–104
sucrose synthetase in, 84–86, 93
resistance mechanism, 173
Sweet sorghum, sucrose synthetase in, 84
Sycamore
antifungal compound in, 164
glycoprotein, 187, 189, 198
sucrose and, 112
tissue differentiation in, 112
Symphytum officinale, plastocyanin, 10
Symphytum X *uplandicum,* plastocyanin, 5

D-Tagatose, substrate, of sucrose synthetase, 89
Takadiatase, 189
Tapioca, sucrose synthetase in, 84–86, 93
Taraxacum officinalis, plastocyanin, 5
Taxus baccata, plastocyanin, 4
Tetrahymena pyriformis, osmotic regulation and, 135
N,N,N′,N′-Tetramethyl-*p*-phenylenediamine, 69
Thiobarbituric acid, sucrose synthetase assay and, 84, 98
β-D-Thioglucose, effect on sucrose synthetase activity, 96
D-Threochloramphenicol, 31
Thymidine 5′-diphosphate-glucose, substrate, of sucrose synthetase, 88
Tobacco, *see Nicotiana*
Tolypella intricata, chloride flux, 221
Tomato
antifungal compound in, 164
glycoprotein, 185, 189
plastocyanin, 5
sucrose synthetase in, 84, 93
Toxin
host-specific, 152–153
definition, 151
nonspecific, 153–155
definition, 151
Tradescantia virginiana, guard cell closure and, 234
Tragopogon porrifolius, plastocyanin, 5
Trehalose, osmotic agent, 124, 135
Trifolium medium, see Clover
Trimethylbenzoquinone, midpoint oxidation-reduction potentials, 66
Trimethylquinol, in model system, 59
Triticum aestivum, see Wheat
Triticum sp., plastocyanin, 10
Triton X-100, 30, 46, 49
Trypsin, 88, 189, 191
effect on photosystem II, 54
Tryptophan, in plastocyanin, lack of, 8–9, 12, 19
TSF-1 particles, 61
TSF-2a particles, 49
Tulip, antifungal compounds in, 161
Tuliposide, 161
Tunneling, 45, 48, 53
Tussilago farfara, see Coltsfoot
Type I copper protein
copper ligands in, 25
oxidation-reduction potential, 9
spectral properties, 12–13, 14
structural homologies, 27–28
Type II copper protein, spectral properties, 14
Tyrosine, in plastocyanin, 8, 26

Ubisemiquinone, 44
Ubiquinone, 44
Umecyanin, 28
spectral properties, 12–13

Uridine 5′-diphosphate
inhibitor of sucrose phosphate synthetase, 104
of sucrose synthetase, 90–91, 95
Uridine 5′-diphosphate-glucose
substrate, of sucrose synthetase, 83, 88, 90–99
Uridine 5′-triphosphate
effect on sucrose phosphate synthetase activity, 104
on sucrose synthetase activity, 95
Uromyces appendiculatur, spore germination, 161
Uromyces fabae, effect on host metabolism, 155, 157–158
Uromyces phaseoli
effect on host metabolism, 156
uredospore, 143
Uromyces viciae-fabae, spore germination, 161
Uronic acid, in plant glycoprotein, 185
Urtica dioica, plastocyanin, 6, 10
Utricularia, 240–241

Vacuole, salt accumulation in, 218–223
Valine, in plastocyanin, 19
Valonia sp., osmotic regulation and, 124–125, 224
Valonia ventricosa, osmotic regulation in, 223–224
Vanillic acid, role in resistance, 172
Venus-flytrap, 239–240
Verticillium alboatrum
cell wall-degrading enzymes, 149
plant inhibitor of, 162
Viburnum tinus, plastocyanin, 4
Vicia faba
fungal infection of, 143, 155
glycoprotein, 188
invertase in, 110
phytoalexin in, 165–166, 170
plastocyanin, 3, 5, 20, 33
stomatal guard cell movement in, 226–230
sucrose phosphate phosphatase in, 105–106, 110
sucrose phosphate synthetase in, 99–100, 103–104, 110
sucrose synthetase in, 84, 91, 110
Vicia sp., osmotic regulation and, 124
Vicilin, 194

Water, physical parameters, 82
Water oxidation, site of, 52, 54
Wheat
antifungal compounds in, 161, 174–175
fungal infections of, 144, 156–158
glycoprotein, 187–188, 190, 193
absence of, 199–200
plastocyanin, 15–19
sucrose phosphate phosphatase in, 105
sucrose phosphate synthetase in, 99–103
sucrose synthetase in, 84–86, 93
Wood, antifungal compounds in, 162
Wyerone, 165–166
Wyerone acid
accumulation, 170
phytoaxelin, 144, 165
structure, 166

X-320, 51
Xanthophyll, 50
Xylanase, 147–149, 193
Xylose, in plant glycoprotein, 186, 193
D-Xylsucrose, 89

Zea mays
stomatal guard cell movement, 226, 228–229, 232
sucrose synthetase in, 95
Zea sp.
osmotic regulation and, 124
plastocyanin, 10
Z scheme, 41–43